THE HISTORY, ECOLOGY AND MANAGEMENT OF THE

ROCKY MOUNTAIN POPULATION OF TRUMPETER SWANS

(1931-86)

RUTH E. SHEA, EDWARD O. GARTON, AND I. J. BALL
2nd Edition

2nd Edition July 2013
Suggested citation: Shea, R. E., E. O. Garton, and I. J. Ball. 2013. The History, Ecology, and Management of the Rocky Mountain Population of Trumpeter Swans (1931-86). North American Swans 34(1). The Trumpeter Swan Society, Plymouth, MN.

1st Edition August 1987
Suggested citation: Gale, R. S., E. O. Garton, and I. J. Ball. 1987. The History, Ecology and Management of the Rocky Mountain Population of Trumpeter Swans. Montana Cooperative Wildlife Research Unit, University of Montana, Missoula.

DEDICATION

George Melendez Wright
1904-1936

We dedicate this work to the memory of George Melendez Wright, the visionary young National Park Service scientist who labored with unshakable determination in the early 1930s to save the last known Trumpeter Swans.

George Wright led and personally funded the first scientific studies of wildlife and their management problems in the western National Parks. His research on Trumpeter Swans in the Yellowstone region was the first biological study of the species. As the first Chief of the U. S. National Park Service's Wildlife Division, George Wright focused public attention on the illegal shooting that imperiled the swans and the possibility of their imminent extinction.

Saving the swans was his passion. Through his science, eloquence and determination, George Wright organized a successful public campaign to establish Red Rock Lakes National Wildlife Refuge, Montana, in 1935 to protect the swans.

Although his remarkable career was cut short by a tragic accident, George Wright's many accomplishments have gained him recognition as one of North America's wildlife conservation giants. His ecological thinking continues to shape resource management in our National Parks today. George Wright's legacy to future generations is profound.

May those who cherish wildlife never forget that George Wright's unwavering campaign to establish Red Rock Lakes National Wildlife Refuge led to permanent protection of this crucial habitat and saved the last Trumpeter Swans in the Lower 48 States.

PREFACE TO 2ND EDITION

When this report was first distributed in 1987, only about 1,100 adult Trumpeter Swans were known to exist in the wild outside of Alaska. Some 400 adults summered in western Canada and ±400 summered in the Tri-state area of southwest Montana, northwest Wyoming, and eastern Idaho, in and near Yellowstone National Park. About 300 adults existed in small restoration flocks scattered from Oregon to Ontario. The species' recovery from near extinction was well underway and its secure restoration was a high management priority.

Over half of the Tri-state adults resided at Red Rock Lakes National Wildlife Refuge (RRLNWR), Montana, where supplemental winter feeding had occurred since 1935. By the early 1980s, declines in cygnet production and adult numbers, centered at RRLNWR, triggered great federal and state management concern. That concern resulted in a 3-year effort involving most of the researchers and managers who had worked with the population since the 1950s. We compiled all the available information into long-term data sets and analyzed population-habitat relationships with funding provided primarily by Region 6 of the U. S. Fish and Wildlife Service. We identified several factors that appeared to be contributing to the declines and presented short-term and long-term management recommendations.

During the subsequent 25 years, significant management actions occurred based in large part on the recommendations of the 1987 report. These actions included termination of winter feeding and a massive program to reduce population vulnerability by expanding winter and summer distribution.

The 1987 report was distributed primarily to the involved management agencies. The very detailed historic information and long-term data sets, particularly with respect to Red Rock Lakes NWR were intended to serve as a reference document for current and future managers and researchers. Never widely distributed beyond the swan management community, few copies remain available today. Therefore, we have reprinted this 2nd Edition of the 1987 report with the hope that it will be useful to those who continue to work to make this population secure.

All population terminology, data, analyses, discussion and conclusions from the 1987 report are presented unchanged in this 2nd Edition. Changes were limited primarily to reformatting, indexing, and converting 1980s text, Tables and Appendices into modern software. Most Figures could not be converted; the originals were therefore scanned and rotated, and captions were reformatted when possible. In addition, we have added an ***Executive Summary***, comprised of the individual chapter summaries and the final Summary Chapter from the 1st Edition. This will enable readers to gain a broad overview of the report's content and then delve further into the full report to explore specific topics and all supporting data.

We thank The Summerlee Foundation and The Trumpeter Swan Society for funding the editing and production of this 2nd Edition.

About the Authors

Ruth E. Shea

The Trumpeter Swan Society, 12615 County Road 9 – Suite 100, Plymouth, Minnesota 55441
rshea@trumpeterswansociety.org

In 1979, Ruth Shea completed her thesis on "The Ecology of Trumpeter Swan in Yellowstone National Park and Vicinity" and received an M.S. in Wildlife Biology from the University of Montana, Missoula. In 1984-86 she worked for the U.S. Fish and Wildlife Service, Cooperative Wildlife Research Unit, Missoula, to compile the long-term data sets and historical information, and write much of the text in the 1st Edition of this publication. Since the 1970s, Ruth has worked in various aspects of wildlife management and research for the U.S. Fish and Wildlife Service, U.S. National Park Service, U.S. Forest Service, and Idaho Department of Fish and Game, with continued focus on Trumpeter Swan management in the Yellowstone region. Currently she serves as The Trumpeter Swan Society's Greater Yellowstone Trumpeter Swan Coordinator.

Edward O. Garton

Emeritus Professor of Fish and Wildlife Ecology and Statistics, Department of Fish and Wildlife Sciences, College of Natural Resources, University of Idaho, Moscow, Idaho 83844-1136
ogarton@uidaho.edu

Oz Garton completed his Ph.D. in Ecology at the University of California, Davis in 1976 and joined the faculty in wildlife at the University of Idaho in 1977. He took primary responsibility for statistical analysis of the long-term data sets on Trumpeter Swan populations of the Yellowstone region and assisted with writing. He retired in 2011 after receiving 9 awards for excellence in teaching and research, advising 20 Ph.D. and 25 M.S. students to completion of their degrees, publication of 87 papers on population ecology and wildlife statistics in refereed journals and completion of $4.7 M in projects funded by external grants. As emeritus faculty member he continues to teach workshops, assist biologists and managers with assessments of population viability of rare and endangered species and publish research in national and international journals.

I. J. Ball

Dr. Joe Ball administered funding and provided editorial support for the original project as Assistant Leader of the Montana Cooperative Wildlife Research Unit at the University of Montana. He retired as Unit Leader in 2004 and currently divides residence seasonally between western Montana and western Oklahoma.

TABLE OF CONTENTS

EXECUTIVE SUMMARY

CHAPTER 1. DISTRIBUTION AND ABUNDANCE

Summary

Historically the Trumpeter Swan was a wide-ranging, migratory species which nested throughout much of Canada and the United States. Trumpeters probably wintered wherever ice-free water was available. Major concentrations of wintering swans occurred in the Mississippi Valley and the coastal estuaries of the Atlantic, Pacific, and Gulf of Mexico. Outside of Alaska, by the 1900s the species was reduced to near extinction due to the commercial swan skin trade, habitat destruction, and over-harvest by early settlers and Indians with firearms.

The only wild trumpeters known to have survived outside of Alaska were a few families that shared the tradition of wintering in perhaps some of the most isolated ice-free waters in the United States, the geothermally warmed waters of Yellowstone National Park (YNP) and the nearby Centennial Valley of Montana. These survivors included two distinct groups, about 50 primarily nonmigratory adults, and about 80 migratory adults that summered near Grande Prairie, Alberta.

The descendants of these two remnants groups increased to a Tri-state wintering total of some 1,600 swans by February 1987. Interior Canadian trumpeters now nest in Alberta, British Columbia, Saskatchewan, the Yukon, and the Northwest Territories. In the Tri-state Area, nesting is concentrated in the marshes and ponds of the Centennial Valley of Montana, including Red Rock Lakes NWR (RRLNWR). Other scattered pairs nest throughout suitable lake and river habitat in northwestern Wyoming, northeastern Idaho, and southwestern Montana. A very small, isolated group nests in central Montana, southwest of Augusta.

Although the Rocky Mountain Population (RMP) has increased in numbers, it has shown little tendency until recently to pioneer other wintering areas. A few trumpeters occasionally migrated to wintering sites outside of the Tri-state Area, but no regular use of these areas has been documented. The high mortality rates of these few migrants have likely slowed the reestablishment of migratory traditions. Recent scattered observations suggest that migration outside the Tri-state wintering area is increasing, or at least becoming more apparent, particularly through the Coeur d'Alene area of northern Idaho on a route that may lead southwesterly to wintering areas in California.

CHAPTER 2. POPULATION TRENDS

Summary

The most thorough surveys of the entire RMP are those made on the Tri-state winter range when both the Tri-state and Interior Canadian subpopulations are present. Prior to 1974, partial surveys of the Tri-state region were made by the states, Red Rock Lakes NWR, and the National Park Service. Since 1974, most of the region has been surveyed, primarily from the air, in early

February. Survey reliability was reduced in some years by storms that prolonged the survey period and by the presence of tundra swans.

Prior to 1974, the number of wintering trumpeters censused in the Tri-state Area approximately equaled the number of swans counted during the late summer Tri-state surveys. There was no indication of the presence of large numbers of migrants, although marked Grande Prairie swans were observed wintering in the Tri-state Area in the 1950s. Wintering swans increased from about 700 in 1974 to 1,600 in 1986. This increase occurred within the migratory Interior Canada Subpopulation (ICSP); during the same period the Tri-state Subpopulation (TSP) declined.

The adult segment of the Tri-state Subpopulation grew from about 50 in 1935, to 548 in 1954. During the subsequent decade their numbers fluctuated. Adult numbers peaked at 554 in 1964 and remained above 500 until 1968. Between September 1967 and September 1968, 17% of the adults were lost from the Subpopulation. These reduced numbers then remained fairly stable over the next decade. Between 1980 and 1986 adult numbers declined again, from 462 to 331.

The 1967-68 decline in the Tri-state Subpopulation was due to a 28% decline in the Centennial Valley (CV) flock. Similar declines followed within three years in the Yellowstone and Idaho flocks. In the 1980s declines have occurred in the Centennial Valley and Yellowstone flocks, but concurrently the Idaho and lower elevation Wyoming flocks have increased or remained stable. Most flocks in Canada, except the Grande Prairie flock, have only been censused once or twice. Between 1980 and 1986, the trumpeters in the Toobally Lakes area, Yukon, declined while the Grande Prairie flock increased. The Grande Prairie flock showed no net growth between 1959 and 1977. Since 1978, the number of adults increased from 88 to 251. Both the area surveyed and the actual numbers of swans have increased.

The Nahanni, NWT flock has increased markedly since one swan was observed in 1970. Although increased survey effort may account for much of the apparent increase between 1980 and 1985, a real increase in adult numbers must have occurred in 1986.

The survey of all Canadian flocks during the summer of 1985 located a total of 418 adults and 190 cygnets. Added to the total of 368 adults and 139 cygnets in the Tri-state Subpopulation in summer 1985, this gave an estimated total size for the Rocky Mountain Population of 786 adults and 329 cygnets. However, the number of swans wintering in the Tri-state Area in February 1986 was 1,304 adults and 299 cygnets. The summer range of nearly 1/3 of the 1,600 swans wintering in the Tri-state Area is unknown.

CHAPTER 3. NESTING AND CYGNET PRODUCTION

Summary

The Rocky Mountain Population includes about 78 nesting pairs in the Tri-state Subpopulation and 145 nesting pairs in the Interior Canada Subpopulation. The number of nests at Red Rock Lakes NWR and YNP declined in the late 1970s and 1980s. Overall the Tri-state Subpopulation has lost about 12 nests since 1978.

The decline in nesting effort at Red Rock Lakes NWR has been greatest on Lower Lake.

Nest success was also lowest on Lower Lake, due at least in part to flooding. The current number of nests at Red Rock Lakes NWR is about 50% of peak numbers.

Very few trumpeter territories regularly fledged cygnets; productivity is sporadic and strongly related to spring weather and water levels. Even during a period of flock increase at Grande Prairie less than 40% of 51 nesting territories fledged cygnets in five or more years out of ten.

Clutch size, percent hatch of eggs, nest success, and cygnet production per nest are higher in the Grande Prairie flock than in the Tri-state flocks. Within the Tri-state Area, these parameters are highest in the lower elevation Idaho and Wyoming flocks and lowest at Red Rock Lakes NWR and YNP. Clutch size is larger in captive trumpeters than in wild trumpeters and increases with the age of the breeding female.

Cygnet production at Red Rock Lakes NWR has been characterized by wide annual fluctuations, with several extremely poor production years in the 1980s. Declining cygnet production has resulted from the failure of most pairs to raise any cygnets at all, rather than uniform mortality among broods. This is evidenced by the reduction in the mean number of broods fledged at Red Rock Lakes NWR from 17.6 in the 1960s to 6.8 in the 1980s, while the number of cygnets per brood has shown little change.

Prior to about 1960, cygnet production per nest at Red Rock Lakes NWR was inversely related to the number of nests. Since 1960, both the number of nests and the cygnet production per nest have declined.

Changes in cygnet production at Red Rock Lakes NWR show a highly significant positive correlation with variations in hatching and fledging success. Variations in clutch size and the proportion of eggs remaining after removals also showed significant positive correlations with cygnet production. We conclude that cygnet production is most closely correlated with environmental factors that influence clutch size, hatching success, and survival to fledging.

Clutch size was positively correlated with increasing May temperatures and negatively associated with April precipitation, water levels during the entire April to July period, and the amount of grain fed per swan during the previous winter.

Hatching success at the Refuge showed a significant negative correlation with median water level in June, and negative associations with increasing rain in May, June, and July. Hatching success showed a positive correlation with the amount of grain fed per swan during the previous winter.

Survival from hatching to fledging at Red Rock Lakes NWR was positively correlated with the mean minimum temperature in June and negatively correlated with July water levels.

Both the instantaneous birth rate and the number of cygnets fledged/breeder at Red Rock Lakes showed highly significant negative correlations with the size of the spring flock, the number of breeding pairs, and the mean number of swans spending the previous winter at Red Rock Lakes NWR. These findings support the hypothesis that complex negative relationships exist between population size and cygnet production. These relationships could result from changing population age structure, age-specific differences in productivity, and/or habitat constraints.

We found significant negative correlations between both measures of birth rate at Red Rock Lakes NWR and both July precipitation and date of peak hatch. Nest initiations, and thus date of peak hatch, are delayed in late cold springs. A less significant negative correlation exists between cygnets per breeder and median water level in July, and a positive correlation exists between cygnets per breeder and mean minimum temperature in May.

In high water years, cygnet production has been reduced by about 50%. When July rainfall exceeded about 1.6 in. the birthrate declined abruptly. During the period of population decline since 1954, the trumpeters have had to cope with an increased frequency of adverse weather conditions. Since 1954, 23% of the years have had both high water levels and high July rainfall, while none of the years before 1954 had such a combination of events.

The correlations between environmental factors and cygnet production at Red Rock Lakes NWR are consistent with the theory that clutch size, hatching success, and cygnet survival are strongly related to the opportunity of the breeding female to obtain adequate prebreeding nutrition, and to the availability of foods to the newly hatched cygnets. Factors which reduce water temperatures and delay the growth of spring foods (heavy April precipitation, cold May temperatures, high spring water levels) are associated with reduced cygnet production. High elevation flocks in the Tri-state Area are particularly vulnerable to these factors. Hatching success is reduced by flooding and cold rains. Young cygnets are particularly vulnerable to cold temperatures and heavy precipitation, and their survival is reduced when water levels are unusually high. Cygnet production to late summer is reduced in years when the peak of hatch is late, and late hatching cygnets may be unable to fledge before autumn freeze-up occurs.

CHAPTER 4. REPRODUCTIVE BEHAVIOR

Summary

Few details regarding the timing and mechanisms involved in the pairing of RMP trumpeters have been studied. Although the subpopulations winter together, to date marking studies have provided no evidence that bonding occurs between individuals from different subpopulations. If the ICSP and TSP do not pair bond and cross mate, they should be regarded and managed as separate populations rather than subpopulations.

Wild trumpeters may delay first breeding attempts until at least six years of age, although some individuals of both sexes are capable of breeding successfully at age three. Captive trumpeters have been observed to pair as early as nine months and lay eggs at 21 months.

Territories at Red Rock Lakes NWR were less than 40 ha in size and averaged about 13 ha. Elsewhere in the Tri-state Area, territory size approximated the size of the nesting lake and averaged 10-15 ha. In Grande Prairie, territory size averaged 127 ha. While most breeding trumpeters exclude other swans from their territory during nesting and brood rearing, others nest successfully within sight of neighboring pairs and nonbreeders. Pairs often tolerate their own subadult offspring in their territory after incubation ends.

Trumpeters go through a lengthy pre-laying period in which feeding is the main activity. During this critical period the female acquires the energy reserves needed for egg-laying and incubation.

Captive trumpeters and wild Grande Prairie trumpeters laid larger clutches and incubated with higher constancies than Idaho or YNP swans. Grande Prairie swans also had higher nest success, hatching success, and cygnet survival than YNP swans. In YNP, pairs that laid larger than average clutches also incubated with higher constancy and with fewer recesses, and had higher hatching success and prefledging cygnet survival than pairs that laid smaller clutches.

Human disturbance occasionally caused direct nest destruction or abandonment. Incubating females that are disturbed by humans leave their eggs uncovered and unprotected, and will temporarily abandon small cygnets. On the average, trumpeters which could accommodate to human disturbance and chose to occupy disturbed territories showed no detectable reduction in productivity. However, disturbed territories were less likely to be occupied than undisturbed territories and therefore total nesting effort was reduced.

Since 1952, frequent airboat use on Red Rock Lakes NWR during the incubation and brood rearing periods caused obvious stress to swans and may have contributed to nest failure and early cygnet mortality. Recently, however, airboat use and other human disturbance at the Refuge have been greatly reduced.

Adult/cygnet bonds are strong and long enduring. Adult feeding activity provides a major source of cygnet food in the first week of life. Adult/cygnet bonds usually persist through the first winter and occasionally longer. Sibling associations often persist for two or three years and subadult offspring may reassociate with their parents.

CHAPTER 5. MORTALITY

Summary

The combined annual survival rate of yearling and older trumpeters was estimated from band recoveries to be at least 80-88% in the CV flock. Survival of trumpeters in age classes 3+ was estimated to be 82% in the Grande Prairie flock and 93% in the Wyoming flock. Due to neck-band loss, neck-band induced mortality, and emigration, these are minimal estimates, particularly for the CV and Grande Prairie flocks. Total post-fledging annual survival in the Tri-state Subpopulation has averaged 79-84% in the decades since the 1940s. During the worst years, the Subpopulation has experienced survival of 65-75%. Although some portion of the annual losses may be due to dispersal of trumpeters beyond the Tri-state range, resightings and/or band returns derived from the marking of over 1,050 CV trumpeters gave evidence of dispersal by only a few individuals.

Survival of first-winter cygnets and yearlings is lower than that of older birds. Cygnet survival from fledging to the following June is about 60% in the Wyoming flock and 43% in the Grande Prairie flock, and probably similar or lower in the CV flock. Yearling survival is about 66-71%.

Factors influencing the loss of swans from Red Rock Lakes NWR were analyzed using a mortality rate estimate (includes dispersal) based upon the loss of swans from the Refuge flock between consecutive annual late summer surveys. This mortality rate was positively correlated with the duration of low water releases at Island Park Dam ($P = 0.01$) and the proportion of cygnets in the fall count ($P = 0.004$). The mortality rate was negatively correlated with the bushels of grain fed per swan ($P = 0.02$) and total bushels of grain fed per winter ($P = 0.04$) at Red Rock Lakes NWR.

The detrimental effect of low flows in the Henry's Fork on Red Rock Lakes NWR trumpeters can be partially mitigated by increased feeding of grain at the Refuge. There was no evidence of density-dependent effects on winter mortality at the Refuge, other than if the amount of grain was held constant as the number of wintering swans increased, making less grain available per bird. A regression model predicted that increasing duration of low water flows in the Henry's Fork, less grain fed per bird at Red Rock Lakes, and increasing winter severity result in higher winter mortality in the Refuge flock. Analysis of the mortality rate of the CV flock showed similar relationships.

The mortality rate of the non-Centennial Valley portion of the Tri-state Subpopulation showed no significant correlations with any factors except the proportion of cygnets in the fall population. These swans winter in a wide variety of locations that are not adequately described by the available data. The mortality rate of the Grande Prairie flock was positively correlated with Tri-state winter severity ($P= 0.0003$).

Low prefledging cygnet survival has been chronic in the Tri-state Area, particularly at Red Rock Lakes NWR and in Yellowstone, and to a lesser extent in Idaho and lower elevation Wyoming habitats. Since 1964, cygnet survival at the Refuge has averaged somewhat less than 46%. Despite above-average cygnet survival in 1985 and 1986, cygnet survival in the 1980s at Red Rock Lakes NWR has averaged somewhat below 38%. Declines in survival rates have occurred in all parts of the Refuge except Upper Lake and the ponds. These declines in cygnet survival have been compounded by reductions in the total number of nesting pairs and low hatching success.

In Yellowstone and the Targhee NF early cygnet mortality averaged 75%, typically involved entire broods, and was strongly specific to sites and/or pairs. Cygnet mortality was significantly higher among cygnets from clutches containing four or fewer eggs, and was probably related to factors that limited the food supplies available both to prenesting adults and to newly hatched cygnets.

Heavy parasitism has been repeatedly documented at Red Rock Lakes NWR, however it may be the result, rather than the cause, of the cygnets' weakened condition. Emaciation/parasitism is commonly involved in the deaths of prefledging cygnets and wintering birds of all ages.

The most frequently diagnosed mortality factors in the Tri-state Area were emaciation, parasites, lead poisoning, disease, and accidents. Approximately 50 trumpeters were known to have died in Idaho and Yellowstone during the winter of 1984-85. Of 18 trumpeters necropsied, most were severely emaciated and suffered from high parasite loads. Toxic levels of lead were found in five carcasses and sublethal lead toxicosis was involved in the death of a sixth bird. Lead poisoning has been diagnosed as the cause of death of 11 of 34 trumpeters found dead in the Tri-state Area and necropsied since 1980. Elevated blood lead levels were detected in 3 of 10 live Wyoming trumpeters and 5 of 22 live Centennial Valley trumpeters.

Attempts to diagnose diseases have been hampered by postmortem contamination of specimens. Only isolated instances of disease related mortality have been detected, including cases of fowl cholera, avian tuberculosis, aspergillosis, generalized peritonitis/pericarditis, and systemic infections by *Pseudomonas sp.* and *E. coli.*

The retarded development of cygnets at Red Rock Lakes NWR and Yellowstone, with cygnets still flightless in late October and November, has been reported on several occasions. Eighteen percent of the cygnets that survived to September in Yellowstone in 1977-79 were stunted. Weakness at hatching and leg abnormalities have also been observed among Red Rock Lakes NWR and Yellowstone cygnets several times since the 1940s.

CHAPTER 6. SUMMER HABITAT
Ruth E. Shea, Leonard J. Shandruk, Kevin J. McCormick

Summary

Red Rock Lakes NWR provides the most important Trumpeter Swan nesting habitat in the Tri-state Area. Although quantitative data are lacking, the productivity of the Refuge marshes has likely declined since the 1930s. Water control structures have increased the water depths on the Refuge since at least 1958, and most likely since the 1930s. In addition to increasing the water depths, the water control structures eliminated the natural water level fluctuations that would have exposed the marsh soils to air during drought cycles. It is likely that the relatively stable, high water levels have decreased the rates of nutrient recycling and vegetation production on the Refuge, and allowed parasites to multiply without interruption to their life cycles. High water levels have also caused the direct destruction of swan nests, particularly on Lower Lake. Reductions in cattle grazing in the marsh may also have reduced rates of nutrient recycling.

Erosion problems in the late 1950s and 1960s noticeably accelerated the deposition of silt into Upper and Lower Lakes. The increased erosion resulted primarily from phosphate mining activities in the Centennial Mountains and from overgrazing. The construction of the 1957 water control structure and the 1959 Hebgen Lake earthquake also probably altered patterns of erosion and deposition.

Nesting habitat in YNP is marginal compared to Red Rock Lakes NWR due to its higher elevations, smaller territories, and deeper waters. Nesting lakes in Yellowstone were located on the Park's most productive soil types. At least three of Yellowstone's historically-used nesting lakes were no longer occupied by nesting swans in the late 1970s due to excessive human disturbance. In 1985, the Park Service began to mitigate for the detrimental effects of human disturbance by providing artificial offshore nest sites in historical nesting lakes where human disturbance prevented the use of traditional shoreline nest sites.

In Idaho, nesting trumpeters selected the more eutrophic lakes which had greater shoreline development and contained high plant and invertebrate diversity. In Grande Prairie, trumpeters used nesting lakes which had above average amounts of emergent vegetation to provide potential nest sites. Lake size, and territory size is much greater in Grande Prairie than elsewhere in the RMP range. Studies of the key features of breeding habitat in the lower elevations of Wyoming are currently in progress; preliminary results have indicated the importance of adequate food resources for prenesting adults and for cygnets in the first weeks after hatching, and emphasized the need for protection from human disturbance.

Although the Tri-state Area is at the southern edge of the RMP's current breeding range, parts of the region are colder than the Canadian habitats. The coldest RMP nesting habitat is in YNP, followed by Red Rock Lakes NWR and the Yukon, which experience comparable temperatures. Milder May temperatures occur in the lower elevations of the Tri-state Area, the Northwest Territories, and Grande Prairie. Flocks which nest in the milder habitats have increased or remained stable in recent years, while the flocks which nest in the colder habitats have declined.

CHAPTER 7. AQUATIC VEGETATION AT RED ROCK LAKES NWR
David G. Paullin, Edward O. Garton, and Ruth E. Shea

Summary and Conclusions

Although quantitative surveys of the aquatic vegetation at RRLNWR have been conducted since 1956, the data do not allow determination of annual changes in species biomass or relative abundance. Use of the survey data is limited by the very large variance of the annual measurements, which resulted from inadequacies of the sampling techniques. However, after examining the long- and short-term changes in macrophyte communities documented in the literature and comparing them with the data in Appendix XV, generalizations can be made that allow some predictive capability. Much of the earlier discussions on individual species has been summarized in Table 29.

Davis and Brinson (1980:47) provide an excellent conceptual model that summarizes how environmental factors affect submersed macrophytes. The three main forces are light attenuation (suspended sediments, eutrophication), toxicity (herbicides), and biomass removal (waves, currents, grazing, and sedimentation). The authors summarized various types of environmental alterations and how these perturbations affect submersed macrophytes. We have excerpted those alterations most likely to be found in the Centennial Valley and have summarized them in Table 30.

The distribution and abundance of aquatic macrophytes in nature reflect the totality of the environmental factors acting upon them. Numerous studies have shown dramatic changes in total biomass and species composition from one year to the next. The situation at Red Rock Lakes is no exception. Annual variations can be caused by cloud cover, nutrient runoff, water temperatures and water depth among others. There are many examples in the literature of waters that once supported luxuriant growths of macrophytes that are now devoid of these plants. The causes for decreases in abundance vary, but most are associated with increased turbidity (Davis and Brinson 1980).

Because turbidity is such an important limiting factor, Davis and Brinson developed a turbidity tolerance index for several species, many of which occur at RRLNWR. The index has been included here in its entirety (Table 31) because it provides much insight into the sensitivities of many species found on the Refuge. Simply stated, species with higher turbidity tolerance indices are better adapted for survival under conditions of low light transmission.

Also, to summarize the long-term changes in submersed macrophyte communities, Davis and Brinson (1980) developed a species survival index (Table 31) which is simply the ratio of the number of lakes in which a species was reported in earlier surveys to the number of lakes in which the species was present when last studied.

The capacity to predict changes in macrophyte communities resulting from natural or human-induced environmental alterations is low. However, Table 29 and Table 30 can be used in combination to make some predictions relative to disturbances and changes in plant abundance or distribution. For example, Table 30 predicts that stream channelization (e.g. Upper Red Rock Creek) increases suspended sediments and eutrophication. Table 29 predicts that *Ceratophyllum demersum*, *P. pectinatus*, *P. richardsonii*, and *Zannichellia palustris* would be the most tolerant of such an alteration whereas *P. praelongus* and the other "northern pondweeds" would be least tolerant. These tables can be used in various combinations to predict species responses to environmental alterations.

Despite the shortcomings of the data in Appendix XV from a statistical standpoint, some major trends are evident, including the decline of *Elodea* on both Upper and Lower lakes, and a refuge-wide increase in species diversity.

The increased turbidity of Upper Lake in the late 1950s and 1960s was detrimental to the luxuriant beds of *Elodea* that dominated the lake in the mid-1950s. *Elodea* is an indicator of low turbidity, and is especially sensitive to suspended sediment early in the growing season.

Water temperatures at RRLNWR are suboptimal for the growth of most species of aquatic plants, particularly during the period of spring runoff. As we discussed in Chapter 6 water depths in Lower Lake were increased by at least 0.3 m (1 ft) by the reconstruction of the lower water control structure in 1957. The increased water depths would have decreased plant productivity by further reducing water temperatures, and also reducing the amount of available light during episodes of high turbidity. Water depths affect the distribution, zonation, and abundance of aquatic plants at RRLNWR, with the maximum species diversity occurring at depths of 0.6-0.9 m (2-3 ft).

Elodea appears to be a pioneer species at RRLNWR; it will invade newly flooded or disturbed sites given the right water and soil conditions. Over time, species diversity increases and overall biomass production may decrease. Heavy waterfowl grazing, as observed at Culver Pond, may maintain *Elodea* in a seral state for over 20 years.

With the exception of Culver Pond, no evidence has been found that indicates that waterfowl grazing at RRLNWR has altered the composition of the aquatic plant communities. The changes which have occurred over the last 30 years are very poorly quantified, and are likely the result of variations in light regime, siltation, wave action, water depths, and water temperatures.

The flora of RRLNWR indicates that the marsh is still very healthy and water quality is near "pristine" conditions as evidenced by the presence of many of the "northern *Potamogetons*". Despite alterations to the watershed caused by mining, grazing, channelization, and installation of the lower water control structure, the Refuge lakes remain healthy, diverse, and productive. Perhaps the most limiting factor on submerged macrophytes is the harsh environment of the Centennial Valley, (including long winters and short, cool, growing seasons, cold waters, ice scouring, wave turbulence, and predominately oligotrophic waters), all of which are suboptimal to maximum standing crop production.

Management Recommendations

1. Much is known about the abundance and distribution of submersed macrophytes at RRLNWR. Unfortunately the key piece of information still lacking is an adequate food habits study for Trumpeter Swans. This should be high on the Refuge's list of priorities.

2. Turbidity and lake levels are critical long-term data needs. Permanent staff gauges should be installed in each lake and pond, and water levels measured at least bi-monthly. The severe ice conditions will probably necessitate annual maintenance to keep gauges functional. Turbidity can be read at the gauging stations. Secchi disc transparencies are helpful but at RRLNWR where visibility to the lake bottom is common, such readings have limited value. Turbidity measurements using a Hach kit or similar equipment are preferable.

3. Consideration should be given to the establishment of permanent plots or transects to periodically sample macrophytes in each lake and pond. Without some known benchmark for reference, year-to-year comparisons are difficult at best, and statistically impossible.

4. Consideration should be given to testing remote sensing (e.g. low level infrared photography as a means of monitoring plant abundance and distribution. Given the clear waters of the Refuge, on a calm day this method may be helpful in determining plant distribution, coverage, zonation, etc.

5. The aquatic plant surveys as they now are conducted will consistently underestimate total biomass of those species that exhibit late season growth, e.g. *C. demersum*, *N. flexilis*, and *M. spicatum*. The accurate estimation of total biomass for all species would require two surveys. This may be worth doing at least once in order to gain insight into the amount of late season production that occurs in the marsh.

CHAPTER 8. WINTER DISTRIBUTION AND HABITATS

Summary

The only known wintering area for the RMP is the Tri-state region where the run-off from geothermal activity creates ice-free winter habitat. Recent sightings of marked trumpeters indicate that some swans from both the rapidly increasing Canadian flocks and the Tri-state Subpopulation are exploring new wintering areas, however no regular use of other areas has yet been documented.

Prior to the impoundment of Culver Springs, little or no dependable ice-free habitat existed in the Centennial Valley and most swans left the Valley at freeze-up. This pattern of movement, however, was deliberately disrupted by the feeding of grain at or before freeze-up. Grain feeding has allowed swans to winter successfully in the Centennial Valley, at least since the mid-1930s. Feeding rates, and the timing and methods of feeding have varied widely and abruptly, effectively varying the winter carrying capacity of the Refuge. The rate of loss of swans from the Centennial flock was highest in years when less grain was fed. In the early 1980s, the amount of grain fed per swan was the lowest in the Refuge's history and was a major factor in the recent decline of the Centennial Valley flock.

The single most important wintering area for RMP trumpeters is the Henry's Fork River below Island Park Dam, at Harriman State Park. This area provides both excellent supplies of aquatic vegetation and protection from waterfowl hunting and other human disturbance. Canadian trumpeters make only limited use of the RRLNWR feeding ponds and are very dependent on the available open water in the Henry's Fork particularly when severe weather freezes the Teton River. Prior to 1968, winter water flows at Island Park Dam were often abruptly and almost completely curtailed, causing substantial portions of the wintering habitat to freeze for weeks at a time. A change in water management at Island Park Dam in 1967-68 provided increased winter flows and was followed by an increase in the number of wintering swans downstream at Harriman State Park. Current management of the River provides no guarantees that sufficient water will be released to prevent the freezing of the Harriman State Park wintering areas in years when water storage supplies are low.

Observations of the composition of the aquatic plant community at Harriman State Park suggest that preferred swan food species have declined in the last decade. Although extensive beds of vegetation remain after the wintering trumpeters leave the Henry's Fork, the carrying capacity of the HSP habitat may have declined. Changes in the plant community could be due to a variety of environmental factors as described in Chapter 7, as well as the effects of herbivory. A better understanding of the ecological relationships on the Henry's Fork is urgently needed, particularly to clarify relationships between sediment deposition, aquatic plant communities, herbivory by swans, and the fishery.

Trumpeters presently appear to have filled the available winter habitat in the lower elevations of Wyoming. Studies currently in progress in Wyoming have described the key characteristics of winter habitat and identified potential areas where winter habitat could be developed. Wintering trumpeters are vulnerable to human disturbance and the eventual development and loss of winter habitat outside Harriman State Park, Red Rock Lakes and Yellowstone NP.

CHAPTER 9. MOVEMENTS OF MARKED TRUMPETERS
David C. Lockman and Ruth E. Shea

Summary

Studies of marked trumpeters of known sex and age have recently produced much information regarding seasonal movements and associations, and clarified similarities between the Grande Prairie and Tri-state trumpeters. Like many other waterfowl species (Hochbaum 1955), trumpeters learn from the behaviors of their experienced companions. The objects, places, or actions thus learned become traditional. With pair bonds that may endure for life and family bonds that endure for several years, family traditions are strong.

Wyoming trumpeters usually remained within the Green River and Snake River drainages of northwestern Wyoming. Movement beyond these drainages included a pre-molt dispersal by yearlings, dispersal after late winter/spring courtship and pair bonding by subadults, and spring foraging in Teton Basin, ID, by one territorial pair. Subadult and yearling dispersal occurred in late spring, and was aided by the territorial defense and aggression of breeding pairs.

Family groups were more sedentary than all other age classes, both in winter and summer. Experienced breeding pairs were the most consistent and traditional in their use of sites and flight paths. Breeding pairs with cygnets segregated from other wintering swans on the National Elk Refuge, and appeared to defend their traditional feeding sites. Breeding pairs were sometimes joined by their yearling and subadult offspring during the winter and permitted them to feed in close association.

Subadults remained on the National Elk Refuge, Wyoming, in April and May. Courtship activity was intense during this period, and pair-bond formation was observed. Breeding pairs left the wintering grounds by about 20 March, and moved to traditional spring feeding sites within 10 km of their territory. Pairs made occasional flights to their territories but did not begin constant territorial occupancy until their lake was at least partially ice-free. The parent-cygnet bond weakened in late winter and early spring. While some cygnets remained on the Refuge when their parents departed in March, others moved to the spring feeding sites with their parents. One brood of cygnets was observed on the nesting territory prior to incubation.

After remaining on their territory throughout the summer, successful breeding pairs occasionally led their newly fledged young to feed in the same sites which the pair had used in the spring. Almost all Wyoming trumpeters staged at the National Elk Refuge annually, between 15 October and 15 November. Nonproductive adults and subadults arrived first, followed by family groups. Swan movements to and from seasonal ranges followed major river courses, and movements over hydrographic divides occurred through natural passes, with gradual relief.

Swans sometimes depleted the available feed at wintering sites. Lack of feed, or its reduced availability due to ice formation during extended cold periods, caused swans to increase their movements to alternate, less preferred feeding sites.

Grande Prairie trumpeters also began to arrive at the wintering areas about 15 October and like the Wyoming trumpeters, nonbreeders were the first to arrive, followed within two to three weeks by family groups. Grande Prairie trumpeters were also quite traditional in their annual movement patterns. Pairs regularly brought their broods to the same wintering sites each year and occasionally associated with their yearling and subadult offspring.

Yellowstone National Park provided important fall habitat for hundreds of migrant trumpeters. Harriman State Park was the most important wintering area and also provided fall habitat for Grande Prairie swans. Heavy use of these two areas, which are closed to waterfowl hunting, protected a large portion of the Canadian trumpeters during the waterfowl hunting season. Trumpeters moved within the Tri-state Area during the winter. Movements were probably influenced by varying weather and ice conditions. Because swans from all the known breeding flocks share the same wintering sites, and movement of swans between these sites occurs throughout the winter, the RMP trumpeters remain quite vulnerable to heavy losses should a disease outbreak occur among wintering waterfowl in the Tri-state Area.

Grande Prairie trumpeters usually left the Tri-state Area about 1-15 March. Earlier departure was observed during unusually mild winters. After a slow migration through southwest Alberta, the breeding pairs arrive in Grande Prairie about 20-25 April. Nonbreeders, including yearlings, arrived by the end of April.

Studies of marked Red Rock Lakes trumpeters have shown movement out of the Centennial Valley by nonbreeders, beginning in August and September. Some family groups also moved out of the Valley in October and November. Swans movements ranged within an 80 km (50 mi) radius of the Refuge. Marked swans moved back to the Refuge in January and February, with a noticeable influx occurring in March. In April the swans dispersed from the feeding ponds, and many nonbreeders moved west along the Red Rock River to Lima Reservoir. Other nonbreeders moved out of the Centennial Valley to Hebgen Lake, the Henry's Fork River, and other nearby sites.

Although most marked Grande Prairie and Tri-state trumpeters wintered in the Tri-state Area, a few swans migrated further south. A marked swan and its mate from Red Rock Lakes wintered near Grande Junction, Colorado, in 1984-85, and another marked Refuge trumpeter was shot at Great Salt Lake in November 1986. In recent years, marked Grande Prairie swans have been seen in Utah and Colorado. Swans marked in the Northwest Territories in 1986 were found during the winter of 1986-87 in northern Idaho, along the Oregon-California border, and in southern Nevada, as well as in the Tri-state Area. If the current increase of the Interior Canada

Subpopulation continues, further exploration of new migration routes and wintering areas is likely. To date, however, this exploration has been primarily by subadults, and mortality has been high. No regular use of any wintering area outside the Tri-state Area has yet been documented.

Observations of marked swans so far indicate that there is little overlap between the ranges of the Pacific Coast Population and the Rocky Mountain Population, although the summer range of some 500 trumpeters that winter in the Tri-state Area is still unknown. Even though the Tri-state and Interior Canada trumpeters share a common wintering area, observations of some 1,300 marked swans have not yet produced any evidence of pair-bonding between members of the two groups. With the lack of evidence to the contrary, there is a real possibility that the subpopulations are reproductively isolated.

CHAPTER 10. BIOENERGETICS AND FOOD HABITS
J. Bradley Bortner and Ruth E. Shea

Summary

Studies of Tundra Swans and arctic nesting geese have shown that weight is lost throughout the winter in an endogenous rhythm and that the energy needed for successful reproduction is gained during spring migration or at the nesting area. Tundra Swans wintering in North Carolina lost up to 25% of their body weight despite abundant food supplies. We conclude that trumpeters likely share a similar annual cycle of nutrient reserve and body weight changes.

We also conclude that use of the Tri-state Area by wintering trumpeters has increased due to the elimination of other migratory traditions, and the "short stopping" of swans by grain feeding and the creation of ice-free habitat downstream from dams. Trumpeters wintering in the Tri-state region endure a long, cold winter with subzero temperatures. The trumpeters' normal endogenous rhythm of winter weight loss, coupled with the marginal climatic conditions of the Tri-state region for wintering and breeding swans, have combined to depress the productivity of nonmigrants. In contrast, the Canadian migrants which leave the Tri-state Area have a greater opportunity to accumulate reserves necessary for breeding.

Few details are known about trumpeter food preferences or the nutritional value of the various food items. Captive trumpeters consumed up to 9 kg (20 lbs) per day of aquatic vegetation. Invertebrates may also be an important part of their diet, particularly during the first weeks of cygnet development.

In summary, numerous observations during the past 20 years support the theory that prenesting food resources are key to successful reproduction by trumpeters. Although the migrant Canadian trumpeters share the same wintering areas with the Tri-state residents, they have access to more varied and milder prenesting habitats as they migrate northward through lower elevations, while the Tri-state swans remain locked in winter. We conclude that these differences in access to transitional spring habitats are a major factor contributing to the superior reproductive performance of the Canadian migrants.

CHAPTER 11. GENETIC CONSIDERATIONS IN TRUMPETER SWAN MANAGEMENT
Peter F. Brussard and Ruth E. Shea

Summary and Management Implications

Judging from considerations of N_e/N ratios in other birds and mammals, there is a good chance that the Centennial Valley flock is suffering from the effects of inbreeding and loss of genetic variation in addition to its many other problems. The genetic effects can be counteracted by implementing one-way gene flow with other trumpeter stocks, preferably through transplants from the relatively genetically diverse Pacific Coast Population. However, there is some chance that this might result in an outbreeding depression in the F1 or F2 generations after the transplants, if the Red Rock Lakes and Alaskan trumpeters represent differently coadapted stocks. The evidence, albeit sketchy, suggests that this is unlikely.

Nevertheless, a higher level of certainty of the degree of differentiation between the extant trumpeter stocks should be obtained prior to making such transplants. Thus, we recommend that a second electrophoretic and morphological study be conducted so that a more adequate assessment of population boundaries can be made. Additional electrophoresis is recommended so that more polymorphic loci can be found. An element of chance exists in such a search; although researchers now routinely resolve at least twice as many loci as Barrett and Vyse (1982) did, the larger samples do not guarantee that any more of these loci will be polymorphic. Thus, we suggest that the first step in a new study should be to survey 40-50 loci in a sample of 50-100 trumpeters, collected from the Pacific Coast Population only. If variation exists in the species, it will be found there. Blood samples, fertile eggs, or freshly dead swans would provide adequate material. If the survey proves successful and additional useful loci are discovered, then the study can be expanded to include samples from the Interior Canada and Tri-state subpopulations. On the other hand, if no more useful loci are found, it may be necessary to consider looking at mitochondrial or ribosomal DNA, a much slower and more expensive procedure.

Additional morphological measurements involving many additional characters from adult birds (e.g., length of fifth primary, forearm, neck, tarsus, bill nail, width of bill and foot web, etc.) should also be taken and analyzed by multivariate techniques. Significant variation between populations and strong concordance between the electrophoretic and morphological data sets would suggest genetic differentiation. Lack of concordance would tend to confirm the results of the earlier studies.

As a final caution, however, we must emphasize that electrophoretic and other molecular techniques assay only a very small proportion of the loci in the swan genome. Furthermore, these loci probably have little, if anything, to do with population fitness. Likewise, one cannot tell how much morphological variation is genetic and how much is environmentally induced without heritability studies. Although reasonable correlations exist between genetic similarity as measured by electrophoresis or DNA analysis, patterns of morphological variation, and overall genetic similarity, absolute proof of potential outbreeding depression can only come from controlled matings between individuals from the populations of interest. Such experiments would be a very tedious and time consuming enterprise in a bird with such a long generation time. However, molecular techniques combined with morphological data can provide an adequate level of precision with which to make informed judgments, and such a data base would provide valuable guidance on how best to incorporate genetic structure into management plans.

CHAPTER 12. POPULATION-HABITAT RELATIONSHIPS

Summary

Banko (1960) concluded that cygnet production in the Tri-state Area was inversely related to population numbers and was depressed at high population levels possibly by mechanisms involving social competition. He also concluded that breeding habitat limitations prevented further growth of the Subpopulation and caused the proportion of nonbreeders to increase because eligible breeding-age swans were prevented from obtaining suitable territories.

Page (1976) suggested that artificially high numbers of swans were maintained on the Refuge after 1954 by supplemental winter feeding. He suspected that overgrazing by swans contributed to the decline of *Elodea* and reduced the carrying capacity of the Refuge. Page suggested that removals were beneficial because they helped to reduce a flock which was damaging its food resources.

Page used a deterministic model of the RRLNWR flock to show that small increases in adult mortality caused a severe decline of the flock. Also, he showed that if a limit were placed on the number of nesting pairs in a simulated population, then the swans would increase and stabilize at an upper maximum that was determined by adult survival rates, and surplus eligible nonbreeders would result.

We found that no shortage of breeding habitat currently exists in the Tri-state Area, except possibly in the lower elevations of Wyoming. We also conclude that breeding habitat was not limiting in the 1950s, because since that time trumpeters have pioneered at least 20 new territories in the Tri-state Area.

Our analysis showed that mortality rates of Centennial Valley trumpeters were not density-dependent, but were significantly correlated with the amount of grain fed per swan at RRLNWR and the reduction in water flows at Harriman State Park due to cutoffs of water released at Island Park Dam. Abrupt variations in the availability of the swans' two key winter food sources have effectively varied the winter carrying capacity of the Tri-state Area and exerted a strong influence on annual mortality rates.

The molting flocks increased in the early 1950s due to several years of high cygnet production and increased overwinter survival made possible by the unusual availability of winter habitat at Harriman State Park in 1950-53. Flocked subadults were also joined in the molt by older failed breeders. The existence of a large surplus of eligible breeding age swans that were assumed to be nonproductive due to a shortage of breeding habitat was never documented.

The leveling off of the growth of the Tri-state Subpopulation after 1954 was caused both by depressed cygnet production, removals, and increased overwinter mortality, due in part to periodic total curtailment of winter water releases from Island Park Dam. Mortality rates also increased as a result of the reduction in the amount of grain fed per bird at RRLNWR after the late 1960s. By the early 1980s, the amount of grain fed per swan was the lowest in the history of the Refuge and likely was a major factor in the flock's continued decline.

Refuge cygnet production was reduced after about 1950 by the combined influence of several factors including: increased human disturbance, the lower productivity of young inexperienced breeders, increased frequency of adverse weather, and increased water levels. The inverse relationships observed between cygnet production and population levels may be due in large part to factors inherent in the shifting age structure and age-specific fecundity of a long-lived species with several nonproductive subadult age classes.

The number of nonmigratory trumpeters in the Tri-state Area has been maintained at an artificially high level by supplemental feeding. However, the productivity of these birds is limited by their ability to obtain adequate spring foods in order to store sufficient energy reserves for successful reproduction. On the other hand, the Tri-state Area could probably support many more breeding trumpeters if they were migrants that imported the energy stores needed for reproduction from more southerly winter and spring habitats.

Nesting swans at RRLNWR and YNP are not fledging enough cygnets/nest attempt to replace themselves and we predict the further decline of these flocks unless productivity or adult survival increases. Yellowstone is now, and probably for many years has been, dependent upon the immigration of Centennial Valley swans to maintain its breeding flock. With the decline of the Centennial Valley flock, the Yellowstone flock has virtually collapsed. At the survival rates estimated from banding data, RRLNWR swans must fledge about one cygnet/nest attempt in order to maintain a stable nesting flock; currently they are only fledging about 0.7 cygnets per nest attempt.

Idaho, lower elevation Wyoming, and the Grande Prairie flocks are producing cygnets at rates that will allow flock expansion if post-fledging survival rates do not decrease.

The Grande Prairie flock was thought to be limited by breeding habitat in the 1950s and 1960s because the flock failed to increase despite high cygnet production. Subsequent analysis by Turner and Mackay (1981) indicated that winter habitat rather than breeding habitat was limiting. Holton (1982) also concluded that breeding habitat did not limit the Grande Prairie flock and suggested that dispersal of swans into expanding flocks in Alberta, B.C., Yukon and NWT may be confounding the estimates of winter mortality.

We found that the mortality rates of Centennial Valley trumpeters, which made regular use of the Henry's Fork and the supplemental grain, were closely linked to the availability of those two food sources but were only slightly influenced by winter severity. In contrast, the mortality rates of the Grande Prairie trumpeters were very highly correlated with winter severity in the Tri-state Area, probably reflecting their dependence upon the availability of natural open water habitats, rather than the Refuge grain. We conclude that the restoration of higher, regular water flows in the Henry's Fork after 1968 increased the availability of high quality winter habitat and resulted in higher winter survival of Grande Prairie trumpeters. We suspect that this increase in winter habitat was a key factor in the expansion and dispersal of the Grande Prairie flock in the 1970s.

CHAPTER 13. SUMMARY

Over the past sixty years, observations of Trumpeter Swans in the wild and in captivity have resulted in the accumulation of a wealth of information about the species. The population and habitat data gathered on the Rocky Mountain Population form a remarkable documentary of the problems and responses of a population as it recovered from near extinction. The biological data only make sense, however, when viewed within an historical perspective. We cannot understand the Rocky Mountain Trumpeter Swans today, without also understanding how their habitats and patterns of habitat use have changed, and why events that happened decades ago still affect the Population.

The task of reconstructing the patterns of habitat use prior to the species' decline is hindered by the lack of historic data. Trumpeters were eliminated from many areas before naturalists were able to identify the species and systematically record their occurrence. Where trumpeters remained, territorial pairs were widely dispersed and difficult to observe from the ground. In the absence of historical data, biologists can either leave important questions unanswered, or they can take what little historic information exists from other locations, and then examine the current behavior and biology of Trumpeter Swans in order to deduce how trumpeters probably functioned prior to their disruption by man. We chose the latter approach. The ideas expressed in the following summary represent a synthesis of information from many sources. Throughout this summary, we will refer to the chapters in which the specific ideas were developed. The reader should refer to the detailed discussions and citations in the preceding chapters in order to trace the origins of each idea.

Prior to the near elimination of the species in the 1700s and 1800s, trumpeters nested widely across most of Canada and the United States. They migrated south to winter on ice-free freshwater habitats scattered across the United States, and to many coastal bays and estuaries (Chapter 1). Then, as now, these long-lived birds likely were highly dependent upon prolonged family group associations and the accumulated experience of family members. Breeding pairs developed traditional annual movement patterns and passed the knowledge of their specific habitats and migration routes to their offspring. Thus, family traditions were passed from one generation to the next. Also as occurs today, dispersal and exploration of new habitat probably occurred mainly among subadults (Chapter 9).

Trumpeter Swans most likely maintained a continuous presence in the Tri-state Area throughout the period of decline, but no data exist to document whether most of the Tri-state trumpeters historically were migrants or year-round residents (Chapter 1). Today, however, both migrant and nonmigrant trumpeters occupy the region. Since the 1930s, observations have revealed that although the majority of trumpeters that summer in the Tri-state Area are nonmigratory, some individuals occasionally migrate to more southerly wintering areas. No evidence of regular migration by Tri-state summer residents has yet been detected however (Appendix I, Chapter 9). Canadian trumpeters, which summer in Alberta, British Columbia, the Yukon, the Northwest Territories, and Saskatchewan have been observed in recent decades to winter primarily in the Tri-state Area. Within the last fifteen years, however, their numbers have grown rapidly and the Canadian trumpeters have shown an increasing tendency to migrate to a variety of other more southerly wintering sites. Spring and fall migrants annually move through the Coeur d'Alene area of northern Idaho, but the winter range and numbers of these migrants are unknown.

The number of trumpeters from both Canada and the Tri-state Area that migrate south of the Tri-state Area probably is higher than records indicate. Most observations of trumpeters from more southerly wintering areas have been reported only because the individuals were marked. Few trumpeters presently are marked and no efforts have been made to search for trumpeters wintering alone or among tundra swans outside of the Tri-state Area. As a result, migrating trumpeters are likely to winter unnoticed due to the inability of most observers to distinguish between tundra swans and Trumpeter Swans (Appendix I, Chapter 9).

The variation shown by the movement patterns of the Rocky Mountain Trumpeter Swans today suggests that the trumpeter families that used the Tri-state Area in pristine times would also have followed a similar variety of movement patterns. As they do today, some families probably wintered in the Tri-state Area and migrated north to summer habitat in Montana and Canada. Other northern families likely passed through the Tri-state Area during migration, possibly spending several weeks at Yellowstone Lake as they do today, before continuing their migration to more southerly wintering sites. Some families probably nested in the Tri-state Area and wintered at a variety of more southerly locations. Still other families remained in the Tri-state Area year-round, finding scattered sites where geothermal activity provided small areas of ice-free winter habitat (Chapter 8).

By the early 1900s, Trumpeter Swans had been eliminated from most of their historic range outside of Alaska. The last survivors in the Rocky Mountains were also reduced to near extinction. Swans were slaughtered when their traditional migration patterns exposed them to the guns of the settlers and Indians, and they were subjected to some 125 years of commercial exploitation by the Hudson's Bay Company on their Canadian breeding grounds. Outside of Alaska and British Columbia, the only trumpeters known to have survived the decline were those that wintered in the Tri-state region. Truly unique on the continent, this remote geothermal area remained virtually unexplored until the 1870s. Despite the region's high elevation and severe winter weather, its isolation from human settlement and the availability of ice-free habitat, created by the runoff from warm springs, provided the last winter refuge for the trumpeters of both the United States and Canada (Chapter 1).

In the 1930s, the first summer surveys of the Tri-state Area found only about 50 adults and their cygnets (Chapter 2). Apparently these few survivors were mostly nonmigrants, whose traditional movements had kept them isolated in the most remote portions of Yellowstone and the Centennial Valley where human-caused mortality was very low. We suspect that trumpeters that nested in the Tri-state Area and migrated south to winter in less remote areas were more vulnerable to human-caused mortality than the nonmigrants and were virtually eliminated (Chapter 1).

Almost all of the trumpeters that nested across the interior of Canada were also eliminated. Only one small flock was known to survive in the vicinity of Grande Prairie, Alberta. When the Grande Prairie flock was first censused in 1946, only 77 adults and their broods were found (Chapter 2). Marking studies in the 1950s revealed that these Canadian survivors also wintered almost exclusively in the Tri-state Area. Like the trumpeters of the United States, any interior Canadian trumpeters that wintered in less remote locations were gradually eliminated (Chapter 9).

The descendants of those two remnant breeding groups now comprise the Tri-state Subpopulation and the Interior Canada Subpopulation; together they are known as the Rocky Mountain Population. Though they still continue to share the same Tri-state wintering range, marking studies have not yet provided evidence that these two groups interbreed. The data gathered so far indicate that these two groups are reproductively isolated and should be regarded as separate populations until evidence to the contrary is found.

The near extinction of the Trumpeter Swans probably caused at least two lasting impacts on the Rocky Mountain Population. First, genetic variability most likely was lost as the number of swans declined to somewhat less than 150 adults. Loss of genetic variation was even more likely to have occurred because trumpeters normally spend much of the year in parent/cygnet or subadult sibling associations, and thus many of the survivors were likely to have been closely related. Although the effects might not be obvious, a loss of genetic variability could render their descendants somewhat less fit to cope with environmental stresses and increase the likelihood of defects due to inbreeding (Chapter 11).

Second, as the vast majority of trumpeters were eliminated most of the Population's cumulative knowledge of traditional migration routes to other winter and spring habitats was also destroyed. The few surviving families retained only two basic annual movement patterns to pass on to their offspring: they either remained in the Tri-state Area year-round, or wintered there and migrated to Canada to nest (Chapter 1). **This loss of knowledge of other migration routes and alternate winter and spring habitats is the underlying cause of several of the problems now facing the Rocky Mountain trumpeters.**

Although management actions were able to increase the numbers of trumpeters, the migratory traditions that were destroyed during the species' decline were not restored. Therefore, as the remnant was protected and their offspring increased, growing numbers of nonmigrants and wintering swans became dependent upon the Tri-state habitat. Managers were able to increase the numbers of nonmigrants and wintering Canadian trumpeters to artificially high levels due to the influence of two factors:

1) the winter carrying capacity of the Tri-state habitat was substantially increased by supplemental feeding which began at Red Rock Lakes NWR in 1935. The feeding program was also intentionally used to hold trumpeters on the Refuge and discourage their traditional fall movements out of the Centennial Valley (Chapter 8).

2) since the 1930s the amount of available winter habitat has also been increased in various locations within the Tri-state region due to man-made warm spring impoundments and the construction of dams. Water stored in reservoirs retains significant amounts of heat and its release during the winter months creates ice-free river habitat downstream. Ice cover in the Harriman State Park area on the Henry's Fork is substantially less under the present water flow regime than it would have been prior to the construction of the Island Park Dam in 1938 (Chapter 8).

One consequence of the loss of traditional movement to more southerly winter and spring habitats has been that the increased numbers of nonmigratory Tri-state trumpeters have access to only very marginal, high elevation spring feeding areas. Recent studies have shown that the reproductive success of other species of swans and geese is directly influenced by the amount of stored energy reserves that the breeding female brings to egg laying and incubation. Energy reserves are catabolized during the winter months, and a female must therefore accumulate sufficient reserves in the weeks prior to egg laying to adequately provision a clutch and to sustain herself during the incubation period. Thus, the quality and quantity of available spring foods are critical to reproductive success (Chapter 10).

The Canadian trumpeters usually depart from the high plateaus of the Tri-state wintering area early in March, and migrate slowly through the lower elevations of Montana and southern Alberta, where they have access to a wide variety of ice-free habitats (Chapter 9). Upon their arrival in Grande Prairie in late April, migrant trumpeters have apparently stored energy reserves sufficient to provision clutches containing an average of 5.6 eggs and to incubate with constancies exceeding 94%. Their average clutch size and constancy are only slightly lower than those of well-fed captive trumpeters (Chapter 4). Their high rate of cygnet productivity has resulted in the rapid expansion of the migrants' numbers and summer distribution (Chapter 12).

In contrast, the Tri-state trumpeters usually face several weeks of winter weather after the migrants depart for lower elevations. Often spring thaw in the higher elevations of the Tri-state Area does not begin until early April. In some years, the lakes at Red Rock Lakes NWR have not thawed until late May (Chapter 6). Access to spring food sources is highly dependent upon spring weather conditions and ice cover. Cold, late springs or unusually high spring runoff delay the development of aquatic plants and invertebrates and reduce their availability, both to the prebreeding adults and to the young cygnets in the critical first weeks of life (Chapter 5, 7). The relatively low availability of spring foods in most years in the Tri-state Area likely limits the amount of energy reserves that the swans can accumulate prior to egg laying in late April or early May and thus reduces their productivity (Chapter 10). Likely as a result of low energy reserves, the mean clutch sizes of the several nonmigratory flocks ranged only from 3.6 to 4.9 eggs (Appendix IX). The Tri-state females incubated with lower constancy and spent more time feeding compared both to the Grande Prairie migrants and to captive trumpeters (Chapter 4).

Several other examples of the relationships between productivity and spring habitat conditions have been shown. Clutch size at Red Rock Lakes NWR was positively correlated with increasing May temperatures ($P = 0.08$) and negatively correlated with water levels during the entire spring and early summer period ($P = 0.08$). Cygnet survival was positively correlated with mean minimum temperatures in June ($P = 0.07$), and negatively correlated with July water levels ($P = 0.06$). The number of cygnets alive in late summer was significantly lower in years when the peak of hatch was delayed ($P = 0.01$); a late peak of hatch was normally associated with adverse spring weather conditions (Chapter 3). Also coinciding with the relative timing of spring thaw and the availability of spring foods, clutch size, hatchability, and cygnet survival are lower in the high elevation Red Rock Lakes and Yellowstone flocks than in the lower elevation Idaho, Wyoming, or Grande Prairie flocks (Appendix IX, Chapter 6).

Another consequence of the population's loss of traditional knowledge of other wintering areas has been its continued high fidelity to the Tri-state wintering area, and thus the dependence of increasing numbers of trumpeters on this single wintering area (Chapter 8). Although the tendency of subadults to explore new wintering habitat is becoming more apparent as the Rocky Mountain Population increases in numbers, over 95% of the trumpeters marked in the Canadian and Tri-state flocks have wintered in the Tri-state Area (Chapter 9). The 1,600 trumpeters that wintered in the Tri-state region in 1986 represented Tri-state residents and all the known Canadian breeding flocks, as well as approximately 500 swans from unknown summer ranges (Chapter 2, Chapter 9). Despite the fact that the main concentrations of swans are scattered between about 8-10 key sites, throughout the winter some trumpeters regularly move from one site to another. If a disease outbreak should occur in the Tri-state region, it could spread rapidly between the wintering sites and be disastrous to the recovery of the Trumpeter Swan both in Canada and the U.S. (Chapter 9).

Very little is known about the relationship between the number of wintering swans and the carrying capacity of the Tri-state winter habitat. Studies of the Wyoming wintering sites suggest that they are presently at their carrying capacity, and this may be the reason that the number of swans wintering in the Jackson area has stabilized. Preliminary data suggest that the carrying capacity of the Henry's Fork has not yet been exceeded, however the vegetation/swan relationships have not been adequately investigated and the potential for the wintering swans to alter aquatic plant communities in the Henry's Fork is unknown.

The future of key wintering habitats is not secure. Private lands along the Teton River in Idaho will likely be developed in the near future and human disturbance is likely to increase in most habitats. The water flows that are essential to maintaining ice-free habitat at Harriman State Park are not guaranteed. Although the Bureau of Reclamation attempts to provide adequate flows when water is plentiful, releases during years of low water storage will likely be insufficient to prevent freezing unless changes in water management are made (Chapter 8).

The artificially high numbers of swans that winter in the Tri-state region must endure some of the most severe winter weather in the lower 48 states, with temperatures regularly declining to -40°C. Winter conditions can last from November until April and cause trumpeters to deplete their energy reserves. In addition to increasing the amount of energy that swans must expend in order to survive, the subzero winter temperatures also create ice cover that severely reduces the amount of available food (Chapter 8). The Canadian migrants appear to be particularly vulnerable to severe winter weather. Mortality rates in the Grande Prairie flock showed a highly significant positive correlation with winter severity in the Tri-state region ($P= 0.0003$) (Chapter 5).

The availability of winter foods also appears to have a large influence on the mortality rates of the Centennial Valley trumpeter flock and was an important factor in the flock's lack of increase after the mid-1950s and decrease since the late 1960s (Chapter 5, 12). Although a few marked Centennial Valley swans have been observed at other Tri-state wintering sites, most of these nonmigratory trumpeters appear to depend on the supplemental grain at Red Rock Lakes NWR and aquatic vegetation at Harriman State Park (Chapter 8, 9). Both of these food supplies have varied widely and somewhat erratically over the last 50 years as a direct result of management actions (Chapter 8). The mortality rate of the Centennial Valley swans increased in years when either one of these food resources was in short supply. The amount of grain fed annually per swan at Red Rock Lakes has varied from about 1-7.5 bu, with a general decline in the 1970s and 1980s to the lowest volumes fed, per swan, in the Refuge's history. A significant correlation between hatching success and the amount of grain fed during the previous winter suggests that the decline in the amount of grain fed in recent years also contributed to decreased productivity (Chapter 3). This decrease in winter feeding was a key factor in the recent decline of the Centennial Valley flock and thus the Tri-state Subpopulation (Chapter 5, 12).

The amount of vegetation available to swans on the Henry's Fork has also varied abruptly, due to the drastic reduction of winter water flows at Island Park Dam during most years between 1938-50, and 1954-67. When winter water releases were curtailed, the resulting low flows at Harriman State Park allowed the river to freeze during periods of normal or subnormal winter temperatures (Chapter 8). Annual mortality rates of the Centennial Valley swans were negatively correlated with the amount of grain fed per swan at Red Rock Lakes NWR ($P= 0.016$), and positively correlated with the duration of reduced water flows on the Henry's Fork ($P= 0.009$).

In effect, the wide, unpredictable fluctuations in these two key winter food sources caused the winter carrying capacity of the Tri-state Area to vary considerably over the last 50 years (Chapter 5, 12). The curtailment of water flows on the Henry's Fork for several weeks or months each winter in most years prior to 1968 apparently also caused high mortality among the wintering Grande Prairie trumpeters, and was an important factor in that flock's failure to increase until the mid-1970s (Chapter 8, 12).

In addition to the underlying influences of spring and winter habitat availability on the Population's productivity and mortality rates, several other factors have contributed to the decline of the Tri-state Subpopulation. Among the most important factors influencing productivity have been habitat changes at Red Rock Lakes National Wildlife Refuge. Increased water levels, due to the influence of the lower water control structure, have directly caused the flooding of numerous nests in the 1980s. Higher, more stable water levels have also probably decreased the productivity of the marsh by reducing the periodic aeration of marsh soils, and reducing average water temperatures, particularly during the critical early spring period. Erosion and the deposition of sediment accelerated in the Refuge lakes in the late 1950s and 1960s due to mining activities, heavy grazing, and increased water levels. The 1959 earthquake probably also altered patterns of erosion and deposition (Chapter 6). Although vegetation changes were poorly quantified, *Elodea* declined dramatically on the Refuge since 1956, and its loss as an important spring food may have further reduced the resources available to the nonmigrants. The decline of *Elodea* most likely was due to the increased turbidity of the refuge waters and other habitat changes, rather than overgrazing by swans (Chapter 7).

Human disturbance also increased substantially at Red Rock Lakes as research and management activities expanded. Use of an airboat in the nesting areas after 1952 caused a dramatic increase in the level of disturbance in the marsh. Several researchers reported that the noise from its engine was very disruptive to the swans and use of the airboat has now been drastically reduced. Although long-term habitat changes in other portions of the trumpeters range have not been studied, increased human disturbance has been found to reduce the rate of occupancy of breeding territories in Yellowstone National Park and Grande Prairie, and to interfere with successful reproduction in the lower elevations of Wyoming (Chapter 4).

Several other factors have contributed to increased adult mortality. Losses of swans to collisions with power lines and fences are the leading cause of death in Wyoming, and may increase as further human development occurs in Trumpeter Swan habitat. The loss of trumpeters to lead poisoning has become more apparent in recent years as managers have made a more determined effort to retrieve carcasses and test for tissue lead levels. Of the 34 swans whose cause of death was determined between 1980 -86, 33% died from lead poisoning. Tests of blood lead levels in 32 nonbreeders found that 23 carried detectable levels of lead. Due to the compounding effect of mortality in the many adult age classes, a slight increase in adult mortality rates due to lead poisoning could easily hasten the decline of the Tri-state Subpopulation. Chronic sublethal lead levels could have long-term detrimental effects both on an individual's reproductive success and its ability to withstand other environmental stresses.

The recovery of Trumpeter Swans is at a crossroads. The Interior Canadian trumpeters are increasing rapidly, and large areas of potential breeding habitat still remain vacant and available in Canada. Their recovery is precarious, however, because of their continued dependence upon the Tri-state wintering area and the vulnerability of all the flocks to high mortality if disease should occur on the wintering ground. An outbreak of avian cholera occurred among Snow Geese at Market Lake in March 1987, less than 75 miles from the Henry's Fork, and was a reminder of the potential for disaster. Even if the trumpeters avoid such an unfortunate event, the restoration of the Canadian flocks will be slowed by occasional high mortality in the Tri-state Area during severe winters, as occurred in 1984-85.

Recent observations indicate that Canadian trumpeters are increasing their exploration of other migration routes and wintering areas as their numbers increase. If this natural tendency can be protected and enhanced by management efforts, Canadian trumpeters have the potential to end their dependency on the Tri-state Area and develop a varied network of winter and spring habitats. A key question will be whether alternate wintering traditions can be restored before high mortality occurs in the Tri-state Area.

The continued survival of the Tri-state breeding population is in doubt. There is currently no evidence that these swans interbreed with the Interior Canada trumpeters. Until evidence of matings between the two groups is found, the Tri-state trumpeters should be viewed as a significant breeding population whose continued existence is threatened, and managed as a threatened population. Unless the current downward trends in total numbers, numbers of nesting pairs, and cygnet production are halted, this breeding population will continue to decline.

Without supplemental feeding, a few pairs of nonmigrants will likely persist in the lower elevations of the Tri-state Area, but their numbers would be inadequate to maintain adequate genetic variation for long-term survival (Chapter 11). Their productivity would be very low and highly vulnerable to harsh spring weather. They would rapidly decline if factors such as lead poisoning or human development increased adult mortality rates (Chapter 12). The past decades of management have increased the number of nonmigratory trumpeters kept alive by supplemental feeding, but managers have not begun to restore the essential patterns of habitat use that could allow the Tri-state trumpeters to truly recover and become self-sustaining.

The immediate problem of stabilizing the Tri-state Subpopulation before further losses occur can be solved. The number of nonmigrant trumpeters can probably be increased fairly quickly if the water management problems and winter feeding program at Red Rock Lakes NWR are corrected, and if other factors, such as habitat loss, human disturbance, and lead poisoning are reduced. But these efforts alone are not sufficient. Increasing the number of grain-fed nonmigrants only buys time, and reduces the rate of loss of genetic variation. The end result however, will be an artificially sustained group of trumpeters, perpetually dependent upon winter feeding. These birds will be continue to have low productivity except in unusually mild years, and population growth will continue to be highly vulnerable to factors that increase adult mortality.

As long as the existence of wild breeding Trumpeter Swan populations in the United States south of Canada depends upon supplemental feeding and the birds are unable to survive on their own, the species should not be considered to be recovered. The true recovery of the Trumpeter Swan as a self-sustaining member of our waterfowl community is possible but it will require the active intervention of managers to actively rebuild broken traditions and enhance the birds' own efforts to migrate to other wintering areas. Migration patterns that were disrupted decades ago must be restored and the Tri-state trumpeters must gain access to less severe and more productive winter and spring habitats. Trumpeters have a high reproductive potential and could increase rapidly if they have access to adequate food supplies. If they can regain the use of a secure foundation of varied winter and spring habitats, the trumpeters' vulnerability to catastrophic loss will be also substantially reduced. With increased spring feeding opportunities and increased productivity, the potential is high that migrant trumpeter could once again return to large portions of their historic range in the United States and Canada.

INTRODUCTION

Trumpeter Swans (*Cygnus buccinator*) once ranged across most of the United States and Canada, but overharvest and habitat destruction brought the species to near extinction by the early 1900s. When Trumpeter Swans were discovered nesting in Yellowstone National Park in 1919, managers and scientists feared them to be the last wild survivors of the species. For over thirty years the isolated remnant flock in the Tri-state region of Idaho/Montana/Wyoming and a small breeding flock near Grande Prairie, Alberta were thought to be the only wild trumpeters remaining. Research at Grande Prairie in the mid-1950s revealed that this last surviving Canadian flock migrated south and shared the Tri-state wintering grounds with the nonmigratory Tri-state residents. The survival of a second Trumpeter Swan population, which nests in Alaska and winters along the Pacific Coast, was not widely recognized until the 1950s.

Believing that the extinction of the species was likely, the National Park Service aggressively protected their few nesting trumpeters. Under the leadership of George Melendez Wright they also initiated studies to identify other trumpeter habitats outside of Yellowstone and to understand the species' life history. These studies led to the recognition of the important breeding marshes in the Centennial Valley of Montana and the creation of Red Rock Lakes National Wildlife Refuge in 1935.

During the next two decades, wildlife managers used supplemental winter feeding, the reduction of illegal shooting, predator control, nest site protection and the reduction of human disturbance to reduce swan mortality and increase cygnet production. The Tri-state trumpeters quickly responded to these management efforts and increased in numbers until the mid-1950s. The swans then entered a period marked by fluctuating adult numbers, high cygnet mortality, and widely fluctuating annual cygnet production, followed by a decline both of total numbers and numbers of nesting pairs after 1968 and extremely poor cygnet production in the 1980s.

In 1960 Winston Banko published a monograph which provided a summary of the history, population dynamics, habits and habitats of the Trumpeter Swan in the most comprehensive treatment to date of the species' ecology and management. Banko concluded that the Tri-state Subpopulation had expanded until it filled the limited available breeding habitat and then ceased to grow further as cygnet production was depressed by the high adult numbers. The Tri-state Trumpeter Swans seemed to be a textbook example of a species that was regulated by the carrying capacity of its breeding habitat.

Although the idea of density-dependent depression of productivity was widespread among managers in the 1960s, concern still surfaced over the very high rates of prefledging cygnet mortality in the Tri-state Area. Studies in the 1960s, 1970s, and 1980s found high levels of parasitism, vulnerability to cold wet weather, and general weakness, but could identify no ultimate cause for what seemed to be the cygnets' low resistance to a variety of proximate factors.

In the 1970s, the theory of density-dependent regulation by breeding habitat limitation became less clear: breeding habitat was obviously vacant as total adult numbers and the number of breeding pairs declined but cygnet production failed to increase as some managers had expected it would. The Tri-state Subpopulation seemed to have entered a downward spiral of low productivity and declining numbers. High cygnet mortality and fluctuating production remained the norm, and reached alarming levels in the early 1980s. By 1986, the Tri-state Subpopulation had declined to its lowest level since 1950. The decline obviously centered at

RRLNWR where losses of nests and cygnets in 1980, 1982, and 1984 resulted in the fledging of only 13 cygnets from 96 nest attempts during those three years.

Concurrent with the poor production and decline of the Tri-state Subpopulation in the 1970s and 1980s, the Canadian trumpeters were highly productive and expanding both in numbers and breeding range. Their recent robust growth was all the more puzzling because they appeared to share the same winter habitat as the Tri-state Subpopulation. Also, the rapid increase of the Canadian trumpeters in the mid-1970s was in marked contrast to their lack of increase during the previous 30 years, despite high cygnet production and management protection.

The unproductive state of the Tri-state Subpopulation in the 1970s was seen by some managers to be symptomatic of a decadent flock, possibly heavy with unproductive older swans. Even though swan numbers had declined since the 1950s and a study at Red Rock Lakes in the 1970s found few if any surplus breeders, the removal of trumpeters was proposed. Using trumpeters for transplants and to satisfy the demands of private propagators was seen as a way to make use of "surplus" swans and to stimulate the flock to be more productive.

By 1983, managers recognized that the Tri-state Subpopulation was in a precarious situation after its third year of poor cygnet production in the 1980s, and an alarming decline in the number of nesting pairs. In addition, the growing realization that virtually all of Canada's breeding trumpeters wintered in the Tri-state Area heightened concern that a single disease outbreak during the winter could severely impact both the Canadian and the Tri-state flocks. Concern also increased that the highly productive, rapidly expanding Canadian flocks would soon compete with the Tri-state swans for winter food resources.

Several graduate students examined various facets of the trumpeters' ecology. Page (1976) studied the ecology of trumpeters at RRLNWR, while Paullin (1973) studied the Refuge vegetation and aquatic environment. Shea (1979) studied the year-round ecology of trumpeters in Yellowstone National Park and adjacent parts of Idaho. Maj (1983) continued studies of Idaho trumpeters and their habitat selection on the Targhee National Forest. Hampton (1981) studied nesting in Idaho, and winter behavior and ecology at Harriman State Park and Yellowstone National Park. Holton (1982) studied habitat use and biology of Grande Prairie trumpeters. Also in 1982, D. Lockman of the Wyoming Game and Fish Department began a long-term study of disease, reproductive success, population dynamics and habitat use, using marked known-age and known-sex birds. The staff of RRLNWR regularly banded swans, monitored nests, surveyed aquatic vegetation, and conducted over a decade of movement studies of marked swans. Marking studies were also conducted in Canada.

In addition, a vast amount of miscellaneous data was accumulated by the various management agencies. Despite over 50 years of data collection, however, understanding of the Trumpeter Swans and their management remained fragmented. Each agency knew a considerable amount about the trumpeters within its own jurisdictional boundaries, but most managers viewed only their piece of the picture, despite notable efforts to coordinate regional census efforts. This wide-ranging population of some 1,600 trumpeters was studied and managed by (at least) the Canadian Wildlife Service, three provinces, two territories, the U.S. Fish and Wildlife Service (2 regions), three states, two National Parks, two National Wildlife Refuges, and five National Forests. As a result of this fragmentation, no single manager possessed an overall understanding of the year-round factors affecting these birds as they moved from one jurisdiction to another and as the various flocks interacted across local and international boundaries.

The decline of the Tri-state Subpopulation in the 1980s and the extremely poor productivity of swans at Red Rock Lakes NWR led to our current research effort. Compared to many wildlife species, the Trumpeter Swan is an easy animal to study because it is relatively sedentary, long-lived, very traditional in its actions, and highly visible. We suspected that if wildlife managers could not understand the Rocky Mountain trumpeters after 50 years of intensive management and research, then perhaps our entire approach needed reassessment. It was not readily apparent that another study of any particular aspect would be fruitful, and the U.S. Fish and Wildlife Service agreed that a different approach was needed. We proposed to reexamine and synthesize the data that already had been collected; to question our basic assumptions; to integrate all the information on population dynamics, behavior, environmental variables, management actions, etc., and then to synthesize the data within an historical perspective. We hoped that perhaps the answers had already been found but were not apparent because the data lay scattered around the country in many filing cabinets.

The volume of data that we located and drew together surpassed our expectations. When finally assembled, the information created a uniquely detailed, 50 year record of the Rocky Mountain Trumpeter Swans, as they literally recovered from the brink of extinction. Taken together, the many pieces provide an understanding of the ecology of Rocky Mountain trumpeters and of the factors, both historic and current, that influence their survival today.

ACKNOWLEDGEMENTS

This project to assemble and synthesize the data on the Rocky Mountain Trumpeter Swan Population was a cooperative effort involving many individuals who have worked with these birds over the years. Many individuals unselfishly made their unpublished data available and contributed countless hours of their time in debates and discussions on the facets of the past, present, and future of Rocky Mountain trumpeters. The continual help and enthusiasm of present and past managers of trumpeters was a source of joy and encouragement throughout the many months of data gathering, analysis, and synthesis. We extend our most sincere thanks to all those friends of the trumpeter, who have worked together over several decades for its successful return and who contributed their knowledge to our project.

Funding for this project was provided primarily by the U.S. Fish and Wildlife Service Refuges Division, Denver. We particularly thank R. Crofts and B. Shranck for initial funding and continued support.

Other financial contributions were made by Idaho Department of Fish and Game, Montana Department of Fish, Wildlife and Parks, Wyoming Game and Fish Department, and The Trumpeter Swan Society. Idaho Department of Fish and Game also provided office space, supplies, and computer access.

Several researchers coauthored various chapters. We extend our sincere thanks to coauthors J. B. Bortner, P. Brussard, D. Lockman, K. McCormick, D. Paullin, and L. Shandruck.

The present and former staff of Red Rock Lakes NWR were essential in helping to locate Refuge data. Special thanks are due to T. McEneaney for his help in reorganizing the Refuge files and uncovering long-buried records, to B. Reiswig, W. Kurtenbaugh, and C. Young for their continual assistance in locating missing pieces of information, and for their helpful insights and questions. Former Refuge staff also were most helpful in their efforts to dig through personal notes to explain and fill in gaps in the written Refuge records. We particularly thank W. Banko, R. Madsen, E. McLaury, R. Sjostrom, E. Stroops, and 0. Vivion.

Most of the data assembled in this report were previously unpublished and represent years of effort by many individuals. We are deeply grateful to the researchers and managers who shared their unpublished data and personal observations so willingly. In particular, we thank D. Lockman for his information on many aspects of the movements, population dynamics, and habitat of Wyoming trumpeters. H. and R. Burgess also made available their information from work at Red Rock Lakes and in Idaho, and their compiled observations of Trumpeter Swans outside of the usual Tri-state range. The constant help and interest of the Canadian researchers was particularly appreciated. Unpublished data and insights were generously provided by G. Holton, K. McCormick, R. McKelvey, L. Shandruk, and B. Turner. These men were very helpful in locating missing data, and in reformatting their information so that comparisons between flocks were simplified. We thank G. Holton for providing us with his unpublished manuscript on the biology of Grande Prairie trumpeters, and we thank B. Turner for providing us with raw data from the observations of marked Grande Prairie trumpeters. We thank others who contributed significant unpublished data: H. Lumsden, M. Maj, T. McEneaney, C. Miller, J. Roscoe, B. Smith, H. Smith, J. Snyder, and G. Worden.

Access to agency information was made much easier by the help of K. Czarnowski and C. McClure of the U.S. National Park Service, J. Naderman and T. Trent of the Idaho Department of Fish and Game, G. Eyraud of Idaho Department of Parks and Recreation, L. Busch of the Bureau of Reclamation, and B. Smith of the National Elk Refuge.

We thank the following individuals for their review of earlier drafts: W. Banko, J. B. Bortner, G. Holton, W. Kurtenbaugh, D. Lockman, T. McEneaney, R. McKelvey, J. Naderman, D. Paullin, B. Reiswig, L. Shandruk, J. Snyder, and B. Turner.

Throughout this project we have deeply appreciated the continued help of G. Holton, D. Lockman, D. Paullin, B. Reiswig, and L. Shandruk. Their comments and discussion have shaped many of the ideas presented. We particularly thank D. Paullin for his literature review and discussion of the vegetation at Red Rock Lakes NWR.

It has been a joy working with you all, and combining our efforts toward our common goal of restoring the Rocky Mountain trumpeters. We hope that you will be pleased with our use of your information, and thank you for sharing so generously.

Ruth Shea Gale
Edward O. Garton
I. J. Ball

July 1, 1987

CHAPTER 1. DISTRIBUTION AND ABUNDANCE

Historic Range

Prior to the settlement of North America by Europeans the Trumpeter Swan was a migratory species which ranged across much of the United States and Canada. By 1832, when Sir John Richardson first described the characteristics of the Trumpeter Swan that distinguish it from the smaller tundra swan (*C. columbianus columbianus*), the trumpeter had already been eliminated from eastern portions of its range. The effort to document the species' early range and abundance has been complicated by confusion between reports of trumpeter and tundra swans and by the rapid elimination of trumpeters as the first settlers and their firearms spread across the continent (Banko 1960:10-14; Rogers and Hammer 1978).

Banko (1960:8-37) provided the first comprehensive summary of the trumpeter's historical status. He concluded that they were once abundant and widespread, with a breeding range that extended from coastal Alaska, east across Canada to the James Bay area, and south through the Rocky Mountains and central plains to Nebraska, northern Missouri, Illinois, and Michigan. In most areas, nesting records were only occasional or from widely separate locations. Banko found that in the lower 48 contiguous states, trumpeters were a common nesting species in at least three areas: the Tri-state Area of southwestern Montana, northeastern Idaho, and northwestern Wyoming; the Flathead Valley in western Montana; and southern Minnesota and northern Iowa. Trumpeters commonly nested on lakes and marshes in the open boreal forest, but near the southern limits of their range in the United States they nested in a variety of forest and prairie life zones. Banko concluded that the prairie potholes and marshes of Canada and the Great Plains supported only a small portion of the total continental breeding population.

More recent research suggests that the trumpeter's historic range was much more extensive than Banko described, particularly in the Midwest and East (Fig. 1). Schorger (1964) concluded that trumpeters nested in Minnesota, Illinois, Indiana, and most probably in Wisconsin. Hansen *et al.* (1971) described the species' distribution in Alaska and suggested a theoretical northern limit of the trumpeter's nesting range based upon the need for a 145 to 150 day ice-free period for cygnet production. Rogers and Hammer (1978) synthesized additional historical and paleontological evidence which led them to conclude that trumpeters had also occurred and possibly nested in northern Florida, the Carolinas, northwestern Ohio, and the lower Mississippi Valley. They concluded that "*C. buccinator probably bred wherever suitable habitat could be found in North America.*" Lumsden (1984) summarized the trumpeters' ancestral range in eastern Canada, and concluded that they occurred through the Hudson's Bay lowlands and other portions of Manitoba, Ontario, and western Quebec, and locally in New Brunswick, Nova Scotia, and Newfoundland.

Trumpeters migrated south in autumn to areas that provided predictable ice-free feeding areas. Banko (1960:26) identified several historic wintering areas including the Pacific Coast from Alaska to Washington, the lower Columbia River, the Sacramento Valley, the Yellowstone National Park area, the Mississippi Valley, the Gulf of Mexico coast and the Atlantic coast from the Chesapeake Bay region south to North Carolina. Rogers and Hammer (1978) suggested that the trumpeters' historic wintering range included the southeastern United States, south to northern Florida.

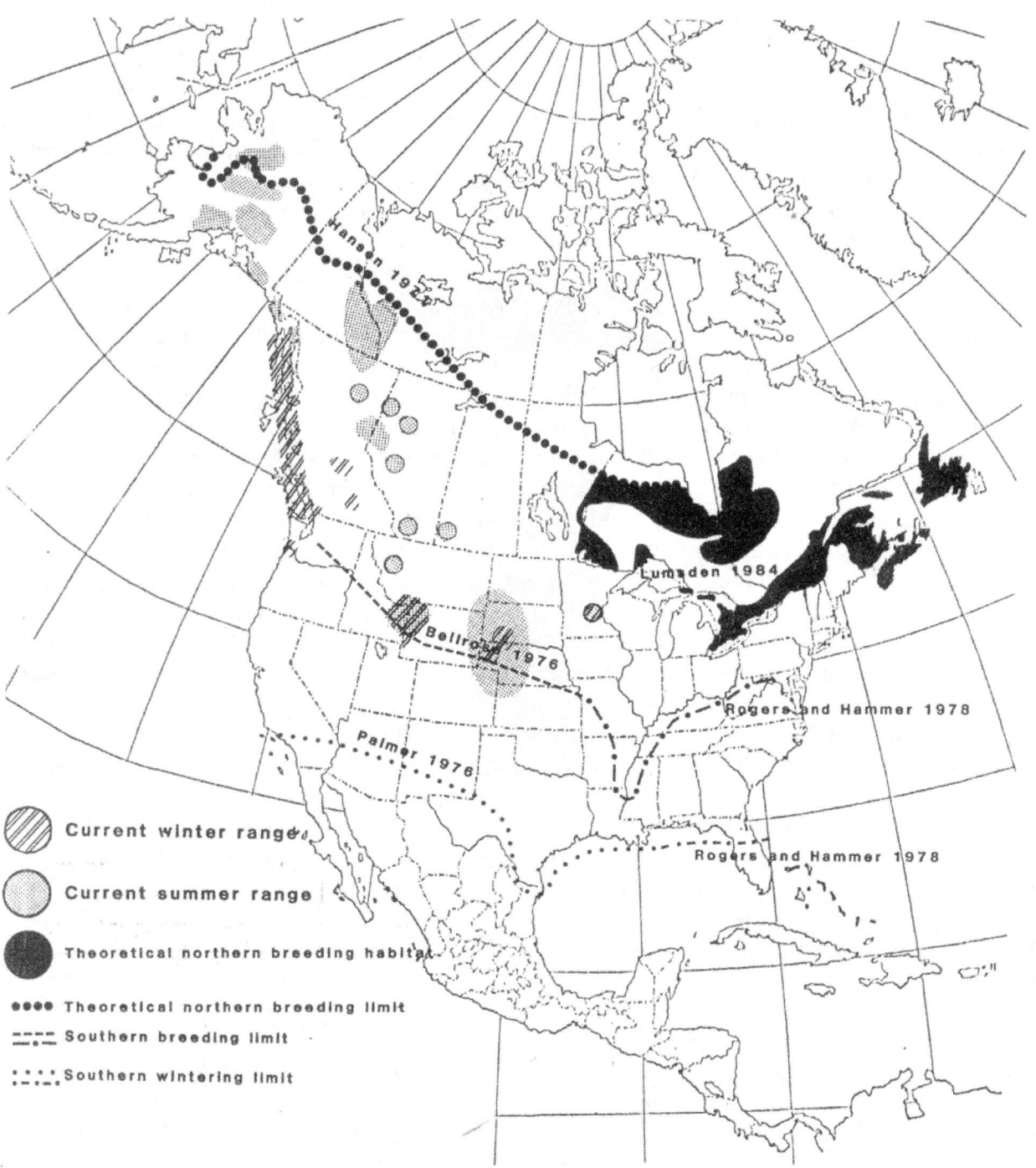

Figure 1. Historic and current range of Trumpeter Swans.

Banko (1960:19, 25) described the trumpeters' decline and attributed their near extinction to exploitation in the north by the Hudson's Bay Company swan skin trade, from about 1772 to 1900, and to overhunting and habitat destruction in the south by settlers. Rogers and Hammer (1978) also identified habitat loss accompanying settlement and overharvest as the main causes of the trumpeter's elimination from the eastern United States by the mid-1800s. Lumsden (1984) concluded that Indians were not capable of killing large numbers of swans with bows and arrows. However, as soon as the Indians acquired guns from the white man, they shot swans by the thousands for food and skins, while settlers slaughtered swans on the southern wintering grounds.

Trumpeters were eliminated from virtually all of their historic range by the early 1900s (Banko 1960:19). Only a few remnant groups survived in the most isolated habitats. In 1939, an estimated 500 trumpeters wintered along the Pacific Coast and central interior of British Columbia (Blackford 1939) and presumably migrated north to Alaskan nesting areas. Though breeding trumpeters persisted in Alaska, their existence was not widely known in the 1930s and extinction of the species was feared imminent. The existence of a substantial breeding population in Alaska was not widely recognized until the mid-1950s (Hansen *et al.* 1971). Except for this population, the trumpeters of the United States and Canada were reduced by 1932 to two small breeding flocks which shared a common wintering area. In the United States, only about 50 adults and subadults, and their cygnets escaped extinction by remaining year-round in the isolation of the Tri-state region. Sparse records from 1874, 1891 and 1902 indicated that a few summering trumpeters had persisted in the Tri-state Area throughout the period of decline (Banko 1960:21, 146). A second surviving group of approximately 77 adults and subadults, and their cygnets summered in the vicinity of Grande Prairie, Alberta, and wintered primarily in the Tri-state Area (Mackay 1978).

These two tiny remnants of a once wide-ranging population of many thousands survived primarily because of their shared tradition of wintering in the Tri-state region. Scattered warm springs provided small pockets of open water even during the coldest winters, and the area was so remote as to be virtually unexplored until the 1870s (Haines 1977). The survivors wintered in isolation from the settlers' guns and the habitat destruction that eliminated all of the trumpeters that followed all other traditional migratory routes to more vulnerable habitats (Hochbaum 1955). As mortality exceeded recruitment in family groups with all other migratory traditions, the groups perished and with them died virtually all traditional knowledge of other wintering areas.

Thus these two remnant flocks, wintering in the remote valleys of the Tri-state region, formed the foundation of the current Rocky Mountain Trumpeter Swan Population. Most of the original migratory strategies were gone by the turn of the century. When efforts to save the last Rocky Mountain trumpeters began, most of the migratory traditions that otherwise could have allowed the species to reoccupy former wintering areas were extinct.

Current Distribution and Abundance

The U.S. Fish and Wildlife Service (USFWS) (1984) recognizes three Trumpeter Swan populations (Fig. 2). Two of these populations developed from the remnant flocks that survived the historic decline. The third population includes several restoration flocks that have been created by transplants and captive propagation efforts.

Because the plumage of yearling and older trumpeters basically appears all white, most surveys can differentiate only between the gray cygnets, and the white adults and subadults. White-plumaged swans will henceforth be referred to as adults except when subadults (age classes 2-5) are specifically discussed.

Pacific Coast Population (PCP)

The Pacific Coast Population nests primarily in interior and south-central Alaska and winters along the Pacific Coast from southern Alaska to Oregon (USFWS 1984). A few swans also winter on open waters in the central interior of British Columbia where they are joined by trumpeters from the south-central and southwestern portions of the Yukon (R. McKelvey, pers. comm.). Recent summer surveys in Alaska have revealed an expanding population. In 1980, 5,259 adults and 2,437 cygnets were counted (King and Conant 1981). Conant *et al.* (1985) censused 7,773 adults and 1,686 cygnets in 1985. A 1986 summer survey of randomly sampled units provided estimates of 7,145 adults and 3,034 cygnets (Hodges *et al.* 1986).

Interior Population (IP)

In an attempt to restore trumpeters to the center of their historic range, several restoration flocks have been created, mainly by transplanting swans from Red Rock Lakes NWR and by captive propagation efforts. Programs at Lacreek NWR, South Dakota, and Hennepin County Park Reserve District, Minnesota, have created flocks which gradually are exploring southward during autumn migration, but are still at least partially dependent on supplemental feeding. The total estimated number of swans in the Interior Population was 245 in December 1985 (Fig. 2, D. Weaver, pers. comm.).

Rocky Mountain Population (RMP)

The Rocky Mountain Population summers along the Rocky Mountain corridor in the United States and Canada, and winters primarily in the Tri-state Area. With nesting areas scattered from Wyoming to the Yukon and Northwest Territories, summer surveys of the RMP have been difficult to coordinate. The best counts of the entire Population have been made in the Tri-state wintering area, where 1,196 adults and 386 cygnets were counted in February 1987 (USFWS Midwinter Trumpeter Swan Survey 1987). Although the extent of intermixing is unknown, two subpopulations and a number of flocks have been described for management purposes (USFWS 1984) (see also Fig. 2). These flocks have often been delineated along state or provincial boundaries which may be biologically artificial, but are useful because they correspond to survey and management jurisdictions.

The RMP includes the predominantly nonmigratory Tri-state Subpopulation (TSP), and the highly migratory Interior Canada Subpopulation (ICSP). Although both subpopulations winter together in the Tri-state region, no pairing between members of the two groups has been documented to date. The existing data suggest that these groups are distinct breeding populations that share a common winter range (see Chapter 9).

Figure 2. Status of North American Trumpeter Swans in 1985 (adults–cygnets)

North American Trumpeter Swans
8,753-2,066

Pacific Coast Population[a] 7,773–1,686		Rocky Mountain Population[b] 786-329		Interior Population[c] 194-51	
Alaska		Tri-state Subpopulation	368-139	Lacreek NWR, SD	144-43
Gulf Coast	1,294-164	Centennial Valley, MT	204- 76	Minnesota	42- 8
Copper River	190- 11	Other Montana	8- 11	Missouri	4- 0
Gulkana	2,474-533	Yellowstone NP	25- 5	Ontario	4- 0
Kenai	137- 51	Lower Wyoming	46- 20		
Cook Inlet	1,320-241	Idaho	83- 27		
Lower Tanana	1,741-503				
Kuskokwim	184- 55	Interior Canada Subpop.	418-190		
Koyukuk	258- 45	Grande Prairie, AB	211-109		
Yukon Flats	10- 3	Other Alberta	31- 14		
Chilkat Valley	24- 16	Ft. Nelson, BC	16- 4		
Upper Tanana	141- 64	Ft. St. John, BC	23- 17		
		Toobally Lakes, YK	82 -20		
		Nahanni, NWT	51- 24		
		Cypress Hills, SK	4- 2		

[a] Data from Conant *et al.* 1985

[b] Data from USFWS 1985 Tri-state Trumpeter Swan Survey, McKelvey (1986) and Shandruk (1986)

[c] Data are December estimates from D. Weaver (pers. comm.)

Tri-state Subpopulation (TSP). Red Rock Lakes NWR and the surrounding Centennial Valley of Montana form the primary nesting area of the Tri-state trumpeters. Other scattered pairs nest on lakes and ponds throughout eastern Idaho and northwestern Wyoming in and adjacent to Yellowstone National Park (Fig. 3). In September 1986, the TSP totaled 333 adults and 63 cygnets (USFWS Tri-state Trumpeter Swan Survey 1986) in the following flocks:

1) Centennial Valley (CV) Flock. Swans that nest in the Centennial Valley of southwestern Montana, including Red Rock Lakes NWR, Bureau of Land Management lands, and private lands are referred to as the Centennial Valley flock. In September 1986, the CV flock included 167 adults and 28 cygnets (USFWS Tri-state Trumpeter Swan Survey 1986). Although swans regularly move across the Refuge boundary to adjacent Centennial Valley habitats, occasionally the Red Rock Lakes flock is referred to as a separate entity in discussions of the history and management of the Refuge. [Although they are on the edge of the Centennial Valley, Elk and Conklin Lakes in the adjacent Gallatin National Forest have most often not been included in summer census compilations for the Centennial Valley flock because of nesting habitat discontinuities between their forest-lake habitat and the Centennial Valley marshlands. These lakes will be included in the census category Other Montana in our tables and appendices.]

2) Yellowstone National Park (YNP) Flock. YNP swans summer in northwestern Wyoming at relatively high elevations (1,770 to 2,515 m) on pond and river habitats administered by the U.S. National Park Service (Shea 1979). In 1986, this flock contained 24 adults and 12 cygnets (USFWS Tri-state Trumpeter Swan Survey 1986).

3) Lower elevation Wyoming Flock. Other trumpeters summer in Grand Teton National Park, the National Elk Refuge, National Forest, and private lands where they are managed by the federal land management agencies as well as by the Wyoming Department of Game and Fish. These swans are grouped together because of similarities in their breeding habitat, which is generally at least 300 m lower in elevation than nesting sites in YNP. In 1986 this flock contained 50 adults and 7 cygnets (USFWS Tri-state Trumpeter Swan Survey 1986).

4) Idaho Flock. Idaho trumpeters nest primarily on small ponds in the Targhee National Forest immediately south and west of Yellowstone National Park, and on adjacent state and private lands. Occasional nest attempts occur at isolated marshes along the desert fringe of the upper Snake River plain, south and west of the primary range. The Idaho habitat is managed primarily by the U.S. Forest Service and the Idaho Department of Parks and Recreation. The swans are managed by the Idaho Department of Fish and Game. This flock contained 83 adults and 14 cygnets in September 1986 (USFWS Tri-state Trumpeter Swan Survey 1986).

5) East Front, Montana Flock. Swans have nested sporadically during the last 20 to 30 years in an area approximately 16 km southwest of Augusta, Montana. One nesting lake near Haystack Butte has been used regularly for at least the last decade and occasional nest attempts by a second pair have occurred in the area (D. Hook, pers. comm.) Only two adults and two cygnets were reported in 1986 (USFWS Tri-state Trumpeter Swan Survey 1986).

Figure 3. Tri-state Area of southwest Montana, eastern Idaho and western Wyoming.

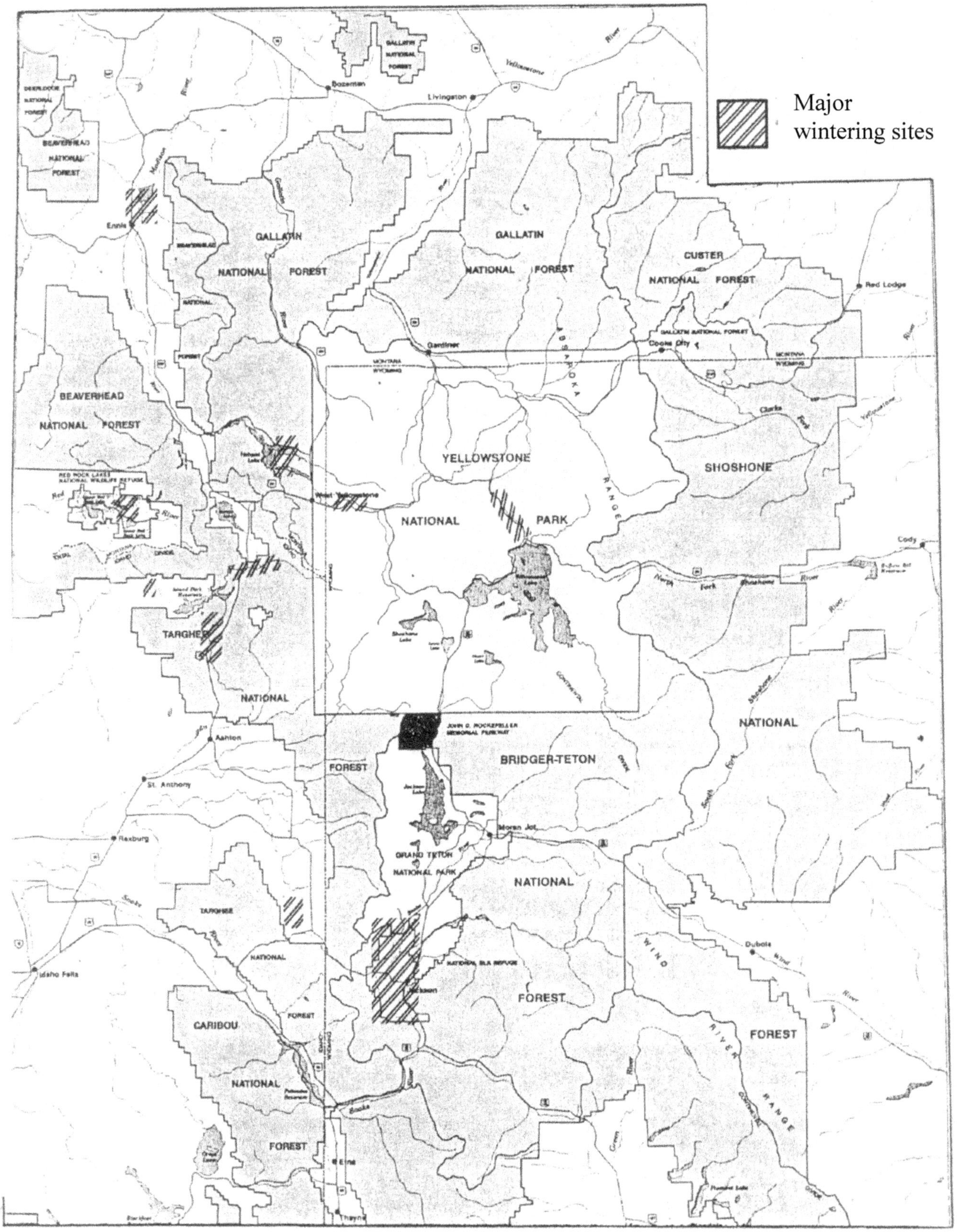

Interior Canada Subpopulation (ICSP). The Interior Canada Subpopulation includes two major concentrations of trumpeters. The Grande Prairie flock nests in the Peace River area of Alberta and British Columbia. Further to the north, trumpeters nest in the southeastern Yukon, with scattered pairs occurring northeastward into the Northwest Territories and southward into northern British Columbia. The first coordinated rangewide survey was flown by federal and provincial biologists during the summer of 1985, and the ICSP at that time was estimated to contain a total of 418 adults and 190 cygnets (Fig. 2).

1) Grande Prairie (GP) Flock. The key breeding habitat of the flock encompasses about 5,700 km^2 of west-central Alberta and east-central British Columbia near the city of Grande Prairie, Alberta (Holton 1982). In September 1985, the flock contained 211 adults and 109 cygnets (L. Shandruk, pers. comm.).

2) Other Alberta Flocks. About 15 adults and their broods summer in the Edson area about 160 km southeast of Grande Prairie. Other scattered nesting pairs have been located near Cardston/Pincher Creek, Otter Lakes, and the Chinchaga River (Shandruk 1986).

3) Toobally Lakes, YK Flock. At least some of the trumpeters that summer in the Toobally Lakes region of southeastern Yukon winter in the Tri-state Area (McKelvey *et al.* 1983). Summer numbers were estimated at 82 adults and 20 cygnets in 1985. Trumpeters nesting further west in the Yukon may belong either to the RMP or the PCP (R. McKelvey, pers. comm.).

4) Nahanni, NWT Flock. Trumpeters have been observed regularly in the southwestern portion of the Northwest Territories since 1977, with the majority located in the Nahanni Butte and Carlson Lake areas (McCormick 1985). The Nahanni, NWT flock contained approximately 51 adults and 24 cygnets in 1985, but a considerable amount of suitable habitat (including Nahanni National Park Reserve) may not have been surveyed (McCormick and Shandruk 1986).

5) Fort Nelson, BC Flock. The 1985 aerial survey found 16 adults and 4 cygnets on about 15 ponds and lakes in the vicinity of Fort Nelson in northeastern BC (McKelvey 1986).

6) Fort St. John, BC Flock. Some 40 adults and 23 cygnets were censused in 1985 in the Ft. St. John/Dawson Creek area northwest of the Grande Prairie flock (McKelvey 1986). Their distribution is virtually continuous with that of trumpeters from the Grande Prairie flock. Due to overlap with the survey route for the Grande Prairie flock, only 23-17 of these swans were added to the total Subpopulation count (Fig. 2) (L. Shandruk, pers. comm.).

7) Cypress Hills, SK Flock. Between one and three pairs have nested annually in the Cypress Hills region of southwest Saskatchewan since at least 1951 (Killaby 1987). In 1985, four adults and two cygnets were counted in the area (Shandruk 1986).

Because the 1985 ICSP surveys were flown by several different observers, at different times and with some overlap of survey areas, the estimated numbers of swans given above differ slightly from data in the original agency survey reports. Small discrepancies were corrected for this account by the original observers.

Observations of RMP Trumpeters Outside of the Tri-state Area

Banko (1960:25-38) described historical and recent migration records. Reports since 1931 indicate that at least a few Tri-state and Interior Canadian trumpeters occasionally migrated beyond the Tri-state wintering area. In recent years a few trumpeters have been observed heading south from the Tri-state Area during the fall Tundra Swan migration (J. Snyder, D. Lockman, pers. comm.). These birds, however, were not likely to be reported unless they were killed or had been marked because no efforts have been made to search for trumpeters in traditional Tundra Swan wintering areas outside of the Tri-state Area.

Unmarked trumpeters are mistaken for Tundra Swans by most observers. The number of migrant trumpeters which have been killed in the Tundra Swan hunt in Utah is unknown. Verbal reports from biologists in Utah, however, suggest that a few trumpeters have been regularly harvested in past years. This mortality on the few pioneering migrants may have been an important factor in preventing the reestablishment of a migratory tradition through Utah. Wintering and migrant marked RMP trumpeters (as well as other unmarked trumpeters) have been reported from northern and southern Idaho, southern Oregon, central and southern Wyoming, Utah, Colorado, Nevada, New Mexico, Nebraska, and Missouri (Fig. 4, Appendix I). Based upon records of marked trumpeters, mortality among these pioneers appears to be high (Chapter 9) and except for northern Idaho, no regular use of these peripheral areas has been documented.

Since about 1980, trumpeters have been observed in the Coeur d'Alene area of northern Idaho during spring and fall migration (F. Bear, pers. comm.). These birds intermingle with migrant Tundra Swans but neither their summer nor winter ranges were known until recently. One clue to the origin of some of these migrants was obtained in November 1986, when a marked trumpeter from Nahanni, NWT, was observed near the south end of Lake Coeur d'Alene (W. Latshaw, pers. comm.). On 15 November 1986, the sighting of a red neck-banded (NWT) trumpeter flying south at Goose Lake, Oregon, also documented that some NWT trumpeters migrate southwesterly, possibly to wintering areas in California. Exploration of new wintering areas by NWT trumpeters was also demonstrated by the January 1987 observation of a pair of marked trumpeters near Alamo, Nevada (L. Shandruk, pers. comm.). It is likely that ICSP trumpeters will increase their exploration of new migration routes and wintering sites, if this Subpopulation continues its recent rate of increase, and if survival rates among migrants are adequate.

Summary

Historically the Trumpeter Swan was a wide-ranging, migratory species which nested throughout much of Canada and the United States. Trumpeters probably wintered wherever ice-free water was available. Major concentrations of wintering swans occurred in the Mississippi Valley and the coastal estuaries of the Atlantic, Pacific, and Gulf of Mexico. Outside of Alaska, by the 1900s the species was reduced to near extinction due to the commercial swan skin trade, habitat destruction, and over-harvest by early settlers and Indians with firearms.

Figure 4. Observations of RMP Trumpeter Swans Outside of Core Use Areas

The only wild trumpeters known to have survived outside of Alaska were a few families that shared the tradition of wintering in perhaps some of the most isolated ice-free waters in the United States, the geothermally warmed waters of Yellowstone National Park and the nearby Centennial Valley of Montana. These survivors included two distinct groups, about 50 primarily nonmigratory adults, and about 80 migratory adults that summered near Grande Prairie, Alberta.

The descendants of these two remnant groups increased to a Tri-state wintering total of some 1,600 swans by February 1987. Interior Canadian trumpeters now nest in Alberta, British Columbia, Saskatchewan, the Yukon, and the Northwest Territories. In the Tri-state Area, nesting is concentrated in the marshes and ponds of the Centennial Valley of Montana, including Red Rock Lakes NWR. Other scattered pairs nest throughout suitable lake and river habitat in northwestern Wyoming, northeastern Idaho, and southwestern Montana. A very small, isolated group nests in central Montana, southwest of Augusta.

Although the Rocky Mountain Population has increased in numbers, it has shown little tendency until recently to pioneer other wintering areas. A few trumpeters occasionally migrated to wintering sites outside of the Tri-state Area, but no regular use of these areas has been documented. The high mortality rates of these few migrants have likely slowed the reestablishment of migratory traditions. Recent scattered observations suggest that migration outside the Tri-state wintering area is increasing, or at least becoming more apparent, particularly through the Coeur d'Alene area of northern Idaho on a route that may lead southwesterly to wintering areas in California.

CHAPTER 2. POPULATION TRENDS

Rocky Mountain Population

Winter Census Effort

The entire RMP can be censused most reliably in winter, when both subpopulations are together in the Tri-state Area (Fig. 3). During the past 35 years, census efforts have varied considerably and this variation has complicated interpretation of the survey data. Prior to 1972, winter swan surveys were flown by the respective state Fish and Game agencies as part of the Pacific Flyway Midwinter Waterfowl Surveys. In addition, Yellowstone National Park employees usually tried to coordinate ground counts in the more accessible areas of the Park with the state aerial surveys. Banko (1960:61) summarized several partial counts of Tri-state wintering swans which were made in the 1950s. The Idaho Department of Fish and Game began aerial counts of swans on the Henry's Fork of the Snake River in 1950 and occasionally supplemented those counts with data gathered incidental to big game survey flights (Salter 1954). In 1966 and 1967, aerial surveys of the Idaho wintering habitats were also flown by the staff of Red Rock Lakes NWR in a special effort to resight neck-banded swans (Research Report RRLNWR-1 in RRLNWR files). The Wyoming Game and Fish Department annually surveyed the Snake River drainage of western Wyoming, beginning in 1955 (Letter from L. Serdiuk to RRLNWR, 1 February 1979).

Because these partial surveys did not provide complete or coordinated coverage of the wintering area, in 1972 the USFWS began its annual Midwinter Trumpeter Swan Survey in an attempt to census all trumpeters wintering in the Tri-state region. Even with a single agency coordinating the survey, however, variation in survey effort still resulted in somewhat ambiguous data.

Since the 1972 and 1973 counts, which included aerial coverage of only Idaho and the Centennial Valley of Montana, the size of the area surveyed has increased substantially. Most of the key wintering sites have been surveyed for at least the last decade, and the surveys have included all the known occupied wintering sites since 1981.

In several years, completion of the survey was delayed by major storm systems and the survey period lasted for three to four weeks. Considerable local movement of swans likely occurred during these extended periods, thus reducing survey accuracy.

Another factor which has reduced survey accuracy has been the presence of varying numbers of wintering Tundra Swans in the Tri-state Area. During the 1950s, up to eight Tundra Swans were regularly observed wintering in the Island Park area but their small numbers did not significantly affect the census data (Banko 1960:60). In some recent winters, however, larger numbers of tundras have been observed. During January 1977, A. Doberstein (pers. comm.) of the Canadian Wildlife Service observed that 40% of the swans at Harriman State Park (HSP) on the Henry's Fork of the Snake River were Tundra Swans. Shea (1979) observed that 50% of the 88 swans observed on the Teton River in January 1978, and about 10% of 200 swans at HSP in February 1979, were tundras.

Since 1978, the USFWS has followed the aerial surveys with ground observations at the major wintering sites to determine the proportion of Tundra Swans included in the Midwinter Survey. Up to 8% (in 1980) of the swans classified in any year have been identified as tundras and estimated Tundra Swan numbers were deleted from the survey totals.

Overall, the Tri-state winter surveys provide a good estimate of the total number of wintering swans and their distribution, but the data are clearly less accurate prior to 1981.

Winter Population Trends

Even though some marked Grande Prairie trumpeters had been observed wintering in the Tri-state Area in the 1950s (Mackay 1957), the winter surveys provided no evidence of a noticeable influx of migrants before 1974. During February 1974, the total number of wintering swans censused (709) was approximately equal to the combined total of estimated swan numbers in the Grande Prairie flock (123) and the Tri-state Subpopulation (530). From 1974-86, the number of wintering swans increased from about 700 to 1,600 (Appendix II). This overall increase was due to a continual increase in migrants; during this same period, the nonmigratory Tri-state Subpopulation declined (Fig. 5).

The increase in swans recorded between 1974 and 1980 undoubtedly was influenced by increased survey coverage, but we strongly suspect that a real population increase occurred as well. During these years the observed ratio of cygnets to adults averaged 22:100. Population modeling by D. Lockman (pers. comm.) predicted that this level of cygnet production would result in growth of the wintering population similar to that actually observed. It is apparent, however, that the unusually high number of swans counted in 1977 was due to the presence of Tundra Swans as previously discussed. The presence of Tundra Swans may also have inflated the count during the very mild winter of 1980-81, when the number of adults increased to 1,000 from 767 in 1980, and then declined to 952 in 1982.

Adult numbers continued to increase through 1985, when they reached a high of 1,326. The lack of increase in adult numbers in 1986 may have been due in part to severe late winter mortality observed during April 1985 (see Chapter 5). The total number of wintering swans increased in 1986, however, due to the high number of cygnets in the population.

The 1987 wintering population contained 1,196 adults and 386 cygnets. The return of only 1,196 white birds in 1987, out of the 1,603 total swans that wintered in the Tri-state Area in 1986, reflected the loss of about 25% of the 1986 population. No unusual mortality was observed within the RMP range during this period, however. The fall of 1986-87 was unusually snow-free and mild, and the dispersal of trumpeters to other wintering areas may have contributed to the low count of adults. The possible dispersal of subadults was also suggested by the high proportion of cygnets to adults that wintered in the Tri-state Area and the sighting of a marked Wyoming subadult male wintering at Lake Powell, Utah (Appendix I).

Tri-state Subpopulation
Census Effort

Systematic efforts to census the Tri-state summer residents and record cygnet production began with surveys conducted by the U. S. National Park Service (USNPS) in 1929. Prior to 1946, coordinated ground surveys were made by USFWS and USNPS observers. The area censused was rather incomplete before 1936 and primarily included the known nesting areas. Aerial surveys by the USFWS began in 1946 and made rather thorough coverage of the Tri-state habitat possible, although occasionally Yellowstone NP was surveyed only by coordinated ground counts (Banko 1960:144-146).

Figure 5. Trends in wintering Rocky Mountain Population. Total swan numbers are from USFWS Midwinter Trumpeter Swan Surveys. Number of nonmigrants equals results of previous summer's Tristate Survey or was estimated from partial surveys. The number of migrants was estimated from total swans minus nonmigrants.

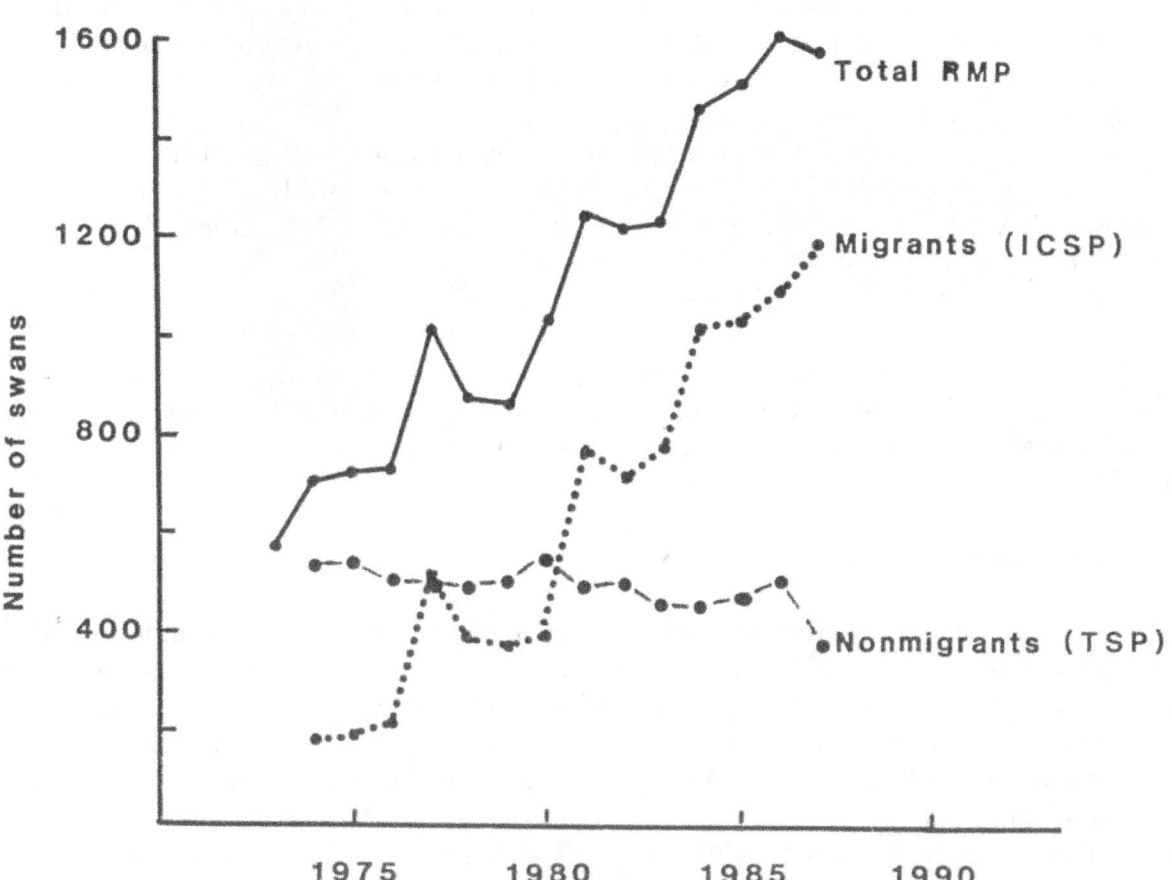

The USFWS Tri-state Trumpeter Swan Surveys provide an extremely valuable long-term record of swan distribution and numbers. The survey has been flown by USFWS personnel from Red Rock Lakes NWR using highly consistent methods and survey routes. The continuity among observers has been exceptional: only two different pilots have flown the surveys since 1953.

Several factors regarding the Tri-state surveys are important to note:

1) Prior to 1954, the annual survey dates varied from 13 July to 31 August. Mid-summer surveys tended to overestimate the number of cygnets fledged because some mortality occurred after the survey but prior to fledging. Since 1954, survey dates have ranged from 20 August to 25 September, and have provided better estimates of cygnet production to fledging.

2) The census data presented in Appendix III differ in numerous instances from the data published previously by Banko (1960:146), the North American Management Plan for Trumpeter Swans (1984), and some of the original USFWS Tri-state Trumpeter Swan Survey reports. Detailed explanations for these discrepancies are provided in Appendix IIIa.

3) In surveys prior to 1972, the USFWS attempted to census all trumpeters in the United States, including those in zoos and in transplant flocks. Thus, the data published in Banko (1960:146) included trumpeters from zoos, Malheur NWR, Ruby Lakes NWR etc., and did not pertain exclusively to the TSP.

4) After 1968, the annual survey was changed to a triennial schedule and census data are therefore lacking for 1969, 1970, 1972, 1973, 1975, 1976, 1978, 1981, and 1982. Partial data for these years were assembled for this report from various Red Rock Lakes NWR files. Annual surveys were resumed by the USFWS in 1983.

5) By themselves, the USFWS Tri-state Surveys do not provide an accurate record of cygnet recruitment at Red Rock Lakes NWR. In some years, cygnets were transferred from the Refuge before the surveys were conducted. In other years, cygnets that were counted in the surveys were later transferred and therefore did not actually recruit into the Subpopulation. Survey results which have been adjusted to account for cygnet removals are presented in Appendix IV.

Subpopulation Trends

Fluctuations in the number of Tri-state cygnets censused annually have been common and have contributed to substantial year-to-year variation in the total number of swans censused. Because of these fluctuations and the high mortality of cygnets during their first winter (see Chapter 5), trends in the number of adults are far more informative than are trends in total numbers. For these reasons the following discussion of trends will deal only with adult numbers unless otherwise specified.

The TSP increased from about 50 adults in the 1930s to a peak of 548 in 1954 (Fig. 6, Appendix III). Numbers then fluctuated (from 359-554) for a decade, reaching another peak of 554 in 1964. Twice, in 1950 and 1957, unusually low counts suggested that from 100-120 swans moved out of the normal survey area but returned before the subsequent year's survey (Banko 1960:146).

Adult numbers remained above 500 from 1964 to 1967 and then declined by 17% between 1967 and 1968. From 1968 to 1980, late summer surveys showed a net change of only 31 swans, from 431-462, although fluctuations during this period may have been less apparent due to the reduced frequency of surveys. Between 1980 and 1986, the number of adults declined from 462 -331, the lowest number censused since 1950.

The decline from 368 adults in 1985 to 331 in 1986 was particularly noteworthy because the 1985 cygnet cohort was the third largest on record, but for undetermined reasons apparently did not contribute many recruits to the 1986 Subpopulation. It is possible that the high number of cygnets produced in 1985 resulted in the dispersal of an unusually large yearling cohort in the early summer of 1986. Observations by B. Reiswig and D. Lockman (pers. comm.) suggest the disappearance of subadults from Wyoming and the Centennial Valley prior to the molt.

Due to the scarcity of marked birds and the lack of surveys of the Tri-state habitats during spring and early summer, however, managers cannot be certain that dispersal beyond the normal survey route, rather than mortality, was responsible for the disappearance of the 1985 production. An unusual hard freeze in early October 1985 (RRLNWR Trumpeter Swan Report 1986) hit the Tri-state region just as many cygnets were fledging and this, or some other factor may have significantly reduced their survival. Until evidence is obtained that documents the dispersal of Tri-state trumpeters to areas outside the summer census route, managers have little choice but to assume that annual losses are due primarily to mortality.

Centennial Valley (CV) Flock The trends shown by the Tri-state Subpopulation in large part mirror trends in the CV flock (Fig. 7). Detailed census data for Red Rock Lakes NWR and the entire CV are given in Appendix V. The CV flock grew from about 30 adults in the 1930s to 401 in 1954. As previously mentioned, an abrupt decline occurred in 1957 due to movement of swans out of the area. However, these birds apparently returned by 1958 (Banko 1960:146). Between 1959-62 the CV flock declined from 363 to 218 adults. This decline was temporarily reversed by record high cygnet recruitment in 1963. Following this one-time increase, however, adult numbers began another decline which culminated in a 28% loss between 1967 and 1968. This unprecedented loss of swans from the CV flock caused the simultaneous, but less extreme, decline in the total Subpopulation.

After the 1967-68 decline, the CV flock increased during the 1970s to 337 adults by 1978. Since 1978, the flock has declined by about 50% to 167 adults in 1986. The 1986 CV flock contained fewer adults than at any time since 1946. The recent decline of the entire Tri-state Subpopulation is basically due to the decline of the CV flock.

The number of adult swans counted on Red Rock Lakes NWR (Fig. 7) also peaked in 1954 when extremely dry weather reduced the availability of off-Refuge habitat and swans moved on to the Refuge (RRLNWR Narrative Report 1954). Numbers fluctuated considerably in the period from 1955-78, with a range of from 137-293 adults. Observations in the Refuge Narrative Reports attribute these fluctuations in large part to the tendency of the swans to concentrate on the Refuge during dry years, particularly when water levels at nearby Lima Reservoir were low. This was the probable reason for the very high number of trumpeters on the Refuge in 1979 (RRLNWR Narrative Report 1979).

Since 1979, the number of adults using the Refuge in late summer continued the downward trend of the late 1960s. The 87 swans censused in 1986 were the fewest counted on the Refuge since 1942. This decline in use of the Refuge has coincided both with the overall decline of adults in the entire CV, and with a reduction in the percentage of CV swans located on the Refuge. From 1944 to 1963, about 74% of the CV flock was censused on the Refuge. After about 1964, there was a shift in swan use to off-Refuge habitat in the Valley, with an average of only 58% of the CV adults censused on the Refuge from 1964-81. This trend continued in 1982-86, when only 50% of the adults in the summering CV flock were found on the Refuge.

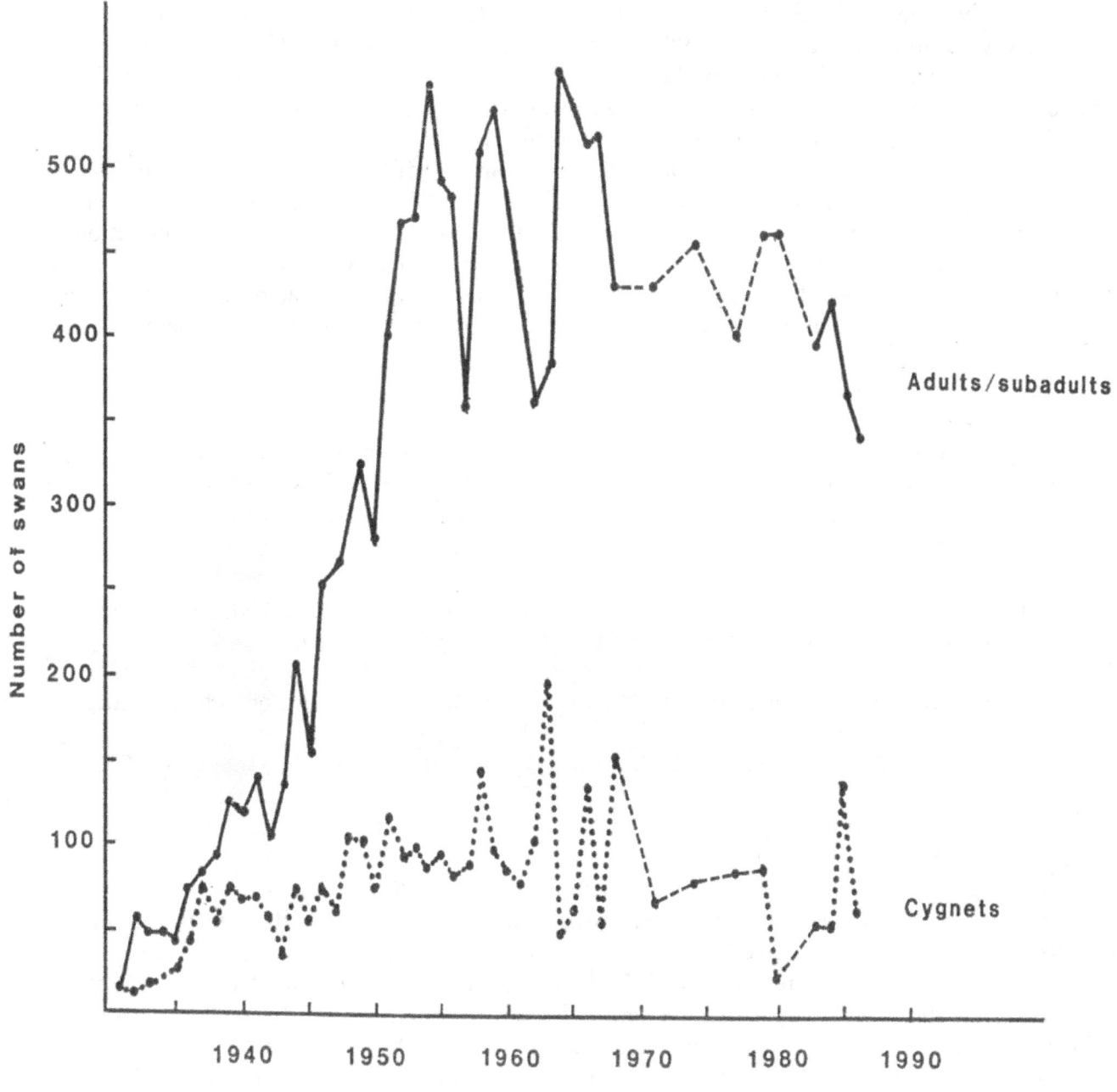

Figure 6. Tri-state Subpopulation trends. Data are from USFWS Tri-state Trumpeter Swan Surveys (late summer).

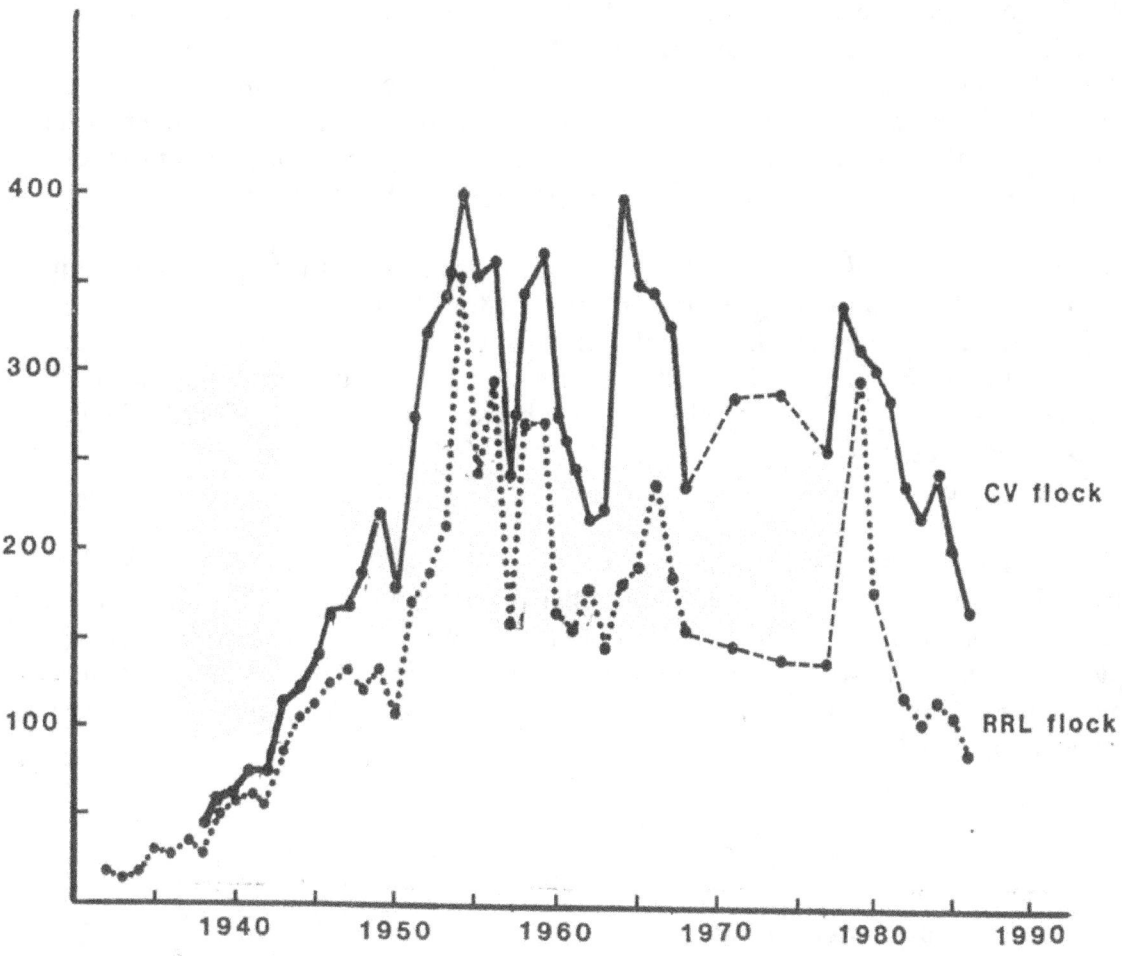

Figure 7. Late summer numbers of adults/subadults in the Centennial Valley (CV) and Red
Rock Lakes (RRL) flocks. Data are from USFWS Tri-state Surveys.
- - - - Missing Data

Yellowstone National Park Flock Changes in the numbers of adults in the YNP flock have roughly paralleled trends in the CV flock (Fig. 8). A core flock of 35-40 adults/subadults in the late 1930s and early 1940s increased to 63 by 1951 and fluctuated between 44 and 69 through 1968. With the change from annual to triennial summer surveys after 1968, annual fluctuations were not documented. However, the surveys show that the YNP adults declined by about 50% between 1968-71, following the decline in the CV flock, and then partially rebounded in the mid-1970s. Since 1976 the YNP flock has shown a rather steady decline to a 50-year low of 24 adults in 1986. The flock fledged 12 cygnets in 1986, its highest production since 1966. Cygnet production in 1986 was increased in part due to the fledging of five cygnets from two territories where artificial nest platforms had been provided (C. McClure, pers. comm.). Neither of these historically productive territories had successfully fledged cygnets in recent years.

Lower Elevation Wyoming Flock Swans in the lower elevations of Wyoming were recorded in the 1931-34 surveys. Their numbers were supplemented by transfers from Red Rock Lakes to the National Elk Refuge in 1938 and 1939. Adult numbers increased to a peak of 69 in the early 1960s and then fluctuated until 1966, when a 29% decline occurred. Numbers have since fluctuated between 25-56, but most frequently remained between 40-50 adults (Fig. 8). This flock did not experience the decline suffered by the CV and YNP flocks in the 1980s, but rather has remained stable at about 50 adults.

Idaho Flock Adult numbers fluctuated with a gradual upward trend to 88 in 1968 (Fig. 8). Confusion by aerial observers has been common in Idaho due to the myriad of unnamed ponds in the dense lodgepole habitat of the Targhee National Forest. Some of the censused fluctuations may be due to survey error; however an overall increase in swans is still apparent. A decline of 32% occurred between 1968-1971, synchronous with the decline in the YNP flock and at least a year after the decline in the CV flock.

Interior Canada Subpopulation

Census Effort and Subpopulation Trends

The scattered distribution of the ICSP and the vast area of potential nesting habitat make a complete summer census of this Subpopulation extremely difficult. Several flocks have only recently been censused for the first time and in several cases consecutive surveys have not covered exactly the same geographic areas. Although some 30 years of survey data have been gathered in the Grande Prairie area, survey effort has varied considerably and this variation makes interpretation of the existing census data quite complicated. Most surveys were flown in early September, shortly before the cygnets attained flight and prior to the arrival of other migratory trumpeter and Tundra Swans.

Grande Prairie (GP) Flock. Early records of trumpeters near Grande Prairie date back at least to 1891 (Mackay 1987) and 1908 (Sheehan 1987). The first aerial survey of this flock was made in 1946 by the USFWS and found 77 adults and 23 cygnets. The Canadian Wildlife Service has flown annual mid-September surveys since 1954 (Mackay 1987). From 1954-77 the surveys covered only the Alberta portion of the habitat, however the number of lakes surveyed rose from

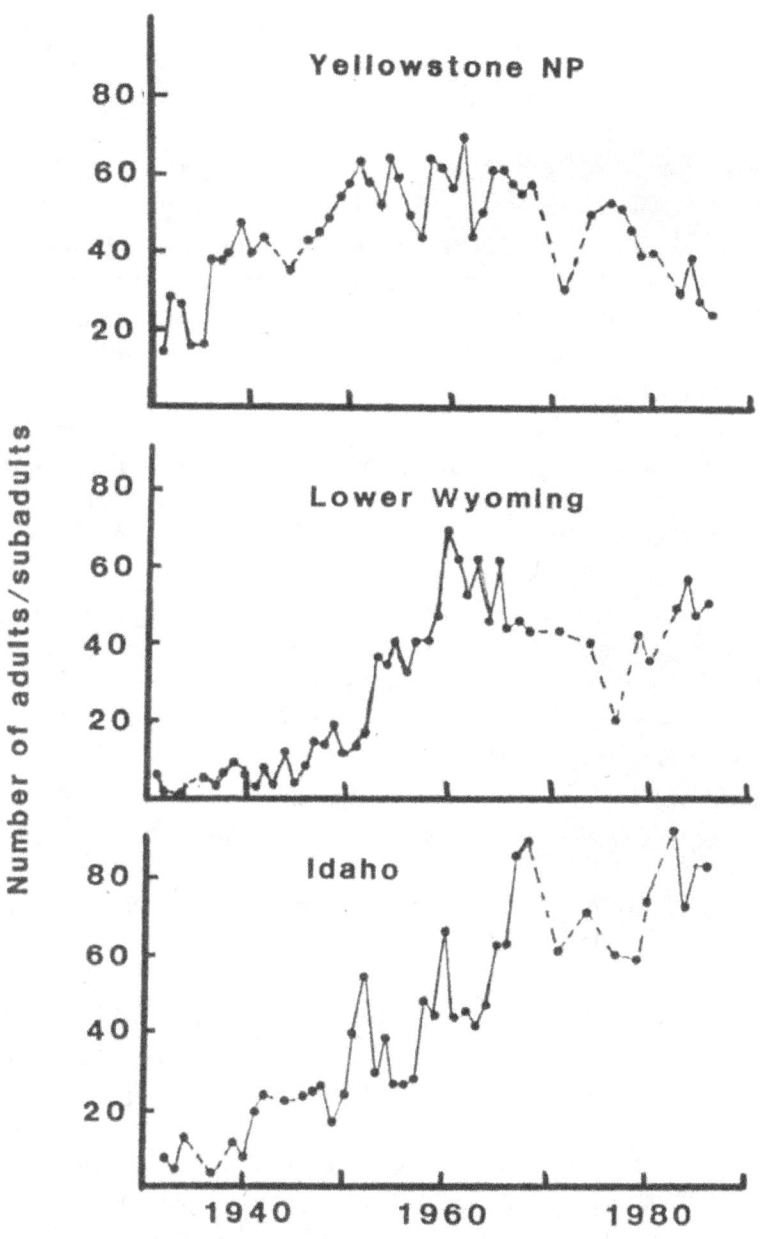

Figure 8. Adults/subadults in the Yellowstone, Lower Wyoming, and Idaho flocks, 1931-86. Data are from the late summer USFWS Tri-state Surveys.

about 36 to 113. Beginning in 1978, lakes in British Columbia were also included in the survey and the number of lakes surveyed in Alberta continued to increase. The 1984 survey covered only Alberta but again in 1985 the effort was expanded to include 30 lakes in British Columbia and 17 additional lakes in Alberta. Survey effort increased substantially in 1986 to include 79 lakes in British Columbia and 192 in Alberta (Holton and Shandruk 1987). Appendix VI presents the Grande Prairie census data, separated by province to aid in year-to-year comparisons.

The annual summer survey data showed that the number of adults fluctuated between 44-98 but made no net change from 1959-77, even though the surveyed number of lakes almost tripled (Fig. 9). In contrast, between 1977-84, the number of adults increased to 225. Although the survey effort was substantially greater in 1985 than in 1984, the number of adults decreased to 211. This decline was likely due to the unusually high mortality noted in the Tri-state Area during the winter of 1984-85 (Chapter 5).

In order to compensate for the effects of increased survey effort, Brechtel (1982) examined data from 55 lakes which were surveyed each year between 1972-81, and from 89 lakes surveyed each year between 1977-81. He then divided annual survey results by the total number of lakes surveyed each year. Brechtel concluded that the territorial component of the Grande Prairie flock had remained very stable during those years; no consistent increase in either the number of pairs or cygnets occurred. He concluded that the increases in the survey totals in the late 1970s were due to increases in the number of flocked and single swans observed.

We note that Brechtel's analysis of the Grande Prairie flock included only those swans occupying habitat in Alberta, and ignored swans occupying contiguous habitat in British Columbia. Brechtel's conclusion that the territorial component had remained relatively stable may correctly describe the situation at the center of the flock's range in Alberta, where territorial pairs could prevent use of their nesting lakes by other swans.

However, his conclusion revealed nothing regarding possible changes in the extent of the flock's total occupied range, nor about nesting activity expansion or contraction at the edge of their distribution. Brechtel also concluded that although increasingly accurate surveys had been flown over the past 20 years, the actual rate of change in the Grande Prairie flock was not calculable because of the changing survey effort.

Holton (pers. comm.) suggested that despite the increased survey effort, most of the censused increase was due to real expansion of the flock, particularly during 1974-1976, and offered the following evidence:

1) Between 1970-73, lakes which were surveyed for the first time contributed only 0-6% of the total swans observed, i.e., expanded survey coverage did not locate significant numbers of previously unsurveyed swans.

2) Between 1974-76, lakes surveyed for the first time contributed 7-21% of the total swans observed, suggesting either that the increasing survey area was "catching up" and including more of the previously occupied area, or that the swans were quickly expanding their area of use.

3) On 32 lakes surveyed for the first time in 1978 by Holton, only 11 swans or 5% of the flock was observed. The expanded effort did not significantly increase the number of swans found. Most of the recorded increase in swan numbers occurred in previously surveyed areas.

4) When Holton returned to Grande Prairie in 1983 swans had begun nesting within the main habitat area on beaver impoundments, which observers previously had considered to be too small to provide suitable nesting habitat. Thus the density of nesting swans in the core habitat had increased.

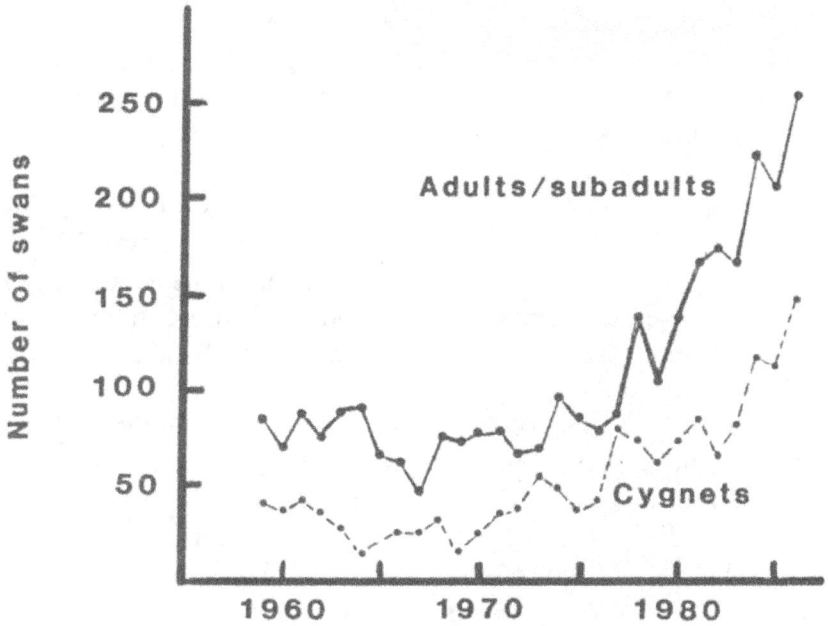

Figure 9. Late summer counts of the Grande Prairie flock by the Canadian Wildlife Service, 1959-86.

Other Alberta Flocks. Ninety-two lakes in the Otter Lakes vicinity were surveyed in 1983 to determine the species identification and population characteristics of swans reported in that area (Holton 1983). Aerial surveys were continued in 1984 and 1985 (Holton 1984, 1985). The 1985 aerial survey was the most complete to date, including the Chinchaga River, Otter Lakes, Edson/Whitecourt, and the Pincher Creek/Cardston area (Shandruk 1986). Otter Lakes trumpeters have remained at 7-9 adults/subadults despite the production of 19 cygnets in 1982 and 1983. In 1986, only 3 adults/subadults and 4 cygnets were observed in the Otter Lakes area, while 19 adults/subadults and 21 cygnets were counted near Chinchaga River (Holton and Shandruk 1987). Swan use of the Edson area increased from 8 adults/subadults in 1983, to 15 in 1985, and 13 in 1986 (Holton 1983, Shandruk 1986).

Yukon Flock. Although records go back only to 1970, McKelvey *et al.* (1983) noted that local knowledge of swans in the Yukon dates to at least 1945. Extensive aerial surveys in 1978 and 1979 revealed a breeding concentration of swans in the southeastern Yukon near Toobally Lakes, and these birds were later confirmed to be trumpeters. In 1980 and 1981, more intensive surveys were made in that area to describe habitat, assess productivity, and to neck-band swans to determine migration routes and wintering sites.

The CWS aerial surveys in 1978, 1979, and 1981 covered much of the southern Yukon in the watersheds of the Big Salmon, Little Salmon, Mugundy, upper Pelly, Ross, and Liard Rivers. Predetermined routes were flown in areas where swans had been previously reported and a total of 32 pairs was estimated for the Toobally Lakes area (McKelvey *et al.* 1983). In July 1985, the CWS made another extensive aerial survey of the southern Yukon which duplicated much of the 1981 coverage. In the Ross/McEvoy and Toobally Lakes areas, which were assumed to be part of the range of the Rocky Mountain Population, fewer swans were found than expected. Only 21 pairs were located in the Toobally area (McKelvey 1986). Recent census data for the Yukon and other Interior Canada flocks are given in Appendix VII.

Nahanni, NWT Flock. General reconnaissance surveys were flown in the mid-1970s in and near what is now Nahanni National Park Reserve to inventory the wildlife resources. The first survey specifically for swans was flown by CWS on 14-15 June 1984 in the Nahanni NPR area (McCormick 1985). On 5-8 August 1985, the CWS made a more extensive survey of areas near the Liard/South Nahanni River junction, North Nahanni River, and Carlson Lake. A portion of the South Nahanni River within Nahanni National Park Reserve was not surveyed in deference to the wishes of Parks Canada. While only 26 adults were reported in 1984, the expanded survey effort in 1985 located 51. A survey of the same area in 1986 found 84 adults. The 1986 increase surpassed that expected from cygnet production and apparently reflected an influx of adults from outside the NWT survey area (L. Shandruk, data presented at The Trumpeter Swan Society 10th Conference, Grande Prairie 1986).

Other British Columbia Flocks. The Fort Nelson area was surveyed for the first time in early summer 1981, and 34 adults were located. A 1985 survey covered a slightly larger area but did not include some lakes where swans were observed in 1981. In the areas surveyed in both years, the number of adults declined from 34 to 16, but cygnets increased from 0 to 4 (McKelvey 1986).

McKelvey (1986) also surveyed the Boudreau Lakes area to the southwest of Fort St. John in 1981, and located 10 adults. He did not cover the Dawson Creek area because it was regularly included in the Grande Prairie flock surveys. On 5 August 1985, McKelvey duplicated his 1981 route and included the Dawson Creek area, finding a total of 43 adults and 23 cygnets. Parts of the Dawson Creek area were also flown during the 1985 Grande Prairie survey, resulting in duplicate counting of swans on some lakes. The actual number of swans observed was approximately 23 adults and 17 cygnets (L. Shandruk, pers. comm.).

Cypress Hills, SK Flock. Although the first authenticated record of Trumpeter Swans in Saskatchewan was made in 1914 when one pair was observed in the Cypress Hills, breeding was not reported until 1951 (Nieman and Isbister 1974). Since then, counts have been made annually and the flock has varied from one to three breeding pairs.
Although cygnets have fledged in most years, this flock has failed to expand (Killaby 1987).

A second group of swans discovered in August 1973 between Meadow Lake and North Battleford in west-central Saskatchewan and thought to be trumpeters (Nieman and Isbister 1974) were later determined to be Tundra Swans (Brechtel 1982).

Total Interior Canada Subpopulation. Using the available census data, there are two ways to estimate the total number of trumpeters in the Interior Canada Subpopulation. Most simply, the sum of all swans counted in the ICSP flocks during the 1985 summer rangewide census gives an estimated total of 418 adults and 190 cygnets (Fig. 2). This survey covered all areas of suspected swan activity within the Subpopulation's known range.

Another method of estimating the size of the ICSP is to estimate the numbers of migrant trumpeters which winter in the Tri-state Area. This estimate is made by subtracting the number of swans counted in the late summer Tri-state Subpopulation, from the total number of RMP swans censused during the subsequent winter. For the winter of 1985-86 this calculation is as follows:

	Adults	Cygnets
February 1986 RMP winter count	1,304	299
- September 1985 Tri-state count	368	139
= estimated number of wintering northern migrants	936	160

We recognize that this crude estimate ignores autumn mortality in the Tri-state Subpopulation and assumes that all Tri-state swans were located during the summer count. Nevertheless, this estimate is strikingly larger than the sum of the individual ICSP flock counts. The estimated number of wintering migrant adults in the Tri-state Area in 1985-86 was 518 more than the number (418) of adults actually censused in Canada in summer 1985.

At first glance the estimated number of migrant cygnets (160) appears close to the number of ICSP cygnets censused during summer (190). However, neck-banding studies (Chapter 5) indicate that cygnets suffer about 30% mortality between fledging and midwinter. Thus, the 329 RMP cygnets that were censused in summer 1985 should have been reduced to about 228 by February 1986; about 70 extra cygnets wintered in the Tri-state Areas in 1986.

This large discrepancy between the censused summer and winter numbers, particularly in the adult age classes is puzzling and suggests several possibilities, none of which are mutually exclusive:

1) More Tundra Swans winter in the Tri-state Area than have been recognized to date.

Observers may successfully detect Tundra Swan family groups and remove them from the census totals, but fail to detect substantial numbers of nonbreeding tundras intermingled with trumpeters.

2) The 1985 summer censuses in Canada overlooked more than 500 trumpeters.

3) Substantial numbers of RMP nonbreeders move outside of the Population's known summer range in a molt migration.

4) A substantial number of Alaskan trumpeters migrated into the Tri-state wintering area as suggested by McEneaney (USFWS Midwinter Trumpeter Swan Survey 1984).

For whatever reasons, about 40% of the adult swans which wintered in the Tri-state Area in 1985-86 could not be located on the known summer ranges in 1985. As discussed earlier in "Winter Population Trends", this high number of northern migrants was not unique in the winter of 1985-86, but has been increasing steadily since the mid-1970s (Table 1). It was only the completion of the first rangewide summer survey in 1985 that made the discrepancy between censused summer and winter numbers quite obvious.

Summary

The most thorough surveys of the entire RMP are those made on the Tri-state winter range when both the Tri-state and Interior Canadian subpopulations are present. Prior to 1974, partial surveys of the Tri-state region were made by the states, Red Rock Lakes NWR, and the National Park Service. Since 1974, most of the region has been surveyed, primarily from the air, in early February. Survey reliability was reduced in some years by storms that prolonged the survey period and by the presence of Tundra Swans.

Prior to 1974, the number of wintering trumpeters censused in the Tri-state Area approximately equaled the number of swans counted during the late summer Tri-state surveys. There was no indication of the presence of large numbers of migrants, although marked Grande Prairie swans were observed wintering in the Tri-state Area in the 1950s. Wintering swans increased from about 700 in 1974, to 1,600 in 1986. This increase occurred within the migratory Interior Canada Subpopulation; during the same period the Tri-state Subpopulation declined.

The adult segment of the Tri-state Subpopulation grew from about 50 in 1935, to 548 in 1954. During the subsequent decade their numbers fluctuated. Adult numbers peaked at 554 in 1964 and remained above 500 until 1968. Between September 1967 and September 1968, 17% of the adults were lost from the Subpopulation. These reduced numbers then remained fairly stable over the next decade. Between 1980 and 1986, adult numbers declined again, from 462 to 331.

The 1967-68 decline in the Tri-state Subpopulation was due to a 28% decline in the Centennial Valley flock. Similar declines followed within three years in the Yellowstone and Idaho flocks. In the 1980s, declines have occurred in the Centennial Valley and Yellowstone flocks, but concurrently the Idaho and lower elevation Wyoming flocks have increased or remained stable. Most flocks in Canada, except the Grande Prairie flock, have only been censused once or twice. Between 1980-86, the trumpeters in the Toobally Lakes area, Yukon, declined while the Grande Prairie flock increased. The Grande Prairie flock showed no net growth between 1959-77. Since 1978, the number of adults increased from 88 to 251. Both the area surveyed and the actual numbers of swans have increased.

The Nahanni, NWT flock has increased markedly since one swan was observed in 1970. Although increased survey effort may account for much of the apparent increase between 1980 and 1985, a real increase in adult numbers must have occurred in 1986.

The survey of all Canadian flocks during the summer of 1985 located a total of 418 adults and 190 cygnets. Added to the total of 368 adults and 139 cygnets in the Tri-state Subpopulation in summer 1985, this gave an estimated total size for the Rocky Mountain Population of 786 adults and 329 cygnets. However, the number of swans wintering in the Tri-state Area in February 1986 was 1,304 adults and 299 cygnets. The summer range of nearly 1/3 of the 1,600 swans wintering in the Tri-state Area is unknown.

Table 1. Winter counts (adults-cygnets) of the Rocky Mountain Population (RMP) and
estimated size of the Tri-state (TSP) and Interior Canada (ICSP) Subpopulations,
1973-1986.

Winter	Total RMP[a]	TSP[b]	ICSP[c]
1972-73	499- 86	No data	No data
1973-74	553-156	450- 80[d]	83- 76
1974-75	595-128	457- 80	138- 48
1975-76	623-102	440- 58	183- 44
1976-77	839-178[e]	420- 80	419- 98
1977-78	695-179	403- 86	292- 93
1978-79	743-123	430- 67[d]	303- 56
1979-80	767-172	462- 87	305- 85
1980-81	1,000-247	462- 23	538-224
1981-82	952-266	440- 65[d]	512-201
1982-83	1,025-207	420- 40[d]	605-167
1983-84	1,128-332	398- 54	730-278
1984-85	1,326-190	424- 53	902-137
1985-86	1,304-299	368-139	936-160
1986-87	1,196-386	331- 61	865-325

[a] Data from USFWS Midwinter Tri-state Trumpeter Swan Survey.

[b] Data from previous late summer USFWS Tri-state Trumpeter Swan Survey.

[c] ICSP = RMP minus TSP

[d] Estimated from partial ground census data; no complete aerial survey made.

[e] High proportion of Tundra Swans in wintering area.

CHAPTER 3. NESTING AND CYGNET PRODUCTION

Much of the available information regarding the nesting and cygnet production of RMP trumpeters comes from observations of nesting efforts at Red Rock Lakes NWR. These records are unique in that they are based on relatively large numbers of nest attempts (24-78 per year), and the observations have been reasonably consistent over the past five decades. Over the last 30 years, 42% of the cygnets fledged in the Tri-state Subpopulation came from the Refuge. Because of the Refuge's important contribution to the productivity of the entire TSP and the wealth of Refuge data, this chapter focuses primarily on the Refuge with comparisons to other flocks where information exists.

Widely fluctuating annual cygnet production, characterized by high prefledging mortality, has worried Refuge managers since the 1950s (Banko 1960:131). In order to analyze some of the factors which affect cygnet production on the Refuge, we searched the entire Refuge files and assembled data sets on a variety of population and environmental variables for the 1935-86 period. These variables included reproductive parameters such as: number of nests, clutch size, percent hatch, number of eggs and cygnets removed for transplants and propagators, nest success, cygnet survival, and number of cygnets fledged. Habitat variables included: water levels, monthly temperature and precipitation, snowfall, length of ice-free period, winter and spring severity indices, and amount of grain provided during the previous winter.

Another important variable, human disturbance, could not be quantified, however narrative comments in Refuge reports provided a substantial amount of information. For most variables, although the original annual Refuge records still existed, the information had never been assembled to form a useful record; data were scattered in a 50-year accumulation of assorted files and reports. Using the resulting data sets, we analyzed the relationships between annual cygnet production at Red Rock Lakes NWR and various environmental and population variables.

Distribution of Nesting Pairs

Tri-state Subpopulation

The total number of nesting pairs has declined from over 90 in 1978 (Shea 1979) to about 78 in 1986 (Table 2). During this period the Yellowstone flock declined from 18 to about 7 nesting pairs. Nesting pairs also declined at RRLNWR, however increased nesting in the CV west of the Refuge somewhat offset this loss. In the Idaho and lower elevation Wyoming flocks, the number of known nesting pairs increased slightly, although this may have been due in part to the increased effort to locate nests.

Red Rock Lakes National Wildlife Refuge. Records of nest locations have been kept regularly since about 1958, and occasionally in earlier years. For years in which the number of nests was not reported, we made a minimum estimate based upon the number of broods observed. The assembled record of annual nest totals on the Refuge, by lake, appears in Appendix VIII. Figure 10 includes a graph of the annual total number of nests on the Refuge. Records regarding the number of nests in the CV downstream from the Refuge are sporadic and reflect widely varying amounts of survey effort. The increase in off-Refuge nesting to 18 nests by 1985, however,

Table 2. Distribution of nesting pairs in the Tri-state Subpopulation in 1978 and 1986 [a].

| | Number of nesting pairs | |
	1978	1986
Red Rock Lakes NWR, MT	43	25
Other Montana	[b]	19
Targhee NF, ID	8	9
Other Idaho	[b]	8
Yellowstone NP, WY	18	7
Other Wyoming	[b]	10
Total Subpopulation	90	78

[a] 1978 data from Shea (1979). 1986 data from IDFG files, brood locations from September 1986 Tri-state Trumpeter Swan Survey, YNP aerial survey of 28 May 1986, and W. Kurtenbaugh and D. Lockman, pers. comm.

[b] A total of at least 21 nests was known to exist in the Centennial Valley outside of Red Rock Lakes NWR, on State and private lands in Idaho, and in the lower elevations of Wyoming (Shea 1979).

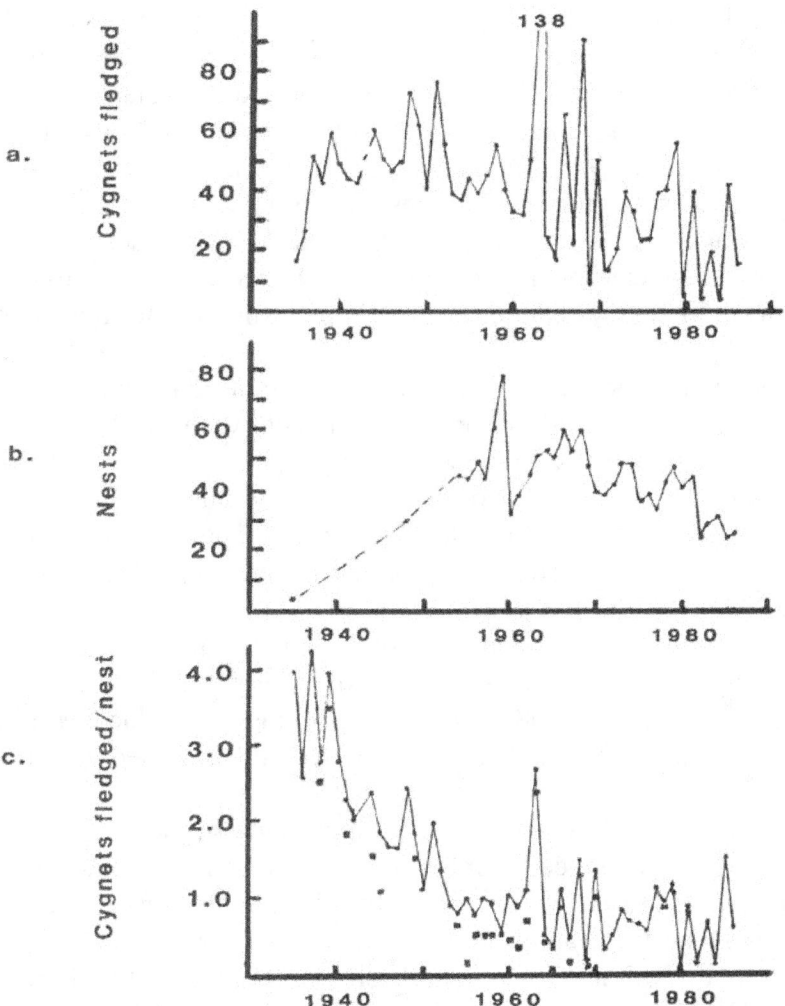

Figure 10. Annual cygnet production (adjusted for removals), number of nests, and
number of cygnets fledged per nest. The effective decrease in cygnets
fledged/nest resulting from removals is shown by ■. All data are from
Red Rock Lakes NWR files, and are given in Appendices IV and X.

appeared to result from a real increase in nesting effort in the western, lower elevation portions of the Valley (Pers. comm. from W. Banko and J. Roscoe).

The number of nests in the Refuge gradually increased to a peak of 78 nests in 1959, of which 67 were visited. Aerial surveys of the Refuge in 1960, however, located only 62 pairs of swans and found only 31 active nests. This apparent abrupt decline in nesting effort occurred primarily on the River Marsh and Swan Lake, where nests declined from 52 in 1959 to 6 in 1960 (RRLNWR Ann. Narr. Repts. 1959, 1960; Appendix VIII). Refuge Narrative Reports expressed no particular concern and it appears that changes in personnel and survey methods were partially involved in the apparent decline. We conclude that a substantial reduction in the number of nests occurred but possibly the loss was not as drastic as the 1960 data indicate. In addition to the new biologist's lack of familiarity with the marsh, the only other relevant factor which we could identify was the Hebgen Lake earthquake of August 1959. As we will discuss in Chapter 6, this earthquake caused some temporary changes, and perhaps some long-term changes, in the marsh environment.

After this decline the number of Refuge nests gradually rebounded, increasing to 60 by 1968. Although the CV flock declined by some 95 adults between 1967 and 1968 (Fig. 7), the number of nesting pairs on the Refuge increased slightly between 1967 and 1968. Thus we conclude that the loss of swans in 1967-68 occurred mainly within the flock's nonbreeding subadult component rather than among the established nesting pairs. During the subsequent two nesting seasons, however, the number of nesting pairs on the Refuge declined substantially, from 60 in 1968 to 39 in 1970.

The decline in the number of Refuge nests, following within two years of the 1967-68 decline in adult/subadult numbers, suggests at least two hypotheses:

1) The loss of subadults in 1967-68 resulted in a shortage of replacement breeders in subsequent years; thus suitable breeding habitat on the Refuge remained vacant as the flock declined.

2) As suggested by Page (1976:34), the carrying capacity of the Refuge for breeding pairs declined in the mid-1960s. The reduced number of adequate nesting territories caused the Refuge flock to stabilize at a lower level, with surplus eligible breeders prevented from nesting by the shortage of suitable nesting habitat (see population models in Page 1976:82-85).

It is important to note that the second hypothesis is not consistent with the sequence of events at the Refuge. The loss of swans from the CV flock in 1967-68 occurred rather abruptly and preceded the reduction in nesting effort; thus, the decline of the CV flock could not have been the result of reduced nesting effort. It is more likely that the reduced breeding effort resulted from the loss of the subadults and the resulting reduction in numbers of new recruits into the breeding segment. We will discuss the likely causes of the loss of subadults in Chapter 8.

The number of Refuge nests fluctuated between 34 and 48 in the period from 1970 to 1981, but never rebounded to the numbers found in the 1950s and 1960s. Again, in 1982, the number of Refuge nests declined abruptly, decreasing from 44 in 1981 to 24 in 1982. As previously mentioned, this most recent decline in Refuge nesting was apparently accompanied by an increase in off-Refuge nesting on downstream CV lands. Between 1982-86, the Refuge contained an average of only 27 nests per year, about 50% of the number supported during the peak years.

While the total number of nests has declined by about 50% from peak levels, the distribution of nests within the Refuge has also shifted (Table 3). Lower Lake experienced the greatest relative loss of nests and has supported only about the same number of nests as Upper

Lake since 1969, despite its apparently superior habitat (Page 1976:55).

Reasons for the shifts in nesting effort within the Refuge have not been conclusively determined, although the effects of construction of the lower water control structure in 1958, the Hebgen Lake earthquake of August 1959, changes in water depths, siltation, and vegetation productivity were all likely involved and are discussed in Chapters 6 and 7. The loss of nests due to flooding on Lower Lake has been an obvious factor in the 1980s (RRLNWR Trumpeter Swan Reports 1980, 1982, 1984). Human disturbance from management and research activities was also likely involved. Page (1976:59) concluded that due to disturbance by the airboat during his 1971-73 study, swans relocated their nests to areas that were less accessible.

Interior Canada Subpopulation

Surveys of the ICSP flocks to determine the number of nesting pairs have not been routinely conducted. Therefore, numbers of nesting pairs in the various flocks were estimated from various late spring and summer aerial surveys. In 1986, the ICSP was estimated to contain approximately 145 nesting pairs, distributed as shown in Table 4 (L. Shandruk, pers. comm.).

Arrival at Territory and Nest Initiation

Trumpeters may arrive at their territory before open water is available, and they occasionally initiate nesting before the ice has completely melted. At Red Rock Lakes NWR and in much of YNP, pairs usually begin nest construction between April 20 and May 30. At lower elevation sites in YNP and Idaho, where spring breakup can begin in late March, swans may occupy their territory for two to four weeks prior to initiating nest building. At higher elevations in YNP, nest initiation may begin in May before the territory is ice-free (Shea 1979). Grande Prairie trumpeters begin nest construction soon after their arrival in early April and up to one week before the lakes become completely ice-free (Holton 1985). Arrival and nest initiation are frequently delayed in late cold springs (Page 1976:65; Cooper 1979; Shea 1979; Holton 1985).

Territory Occupancy and Cygnet Production

Red Rock Lakes National Wildlife Refuge

Documentation of the occupancy rates of particular nesting territories at the Refuge has been hampered by difficult viewing conditions, lack of marked nesting swans, and use of alternate nest sites within territories. The determination of long-term occupancy rates and/or cygnet production at specific Refuge territories was not possible from the available data. Page (1976:63) observed that in 1971-73, about 95% of the Refuge nests were reoccupied the following year, and 11 of 17 new territories were occupied for at least one year before swans attempted to nest.

Table 3. Distribution of nests at Red Rock Lakes National Wildlife Refuge during four periods since 1954[a]. Data are presented as mean number of nests (range) and percent change from peak numbers of nests to present (1982-86) numbers.

Years	Upper Lake	Lower Lake	River Marsh	Swan Lake	Refuge Ponds	Refuge Total
1954-59	6 (5- 7)	11 (6-19)	26 (19-38)	10 (7-14)	0	54 (44-78)
1960-68	8 (3-11)	13 (11-18)	18 (4-25)	9 (2-13)	3 (2-4)	49 (31-60)
1969-81	7 (3-10)	6 (4- 8)	15 (10-20)	9 (5-11)	6 (3-7)	42 (34-48)
1982-86	3 (1- 4)	3 (1- 4)	10 (7-12)	7 (5- 8)	4 (3-5)	27 (24-31)
Percent Change[b]	-63%	-77%	-62%	-30%	-33%	-50%

[a] Data were extracted from RRLNWR Annual Narrative Reports and RRLNWR Annual Trumpeter Swan Reports, and represent minimum estimates of the mean number of nests and the range during each period.
[b] Percent change from peak number of nests to present number.

Table 4. Estimated distribution of nesting pairs in the Interior Canada Subpopulation, 1986[a]

Area	Number of nesting pairs
Grande Prairie, AB	51
Other Alberta	13
Ft. St. John, BC	8
Other BC	3
Toobally Lakes, YK	38
Nahanni, NWT	15
Cypress Hills, SK	1
Total Subpopulation	145

[a] Estimated by L. Shandruck, CWS, Edmonton.

Other Areas

Very few territories consistently fledge cygnets year after year; sporadic cygnet production at individual territories is the norm in the Rocky Mountain Population, even in expanding flocks. Based upon data from the late summer USFWS Tri-state Surveys, few lakes in YNP were consistently occupied by nesting pairs between 1931-78. Of 37 lakes where swans nested, only 10 (27%) were occupied by pairs in more than half of the years. Only two lakes were occupied in more than 70% of the years. Cygnet production has been even more sporadic than territory occupancy in YNP; only one lake (Riddle Lake) has fledged cygnets in more than 40% of the years from 1931-78. Approximately two-thirds of the nesting lakes fledged cygnets in less than 10% of the years between 1931-78.

Sporadic cygnet production is not limited to territories in the Yellowstone area. At 12 territories used by the lower elevation Wyoming flock between 1981-85, only three (25%) fledged cygnets in more than half of the years. Four sites (33%) fledged no cygnets during the period (Lockman *et al.* 1986). Brechtel (1982) classified the lakes in the Alberta portion of the Grande Prairie habitat according to the percentage of years between 1972-81 in which they fledged cygnets. Of 51 lakes, only 39% fledged young in more than half the years. Despite this sporadic production, the Grande Prairie flock increased during the period (Appendix VI).

Egg Laying

At Red Rock Lakes NWR, first eggs have been reported as early as 22-26 April (RRLNWR Annual Narr. Repts. 1941, 1966). A clutch of 8 eggs, of which seven hatched on about 31 May 1987 (C. Mitchell, pers. comm.) must have been laid beginning about 15 April 1987. Most egg laying at the Refuge, however, occurs between 1 May and 15 May (Banko 1960:116). Elsewhere, the first eggs observed have been on 23 April in Idaho (Hampton 1981), 30 April in lower Wyoming (Lockman *et al.* 1987), and 10 May in YNP (Shea 1979). Late nesting pairs in the Tri-state Subpopulation have been observed to delay laying until at least 30 May, as evidenced by mid-July hatching dates.

In studies of Grande Prairie nests in 1958-63, Mackay (1987) observed the earliest clutch initiation on 25 April 1956 and the latest on 11 May 1959.

Clutch Size

Comparative data on the reproductive parameters of most RMP flocks are given in Appendix IX, and will be a helpful reference to the following section.

Wild RMP trumpeters rarely lay more than seven eggs in a clutch, although the largest reported clutches contained nine eggs (Banko 1960:114; Mackay 1987). Hull (1939) noted that Red Rock Lake trumpeters laid from four to eight eggs although the latter was unusual. During studies in 1977-80, many clutches in YNP were unusually small. Seventeen of 55 clutches contained less than four eggs. In addition, six pairs of swans attended nests and maintained incubating postures for weeks, apparently without laying any eggs (Shea 1980). In contrast, clutches containing fewer than four eggs were significantly less frequent in Grande Prairie (Table 5).

Mean clutch size has been smaller in the Tri-state flocks than in the Canadian flocks or at Lacreek NWR, SD (Appendix IX). At Red Rock Lakes NWR during the last 20 years, clutch size has averaged 4.6 ($n= 494$) and ranged from 3.41 ($n= 32$) in 1967, to 5.39 ($n= 33$) in 1972 (Appendix X). Clutch size averaged 4.0 in YNP ($n= 56$) (Shea 1980), 4.3 ($n= 21$) in lower elevations of Wyoming (Lockman et $al.$ 1987), and 4.4 ($n= 21$) in Idaho (Maj 1983). In contrast, clutch size averaged 5.6 ($n= 14$) in Grande Prairie (Holton 1985), and 6.4 in Cypress Hills, SK (Nieman 1979). Leach (1977) reported a mean clutch size of 6.2 at Lacreek NWR during the period 1963-76.

In captivity, the mean clutch size of trumpeters is larger than in the wild (Lumsden 1987b). Captive descendants of Red Rock Lakes trumpeters kept at the Brit Spaugh Zoo in Great Bend, Kansas, laid up to 11 eggs and maintained a mean clutch size of 8.4 over a six-year period (Page 1976:106). In 104 clutches laid by 41 captive trumpeters, clutch size increased with the age of the laying female. Clutch sizes averaged 5.0 for 3 year-old birds ($n= 4$), 5.3 for 4 year-olds ($n=15$), 5.7 for 5 year-olds ($n= 20$), 5.9 for 6 to 8 year-olds ($n= 35$), and 6.9 for 9 to 14 year-olds ($n= 19$) (Lumsden 1987b). Lumsden also concluded that clutch size increased with nesting experience. The mean clutch size of 4.5 among first-time breeders of ages 3-7 years was significantly less than the mean clutch size of 6.3 for breeders with previous experience.

Studies based upon marked swans in Wyoming showed a similar trend. In six first nest attempts recorded in the lower elevation Wyoming flock in 1982-85, clutch size averaged 3.8. In 17 clutches laid by pairs with previous nesting experience, mean clutch size was 4.6 (Lockman et $al.$ 1987).

Recent work has shown that captive trumpeters can be made to lay much larger than normal clutches. Three experiments involving the removal of eggs from clutches during the laying period resulted in final clutch sizes of 10, 12, and 17 eggs. Examination of dead females has shown that they may produce many more follicles than the number of eggs actually laid. Their follicles normally are resorbed when the female's body reserves diminish to the level essential to her survival (Lumsden 1987b).

Hatching

In the Tri-state Area, the peak of hatching typically occurs during the third week of June, although the entire hatching period may last from late May to early July (Appendix XI). Low elevation clutches usually hatch before those at higher elevations where nest initiation is often delayed by late ice-out (Shea 1979, Lockman et $al.$ 1987). The peak of hatching occurs later following late, cold springs (Shea 1979, Holton 1985), and as we will discuss later in our analysis, a strong relationship exists at Red Rock Lakes between a late peak of hatch and the reduction in the number of cygnets per breeder in late summer.

The peak of hatching also normally occurs about the third week in June in Grande Prairie (Turner 1982, Holton 1985, Mackay 1987). During the cold late spring of 1979, the median hatch date in Grande Prairie was later than in 1978 or 1980, and the span of the hatching was shorter (Holton 1985). Based upon an aerial assessment of cygnet development, McKelvey et $al.$ (1983) estimated that the peak of hatching in the Yukon occurred about the first week of July. In the Northwest Territories, McCormick (1985) observed one pair of trumpeters accompanied by five cygnets, and two other pairs with clutches of eggs on 14-15 June 1985.

Table 5. Mean clutch sizes and frequency at Grande Prairie, AB, and Yellowstone
National Park, WY.

Location	N	|				Clutch Size					Mean
		1	2	3	4	5	6	7	8	9	
Yellowstone NP (Shea 1980)	55	1	8	8	15	16	7	0	0	0	4.0
Grand Prairie (Holton 1985)	34	0	1	0	5	11	9	8	0	0	5.6
Grand Prairie (Mackay 1987)	53	0	1	2	3	14	10	14	8	1	6.1

Low hatchability of eggs has long been considered a problem at Red Rock Lakes (Banko 1960:130). Recent records from Red Rock Lakes, YNP, and the National Elk Refuge, WY, showed that only about 50-56% of the eggs hatched. Hatching success of 61-91% has been documented in lower elevation nests in Wyoming and in Idaho (Appendix IX). At Red Rock Lakes, hatching success in the best years is typically 70-75% (Appendix X), however it has occasionally been much lower, as when flooding caused the loss of entire clutches in 1980, 1982, and 1984 and reduced hatching success to less than 30% (RRLNWR Ann. Trumpeter Swan Repts. 1980, 1982, 1984). Flooding has also increased hatching failures in YNP (Shea 1979).

Other observed causes of hatching failure in the Tri-state Area are egg infertility, nest abandonment related to human disturbance, and the abandonment of unhatched live eggs after part of the clutch has hatched. RRLNWR Annual Narrative Reports occasionally expressed concern over the seemingly high proportion of unhatched eggs which contained fully developed embryos. Page (1976:119) noted that females took their cygnets into the water shortly after hatching, often abandoning eggs with fully developed embryos that apparently would have hatched had incubation continued. Of 34 unhatched eggs collected by Maj (1983) from Red Rock Lakes NWR in 1981, 18 contained full term embryos while 15 showed no embryo development. Of 31 eggs that failed to hatch in YNP in 1977-78, Shea (1979) found 14 with no embryo development, 10 with partly developed embryos, and 7 with fully developed embryos.

Efforts to determine the proportion of infertile eggs have been made intermittently at Red Rock Lakes. Frequently, 20-30% of the examined eggs were classified as infertile, with a high of 50% in 1964 (data from RRLNWR Ann. Narr. Repts. and Ann. Trumpeter Swan Repts.). These data are misleading, however, because eggs were classified as infertile if no embryo could be seen. Some of these "infertile" eggs may in fact have been fertile eggs in which the embryo died early in its development. Romanoff (1972) reports that when examined histologically, 75% of supposedly infertile hen and turkey eggs were found to have actually had blastodisks which died before the eggs were laid. In YNP, only 9% of 195 eggs checked in 1977-79 showed no embryo development (Shea 1980).

Nest Success

Nest success (the percentage of nests hatching at least one egg) at Red Rock Lakes NWR has fluctuated widely, with occasional large losses due to flooding. In good years, nest success has been 85-95%; in the wet years of 1980, 1982, and 1984, it ranged from 26-35% (Table 6). Studies in 1971-73 found that nest success varied significantly between different areas of the Refuge. During those three years, success was 58% on Lower Lake, 71% on Swan Lake, 78% on River Marsh, 87% on Upper Lake, and 88% on the ponds (Page 1976:68). In studies elsewhere in the Tri-state Area, nest success ranged from 60% in YNP to 100% on the Targhee National Forest (Appendix IX).

Shea (1980) documented the fate of 70 nests in and near YNP during 1977-80 (Table 7). In six cases, failure was apparently due to the inability of the nesting pair to lay any eggs: no evidence of eggs or egg shell fragments was found immediately following the normal incubation period although the swans had appeared to be incubating when observed during several aerial surveys. Nest success did not differ significantly among nest habitat categories (beaver lodges, islands, and mounds of emergent vegetation) in YNP. Nest success in Grande Prairie was substantially higher than at YNP or RRLNWR, with studies based on 53 total nests reporting success of 94-100% (Burgess 1972, Holton 1985).

Brood Size At Hatch

Few data are available from Red Rock Lakes NWR concerning brood size at hatch because the newly hatched cygnets are very difficult to observe in the marsh. Several recent studies have found mean brood size at hatch (or cygnets hatched per successful nest) to range from 2.9-3.9 elsewhere in the Tri-state Area (Appendix IX). Mean brood size at hatch at Grande Prairie ranged from 4.5 (Holton 1985) to 4.9 (Burgess 1972), a result both of larger clutch size and higher hatchability of eggs at Grande Prairie than in the Tri-state Area (Appendix IX).

Brood Size and Number of Broods at Fledging

In most areas of the Tri-state region, average brood size at fledging appeared somewhat higher in the 1930s and 1940s ($\bar{x} = 3.4$), than in later years ($\bar{x} = 2.7$) (Appendix XII), but the decline coincided with later survey dates after 1953. The record seemed quite compatible with an explanation that actual brood size remained unchanged but later survey dates resulted in more accurate (lower) estimates of brood size at fledging.

At Red Rock Lakes NWR, the mean number of broods fledging increased as the number of nests increased, up to 17.6 broods fledged per year in the 1960s, then declined over 60% to 6.8 broods per year in the 1980s (Appendix XII). Obviously, declining recruitment at Red Rock Lakes primarily reflects the failure of many pairs to fledge any young, rather than a uniform decline in fledging success among most broods.

At Grande Prairie, average brood size at fledging remained about stable ($\bar{x} = 3.3$) from the 1960s to the 1980s, while the number of broods fledged per year more than tripled (Appendix XII).

Cygnet Production Rates

Low cygnet production, characterized by the site and/or pair specific loss of entire broods in the first few weeks after hatching, has been found in several studies of Tri-state trumpeters (Page 1976; Shea 1979, 1980; Maj 1983). When annual productivity is calculated as the number of cygnets fledged per 100 adults/subadults, cygnet production was highest in the Tri-state Area in the 1940s and declined by about 50% by the 1980s (Table 8). In contrast, productivity of the Grande Prairie flock has been almost three times greater than in the Tri-state in recent decades. This high productivity may be overestimated somewhat due to emigration of subadults from the Grande Prairie flock, thus reducing the number of nonproductive swans present during the late summer censuses. Whether or not emigration is a factor, Grande Prairie trumpeters surpass Tri-state trumpeters in virtually every reproductive measure (Appendix IX).

Table 6. Nest success at Red Rock Lake National Wildlife Refuge[a].

Year	Number of nests checked	Percent success
1944	23	91
1945	22	77
1946	19	84
1949	12	83
1951	12	92
1958	38	71
1959	74	92
1962	45	73
1966	60	95
1969	25	76
1970	26	81
1971	34	77
1972	33	64
1973	34	88
1974	48	48+[b]
1975	36	67
1976	23	70
1977	34	62+[b]
1978	37	86
1979	42	64
1980	41	27
1981	44	61+[b]
1982	22	35
1983	29	55
1984	31	26
1985	24	75+[b]
1986	25	36

[a] Data from RRLNWR Annual Narrative Reports and RRLNWR Annual Trumpeter Swan Reports.
[b] Minimum estimate based upon number of broods reported.

Table 7. Fate of 70 nests in Yellowstone National Park and vicinity, 1977-79[a].

Fate of nest	n	%
Successful	43	61
Flooded	7	10
Predator loss	1	1
Deserted during incubation	6	9
Fertile eggs, none hatched	4	6
No embryo development	3	4
No evidence of eggs	6	9

[a] Shea (1980)

Cygnet Production at Red Rock Lakes National Wildlife Refuge

Annual cygnet production to fledging on the Refuge has ranged from a very unusual high of 138 in 1963, to a low of four in 1982 and in 1984 (Fig. 10a). The record of actual cygnet production, with adjustments made from previously reported survey totals to account for cygnets transferred from the Refuge, is given in Appendix IV.

Cygnet production per nest was inversely proportional to the number of nests on the Refuge until about 1960. Since the early 1960s, however, this relationship has changed. After 1959 the number of nests fluctuated with a gradual downward trend. At the same time, the cygnet production per nest fluctuated but also showed a slight downward trend, with three record lows occurring in the 1980s.

Figure 10c also shows the effective reduction in the number of cygnets fledged per nest that was due to removals of cygnets. For 12 of 14 years during the period 1954-67, actual cygnet production after removals was less than 1.0 fledged per nest. As discussed in Chapter 12, at the current estimated adult survival rates, cygnet production must average at least 1.0 fledged per nest in order to maintain a stable flock at RRLNWR.

Estimates of cygnet production by lake can be made using Refuge nest records (Appendix VIII) and late summer Tri-state survey data (Appendix V). These estimates were somewhat inaccurate because occasional brood movement occurred between lakes and because the lake origin of cygnets removed for transplant in the 1950s and 1960s could usually not be determined. Nevertheless, these estimates for the 1970s and 1980s showed (Table 9) that compared to most other RMP flocks, which have fledged from 1.2-2.9 cygnets per nest attempt in recent studies (Appendix IX), production in all areas of the Refuge has ranged from 0.5-0.9 cygnets per nest attempt. Only the Yellowstone flock had a lower cygnet production rate. The causes of low cygnet production have been a management concern for decades. We focused our analysis on the productivity of the Refuge flock because it is the only RMP flock for which long-term production, population, and environmental data exist.

Key Factor Analysis

The importance of various factors that might affect the recruitment of cygnets to the fall population was analyzed using key factor analysis (Morris 1963). The number of fall cygnets was determined from the following identity:

Number of Fall Cygnets = (Spring Population Size) x (Proportion Breeders) x (Average Clutch Size) x (Proportion of Eggs Not Removed) x (Proportion Hatching Success) x (Proportion Fledging Success).

This relationship was converted to a key factor model by taking natural logarithms of both sides:

$$K = K_1 + K_2 + K_3 + K_4 + K_5 + K_6,$$

where

$K = ln$ (Number of Fall Cygnets),
$K_1 = ln$ (Spring Population Size),
$K_2 = ln$ (Proportion Breeders),

Table 8. Cygnets fledged per 100 adults/subadults in the Tri-state Subpopulation and Grande Prairie Flock[a]

| Decade | Tri-state Subpopulation | | | Grande Prairie Flock | | |
	Mean	Range	*n*	Mean	Range	*n*
1940s	35	23-57	10		No data	
1950s	21	14-30	10		No data	
1960s	20	9-48	9	35	8-54	10
1970s	18	16-21	4	56	32-91	10
1980s	17	5-38	5	51	39-59	7

[a] Tri-state data were calculated from late summer USFWS Tri-state Trumpeter Swan Surveys (Appendix III); Grand Prairie data were calculated from late summer CWS surveys (Appendix VII).

Table 9. Average annual cygnet production (A) and cygnet production per nest (B) in major nesting areas at Red Rock Lakes National Wildlife Refuge[a]

| | Upper Lake | | Lower Lake | | Swan Lake | | River Marsh | | Ponds | |
	A	B	A	B	A	B	A	B	A	B
1940s	11.1	2.0	21.6	3.2	3.8	1.9	No data		No data	
1950s	10.9	1.8	8.9	0.8	4.4	0.5	18.4	0.7		
1960s	6.9	0.9	12.3	1.0	6.5	0.7	16.9	0.9	4.0	2.0
1970s	5.3	0.8	5.2	0.9	6.4	0.7	11.5	0.8	5.1	0.8
1980s	4.0	0.9	1.6	0.5	3.7	0.5	5.1	0.5	3.6	0.8

[a] Data were extracted from RRLNWR files. Detailed records are in Appendix V and Appendix VIII.

$K_3 = ln$ (Average Clutch Size),
$K_4 = ln$ (Proportion of Eggs Not Removed),
$K_5 = ln$ (Proportion Hatching Success),
$K_6 = ln$ (Proportion Fledging Success).

Across all years of available data (1932-1985), hatching success ($P= 0.0001$) and survival to fledging ($P= 0.0001$) were the factors most highly correlated with cygnet production to fledging (Table 10). Clutch size ($P=0.01$) and the proportion of eggs not removed ($P= 0.02$) were highly significant also. The proportion of adults breeding was weakly related ($P= 0.20$), and size of the adult population was not significant at all ($P=0.75$) (Table 10).

During the period of population increase (1932-54), no data were gathered on hatching or fledging success and few if any eggs were removed. Adult spring population size showed a significant positive correlation with cygnet production ($P= 0.10$). The proportion of breeders in the adult population showed a negative association ($P= 0.28$) with number of cygnets produced during the increase phase, but shifted to a positive, significant correlation ($P= 0.05$) after 1954.

After the population peaked in 1954, adult spring population size was again positively correlated with cygnet production ($P= 0.08$), but the most significant correlations were between cygnet production and hatching success ($P= 0.0001$), and survival to fledging ($P= 0.0001$).

Conclusions could be confused by the effects of correlations between the key factors themselves and thus it is important to examine the correlation matrix between key factors for significant correlations (Table 11). The only highly significant correlation was a negative relationship between spring population size and proportion breeders ($P= 0.0001$). As we would expect for a territorial species with several nonbreeding subadult age classes, the proportion of the population breeding in a given year declined as total population size increased. For several years following years of high cygnet production or high survival, spring population size would be large and contain a high proportion of nonbreeding subadults. Hatching success and fledging success were also positively correlated ($P= 0.08$) (Table 11), suggesting that common factors influence both parameters.

The overall conclusions suggested by these analyses are:
1) Variations in hatching and fledging success have been most correlated with changes in cygnet production since the population peaked, and may both be influenced by common factors.
2) Variations in clutch size also correlate positively with changes in cygnet production.
3) Adult spring population size has shown a positive relationship to cygnet production from 1932-1985, but the slope of the relationship has changed between periods (Figure 11).
4) The relationship between the proportion of breeders in the spring adult population and cygnet production has shifted more dramatically yet, from weakly negative in the early years, to strongly positive after 1954.
5) Conclusions 3 and 4 suggest that relationships between population size and cygnet production are extremely complex.
6) The proportion of eggs not removed correlated positively with cygnet production.

Table 10. Key factor analysis of cygnet production at Red Rock Lakes NWR using correlations between key factors and cygnet production.

Data Set	Adult Spring Pop. Size K_1	Proportion Breeders K_2	Clutch Size K_3	Prop. Eggs Remaining K_4	Hatching Success K_5	Fledging Success K_6
All Years (1932-85)	r=0.05	0.21	0.40	0.33	0.73	0.72
	P=0.75	0.19	0.01	0.02	0.0001	0.0001
	n= 42	42	37	50	29	28
					--	--
Increase Phase (1932-54)	r=0.39	-0.26	0.36	0	--	--
	P=0.10	0.28	0.38	1.0		
	n= 19	19	8	19	2	0
After Peak (1955-85)	r=0.37	0.42	0.30	0.28	0.743	0.72
	P=0.08	0.05	0.11	0.13	0.0001	0.0001
	n= 23	23	29	31	27	28

[a] Estimates of the spring population size in the Centennial Valley flock were obtained from the late summer counts by assuming no loss of adults/subadults during the period from late spring until the late summer count. The proportion of breeders was estimated for the entire Centennial Valley as number of nests divided by the spring population size. Spring population size for the Refuge flock was then calculated from the number of nests on the Refuge divided by the proportion of breeders determined for the entire Centennial Valley. The key factor analysis was performed three times: once for all the years since 1932, once for the increase phase of the population (1932-54), and once for the period since the population peaked (1955-85).

Table 11. Correlation matrix of key factors for cygnet production at Red Rocks Lakes National Wildlife Refuge, 1932-85.

		Spring Population K_1	Proportion Breeders K_2	Mean Clutch Size K_3	Proportion Eggs Remaining K_4	Proportion Hatching K_5	Proportion Fledging K_6
K_1	$r=$	1.00					
	$P=$	0.00					
	$n=$	46					
K_2	$r=$	-0.75	1.00				
	$P=$	0.0001	0.00				
	$n=$	43	43				
K_3	$r=$	-0.10	0.14	1.00			
	$P=$	0.60	0.47	0.00			
	$n=$	30	30	38			
K_4	$r=$	-0.13	0.24	0.18	1.00		
	$P=$	0.42	0.12	0.28	0.00		
	$n=$	43	43	38	51		
K_5	$r=$	0.04	0.26	0.17	0.26	1.00	
	$P=$	0.86	0.25	0.38	0.17	0.00	
	$n=$	22	22	30	30	30	
K_6	$r=$	-0.13	0.14	0.07	-0.06	0.34	1.00
	$P=$	0.57	0.56	0.73	0.75	0.08	0.00
	$n=$	21	21	29	29	28	29

Based on these findings, we suspected that annual cygnet production would be most closely correlated with environmental factors that influenced hatching, survival to fledging, and clutch size. Effects of adult population size also could be important and these effects could result from changing population age structure, differences in age-specific productivity, or habitat constraints. In the next step of our analysis, we examined a large number of environmental factors that might be correlated with these key productivity factors.

Key Factors and Environmental Variables

The preliminary method for identifying environmental variables that may influence cygnet production involved testing for significant correlations between key production factors and these environmental variables. Such simple correlations must be treated with caution, however, due to the possibility of cross correlations with other variables. We calculated simple correlations between key factors K_3 through K_6, and 19 environmental variables (Table 12).

As expected, the proportion of eggs not removed was not significantly correlated with any environmental variables. Clutch size was positively correlated with May temperatures ($P=0.08$) and negatively correlated with April precipitation ($P=0.10$) and water levels ($P=0.08$) during the entire April-July period. The most highly significant correlations would appear to be spurious, unless one is prepared to believe that swans can predict July precipitation and water levels during their laying period in May. Clutch size was also negatively correlated with the amount of grain fed during the previous winter ($P=0.10$). McLandress and Raveling (1981) indicated that succulent wild food items, as opposed to cultivated grain seeds, were of major importance in the process of spring weight gain prior to egg laying. Apparently, winter grain feeding will not increase clutch size in trumpeters, however we do not understand the full implications of this relationship.

We expect that warm May temperatures would increase water temperatures and accelerate the growth of aquatic plant and invertebrate food sources. High April precipitation and high water levels, on the other hand, may retard the development of spring foods, thus reducing the breeding female's opportunity to acquire the energy reserves necessary to provision a clutch of eggs.

The proportion of eggs hatching showed a significant negative correlation ($P=0.03$) with median water level in June, indicative of the effect of high water flooding nests. Hatching success also showed a significant positive correlation with the amount of grain fed per bird during the previous winter ($P=0.09$) and a positive association with mean maximum temperature in June ($P=0.13$). Hatching success was negatively associated with increasing rain in May, June, and July.

Survival from hatching to fledging (Proportion Fledging) was positively correlated with mean minimum temperature in June ($P=0.07$) and negatively correlated with July water levels ($P=0.06$) (Table 12).

We conclude that the annual production of cygnets at Red Rock Lakes NWR is reduced by cold spring temperatures, high precipitation, and high water levels. We suspect that these factors interact in several ways: 1) they delay spring food development and reduce the breeding bird's opportunity to acquire adequate prebreeding nutrition, resulting in smaller clutch sizes and reduced viability of eggs and cygnets; 2) they also reduce the availability of foods to the newly hatched cygnets, and 3) subject the cygnets to increased thermoregulatory stresses.

52

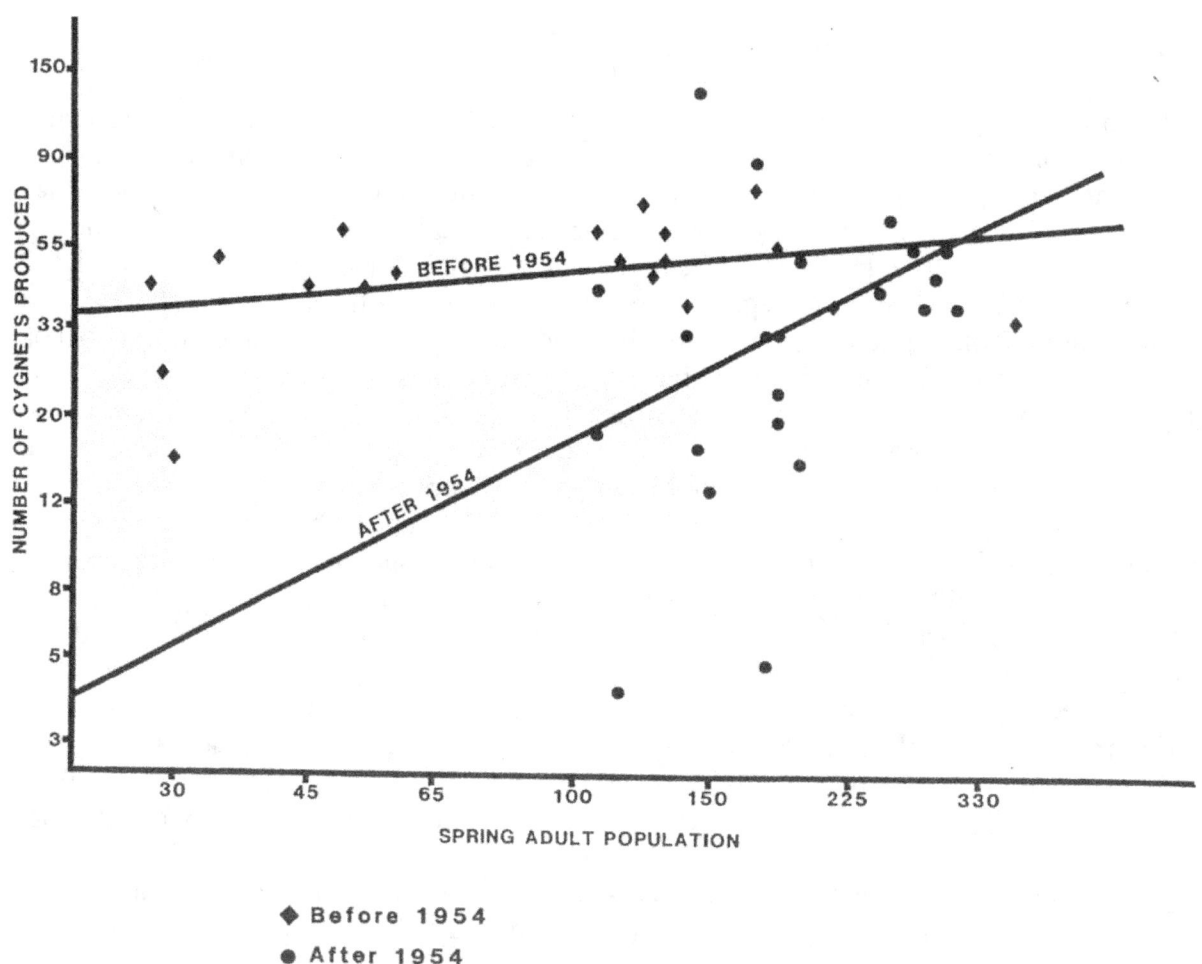

Figure 11. Relationship between spring adult population size at Red Rock Lakes NWR and the number of cygnets produced.

Modeling and Predicting Cygnet Production

The reproductive rate in a swan population is traditionally measured as the number of cygnets produced per breeding pair. This approach is not completely adequate from a population perspective even though it is intuitively appealing. It leaves out the nonbreeding portion of the population and does not lead directly to a population model. An alternative approach can be developed as follows:

A simple model to predict the population size in succeeding years:

$$N_{t+1} = N_t R_t,$$

where N_t = population size in year t,
R_t = finite rate of increase during year t.

We can break the year into two basic periods, the spring-to-fall period of population increase, and the fall-to-spring period of population decline. The finite rate of increase from spring to fall (B_t) is due primarily to recruitment of cygnets to the fall population. The finite rate of decline from fall to spring (S_t) reflects mortality of cygnets and older birds during the winter. Thus $Rt = BtSt$ and $N_{t+1} = NtBtSt$. If we assume no mortality of adults/subadults from the start of the breeding period until the late summer count, the finite birth rate can be estimated as:

$$B_t = \frac{\text{Number of cygnets + adults/subadults in late summer}_t}{\text{Number of adults/subadults in late summer}_t}$$

Likewise, the finite rate of decline can be estimated as:

$$S_t = \frac{\text{Number of adults/subadults in late summer}_{t+1}}{\text{Number of cygnets + adults/subadults in late summer}_t}$$

The sequence of events resulting in cygnet recruitment in late summer (Number of nests times mean clutch size times hatching success times fledging success) suggests that environmental factors influencing recruitment ought to be combined for analysis in a multiplicative manner. This is an intractable process for a large number of variables but can be made tractable by converting the birth rate to an instantaneous rate (b_t) as follows:

$$b_t = ln(B_t) = ln(\text{Late Summer Count}_t \text{ of all age classes}) - ln(\text{Late Summer Adult/subadult Count}_t)$$

In this form, the instantaneous birth rate can be modeled as the result of environmental and other factors combined in an additive manner.

Both the instantaneous birth rate and the number of cygnets fledged per breeder showed highly significant negative correlations ($P=0.0001$) with size of the spring Refuge population, the number of breeding pairs at Red Rock Lakes NWR, and the mean number of swans spending the previous winter at the Refuge (Table 13). These correlations support the hypothesis that there are negative relationships between population size and birth rate. The environmental

Table 12. Correlations between environmental factors and key factors (all years 1932-1985).

		Clutch Size K_3	Eggs Remaining K_4	Proportion Hatching K_5	Proportion Fledging K_6
Bushels grain fed/bird previous winter	$r=$	-0.31	0.21	0.37	.11
	$P=$	0.10	0.20	0.09	0.65
	$n=$	29	41	22	20
Peak Water Level	$r=$	-0.26	-0.27	-0.10	-0.01
	$P=$	0.18	0.15	0.61	0.98
	$n=$	29	31	26	27
Median Water Level April	$r=$	-0.41	-0.19	-0.18	0.01
	$P=$	0.07	0.40	0.49	0.97
	$n=$	20	21	18	19
May	$r=$	-0.33	-0.22	0.18	-0.08
	$P=$	0.08	0.24	0.37	0.69
	$n=$	30	32	27	28
June	$r=$	-0.39	-0.16	-0.44	-0.18
	$P=$	0.04	0.39	0.03	0.37
	$n=$	28	30	25	26
July	$r=$	-0.48	-0.25	-0.31	-0.36
	$P=$	0.01	0.16	0.12	0.06
	$n=$	29	32	26	27
Precipitation: April	$r=$	-0.27	-0.05	-0.04	-0.09
	$P=$	0.10	0.72	0.82	0.66
	$n=$	38	46	30	29
May	$r=$	0.18	0.03	-0.23	0.11
	$P=$	0.29	0.83	0.23	0.58
	$n=$	38	46	30	29
June	$r=$	-0.17	0.19	-0.26	0.10
	$P=$	0.31	0.20	0.17	0.61
	$n=$	38	46	30	29
July	$r=$	-0.51	-0.23	-0.33	-0.12
	$P=$	0.001	0.12	0.08	0.52
	$n=$	38	46	30	29

Table 12, cont.

		Clutch Size K_3	Eggs Remaining K_4	Proportion Hatching K_5	Proportion Fledging K_6
Mean Temperatures					
May	$r=$	0.30	0.19	-0.20	0.03
Minimum	$P=$	0.08	0.25	0.31	0.89
	$n=$	35	39	27	28
May	$r=$	0.30	0.18	-0.03	-0.04
Maximum	$P=$	0.08	0.26	0.86	0.85
	$n=$	36	40	28	29
June	$r=$	-0.12	-0.02	0.06	0.36
Minimum	$P=$	0.48	0.92	0.78	0.07
	$n=$	34	38	26	27
June	$r=$	0.10	-0.23	0.30	0.29
Maximum	$P=$	0.59	0.17	0.13	0.15
	$n=$	34	38	26	27
Number June	$r=$	0.14	-0.20	-0.02	-0.27
of days	$P=$	0.42	0.23	0.92	0.18
with	$n=$	34	38	28	27
freezing					
temper- July	$r=$	0.11	0.20	0.28	-0.04
atures	$P=$	0.54	0.21	0.14	0.83
	$n=$	35	40	29	28
Spring	$r=$	-0.08	-0.02	0.01	-0.11
Severity	$P=$	0.65	0.89	0.97	0.56
Index	$n=$	37	42	29	28
Previous Winter	$r=$	0.12	0.14	-0.05	0.06
Severity	$P=$	0.49	0.36	0.78	0.76
Index	$n=$	38	43	30	29
Length of	$r=$	-0.06	0.08	-0.03	0.03
Growing	$P=$	0.74	0.61	0.87	0.88
Season	$n=$	35	46	27	26

factors which showed strong negative correlations with either measure of birth rate were the amount of precipitation in July ($P= 0.02$), July water levels ($P= 0.08$) and the date of peak hatch ($P= 0.01$). These highly negative correlations suggest that high rainfall and water levels during July decreased the survival of the recently hatched cygnets. Also, in years when the date of peak hatch was late (usually years with cold, late springs) cygnet survival to the late summer survey decreased. In such years, Refuge Narrative Reports have also reported that mortality of late-hatched cygnets occurred after the survey at freeze-up.

The effects of water levels at the Refuge were examined further by comparing the instantaneous birth rate and cygnets per breeder in years which the managers rated as high water years against other normal and low water years. The instantaneous birth rate in high water years (0.182) was only about half as large as in other years (0.334). Likewise, cygnets per breeder in high water years (0.389) was about half that of other years (0.727). The differences between both measures of birth rate were highly significant ($P= 0.02$). This agrees with the managers' perceptions that in high water years, nests were flooded and entire clutches destroyed, and mortality of newly hatched cygnets was high.

Examination of the plot of instantaneous birth rate against July rainfall (Fig. 12) suggests that a threshold exists at about 1.6 inches of rainfall. Above that level, the birth rate drops dramatically. We used this threshold to divide years into groups of high rainfall (> 1.6 inches) and low rainfall (< 1.6 inches) years.

We then asked if the weather has differed recently from what it was like during the period of swan population growth. We examined this question by testing for a difference in the frequency of high water and high rainfall years between the period of population increase (1932-54) and the period since the population peaked (1954-85). High water years have been more frequent since 1954 (35% of the years versus 23% of the years 1932-54), although the difference was not significant ($X^2= 0.73$, $P= 0.39$). Years with high rainfall in July have definitely been more frequent since 1954 (45% of the years versus 6% of the years 1932-54) and the difference is highly significant ($X^2= 7.88$, $P= 0.01$). Obviously, the worst type of year for cygnet production would be one in which there was both high water during nesting and high rainfall in July. No such years occurred in the 17 years for which weather data exist prior to 1955. Since that time, 23% (7 out of 31 years) of the years have been of this worst type. This difference is significant ($X^2= 4.49$, $P= 0.04$). It therefore appears that cygnet production at Red Rock Lakes NWR has in part been reduced by the increased frequency of adverse weather in recent years.

Cygnet Production for Entire Centennial Valley Flock

Repeating all of the analyses above for the entire CV flock leads to the same conclusions as for the analyses of the Refuge flock alone, but this is no doubt because the Refuge flock makes up most of the CV flock.

Cygnet Production for Tri-state Subpopulation Excluding Centennial Valley

We could not identify any significant correlations between the birth rate (cygnets/breeder) of the non-Centennial Valley Tri-state Subpopulation and the number of nesting pairs, previous winter's severity index, spring severity index, July precipitation, annual precipitation, or water levels. The environmental variables calculated for Red Rock Lakes NWR were used in these

Table 13. Correlations of instantaneous birth rate (b_t)[a] and cygnets produced per breeder at Red Rocks Lakes NWR with environmental and other factors potentially influencing cygnet production.

Variable	Instantaneous Birth Rate			Cygnets per Breeder		
	r	P	n	r	P	n
Spring population size	-0.71	0.0001	42	-0.65	0.0001	42
Number of breeding pairs	-0.65	0.0001	42	-0.66	0.0001	50
Mean winter swan count	-0.64	0.0001	41	-0.66	0.0001	49
Bushels grain fed per bird	-0.11	0.53	35	0.13	0.42	40
Peak water level	0.19	0.37	24	-0.04	0.83	31
Median -- April	-0.035	0.93	15	-0.01	0.96	21
Water -- May	0.10	0.63	24	-0.13	0.48	32
Level -- June	-0.01	0.96	23	-0.25	0.18	30
-- July	-0.07	0.75	25	-0.32	0.08	32
April precipitation	-0.001	0.99	37	-0.02	0.87	45
May precipitation	0.07	0.69	37	-0.12	0.43	45
June precipitation	0.07	0.68	37	-0.04	0.79	45
July precipitation	-0.41	0.01	37	-0.36	0.02	45
Mean minimum temperature - May	0.25	0.17	31	0.29	0.07	39
Mean maximum temperature - May	0.14	0.45	31	0.15	0.38	39
Mean minimum temperature - June	0.19	0.32	30	0.14	0.40	38
Mean maximum temperature - June	-0.13	0.49	30	-0.44	0.39	38
Spring severity index	0.01	0.97	33	-0.14	0.40	41
Previous winter severity index	0.107	0.55	34	0.19	0.23	42
Date of peak hatch	-0.713	0.0003	21	-0.47	0.01	28

[a] $b_t = ln (Bt) = ln$ (total count) - ln (white bird count)

58

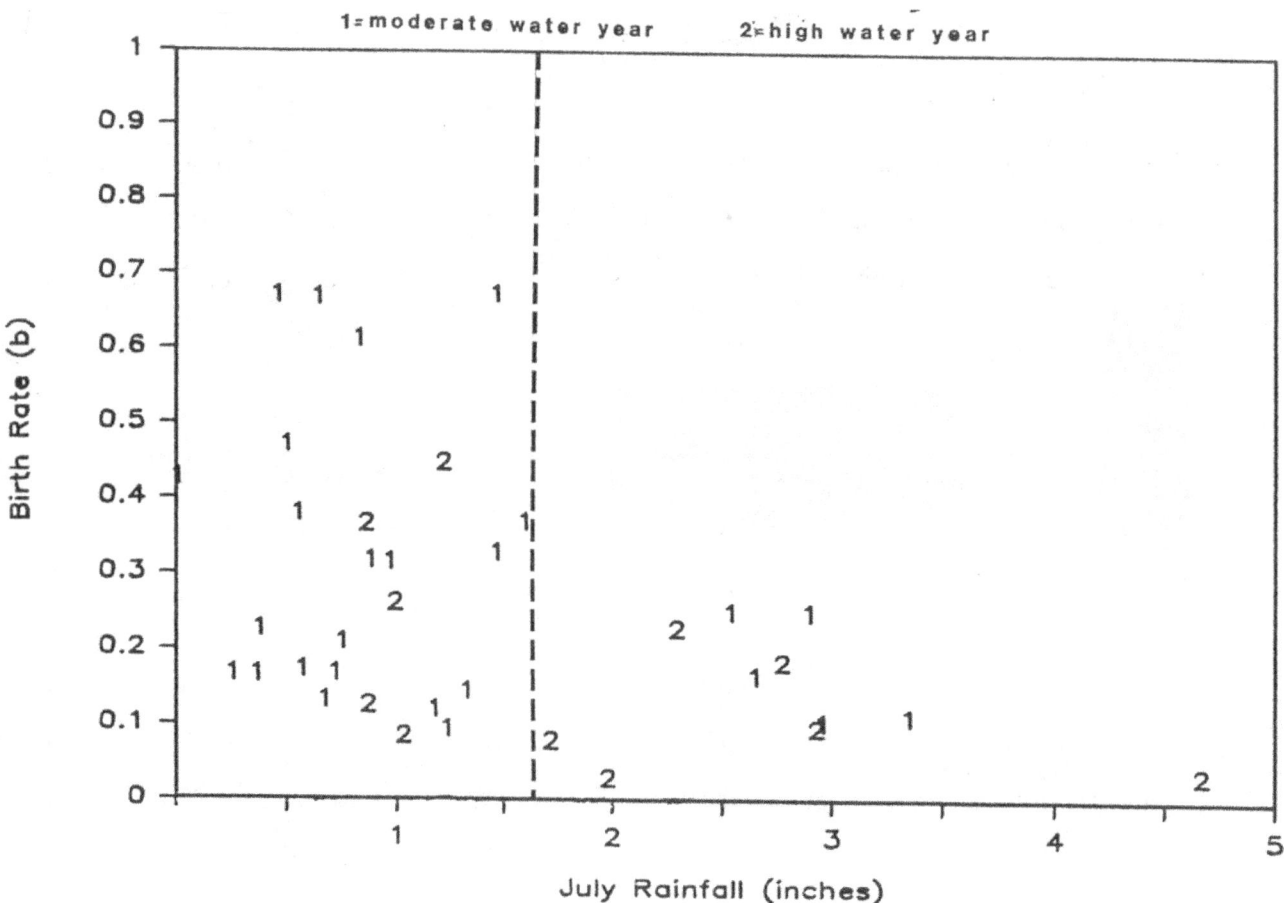

Figure 12. Birth Rate vs. July Rainfall

analyses but they probably do not adequately measure the variation in weather conditions in other parts of the Tri-state region. It might be better to break the Subpopulation into appropriate segments (YNP, other Wyoming, Idaho, etc.) and use weather data from weather stations in each of those regions. However, the flock sizes are so small and the variation in nesting habitats so great, that this was not deemed worthwhile.

Summary

The Rocky Mountain Population includes about 78 nesting pairs in the Tri-state Subpopulation and 145 nesting pairs in the Interior Canada Subpopulation. The number of nests at Red Rock Lakes NWR and YNP declined in the late 1970s and 1980s. Overall the Tri-state Subpopulation has lost about 12 nests since 1978.

The decline in nesting effort at Red Rock Lakes NWR has been greatest on Lower Lake. Nest success was also lowest on Lower Lake, due at least in part to flooding. The current number of nests at Red Rock Lakes NWR is about 50% of peak numbers.

Very few trumpeter territories regularly fledged cygnets; productivity is sporadic and strongly related to spring weather and water levels. Even during a period of flock increase at Grande Prairie less than 40% of 51 nesting territories fledged cygnets in five or more years out of ten.

Clutch size, percent hatch of eggs, nest success, and cygnet production per nest are higher in the Grande Prairie flock than in the Tri-state flocks. Within the Tri-state Area, these parameters are highest in the lower elevation Idaho and Wyoming flocks and lowest at Red Rock Lakes NWR and YNP. Clutch size is larger in captive trumpeters than in wild trumpeters and increases with the age of the breeding female.

Cygnet production at Red Rock Lakes NWR has been characterized by wide annual fluctuations, with several extremely poor production years in the 1980s. Declining cygnet production has resulted from the failure of most pairs to raise any cygnets at all, rather than uniform mortality among broods. This is evidenced by the reduction in the mean number of broods fledged at Red Rock Lakes NWR from 17.6 in the 1960s to 6.8 in the 1980s, while the number of cygnets per brood has shown little if any change.

Prior to about 1960, cygnet production per nest at Red Rock Lakes NWR was inversely related to the number of nests. Since 1960, both the number of nests and the cygnet production per nest have declined.

Changes in cygnet production at Red Rock Lakes NWR show a highly significant positive correlation with variations in hatching and fledging success. Variations in clutch size, and the proportion of eggs remaining after removals also showed significant positive correlations with cygnet production. We conclude that cygnet production is most closely correlated with environmental factors that influence clutch size, hatching success, and survival to fledging (Table 14).

Clutch size was positively correlated with increasing May temperatures and negatively associated with April precipitation, water levels during the entire April to July period, and the amount of grain fed per swan during the previous winter.

Hatching success at the Refuge showed a significant negative correlation with median water level in June, and negative associations with increasing rain in May, June, and July.

Hatching success showed a positive correlation with the amount of grain fed per swan during the previous winter.

Survival from hatching to fledging at Red Rock Lakes NWR was positively correlated with the mean minimum temperature in June and negatively correlated with July water levels.

Both the instantaneous birth rate and the number of cygnets fledged/breeder at Red Rock Lakes showed highly significant negative correlations with the size of the spring flock, the number of breeding pairs, and the mean number of swans spending the previous winter at Red Rock Lakes NWR. These findings support the hypothesis that complex negative relationships exist between population size and cygnet production. These relationships could result from changing population age structure, age-specific differences in productivity, and/or habitat constraints.

We found significant negative correlations between both measures of birth rate at Red Rock Lakes NWR and both July precipitation and date of peak hatch. Nest initiations, and thus date of peak hatch, are delayed in late cold springs. A less significant negative correlation exists between cygnets per breeder and median water level in July, and a positive correlation exists between cygnets per breeder and mean minimum temperature in May.

In high water years, cygnet production has been reduced by about 50%. When July rainfall exceeded about 1.6 in. the birthrate declined abruptly. During the period of population decline since 1954, the trumpeters have had to cope with an increased frequency of adverse weather conditions. Since 1954, 23% of the years have had both high water levels and high July rainfall, while none of the years before 1954 had such a combination of events.

As summarized in Table 14, the correlations between environmental factors and cygnet production at Red Rock Lakes NWR are consistent with the theory that clutch size, hatching success, and cygnet survival are strongly related to the opportunity of the breeding female to obtain adequate prebreeding nutrition, and to the availability of foods to the newly hatched cygnets. Factors which reduce water temperatures and delay the growth of spring foods (heavy April precipitation, cold May temperatures, high spring water levels) are associated with reduced cygnet production. High elevation flocks in the Tri-state Area are particularly vulnerable to these factors. Hatching success is reduced by flooding and cold rains. Young cygnets are particularly vulnerable to cold temperatures and heavy precipitation, and their survival is reduced when water levels are unusually high. Cygnet production to late summer is reduced in years when the peak of hatch is late, and late hatching cygnets may be unable to fledge before autumn freeze-up occurs.

Table 14. Summary of significant ($P < 0.10$) correlations between environmental factors and reproductive parameters.

Factors	Clutch Size	Percent Hatch	Percent Fledging	Instantaneous Birthrate	Cygnets per breeder
Negative correlations					
Median water levels					
April	0.07				
May	0.08				
June	0.04	0.03			
July	0.01		0.06		0.08
Monthly precipitation					
April	0.10				
July		0.08		0.01	0.02
Date of peak hatch				0.0003	0.01
Spring population size				0.0001	0.0001
Number of breeding pairs				0.0001	0.0001
Mean winter swan count				0.0001	0.0001
Bushels of grain fed per swan during previous winter	0.10				
Positive correlations					
Temperature					
May mean max.	0.08				
May mean min.	0.08				0.07
June mean min.			0.07		
Bushels of grain fed per swan during previous winter		0.09			

CHAPTER 4. REPRODUCTIVE BEHAVIOR

Pair Bond Formation

In a long-lived species such as the trumpeter, determining the process involved in the initiation and maintenance of a lasting pair bond requires long-term, detailed observations of marked individuals of known age and sex. Although occasional observations of marked swans have been made, few details of the pair bonding process among Rocky Mountain trumpeters are yet known.

Five female and four male trumpeters, transplanted from Red Rock Lakes NWR to Lacreek NWR as cygnets in 1962, were wing-clipped, neck-banded, and kept together in a holding pen. The birds began courtship displays in mid-January 1964, at 20 months of age, and continued intermittently until about mid-March. A less active period of display occurred during late March and continued until mid-April. After one week of very active courtship display in January, two strong pairs developed. One trio of two males and one female also formed, and two females showed no tendencies to pair. Unfortunately, the neck-bands fell off during the following summer so further study of the permanence of these bonds was not possible (Monnie 1966).

Based upon resightings of marked swans, none of the trumpeters that were marked in Grande Prairie has remained in the Tri-state Area to breed, and none of the swans marked in the Tri-state Area has ever nested in Canada. This lack of observed intermixing led Turner (1987) to suspect that pair bond initiation occurred when the subpopulations were apart, either during migration or on the breeding grounds. Turner suggested that it most likely coincided with fall recrudescence of the gonads. In support of this conclusion, Turner (1987) also reported that marked territorial pairs returning to Grande Prairie in the spring did not include 2-year-old birds, but in two cases 2 year-old birds were observed to be paired in autumn when they reached the wintering area. Holton (1985) recorded no courtship displays while observing summer nonbreeding flocks in Grande Prairie and concluded that pair formation probably occurred after swans departed in the fall and prior to their return in the spring.

Although Delacour and Mayr (1945) stated that "pair formation in all temperate zone swans occurs in the fall", others have observed that pairing can occur in the spring. An adult female which migrated from Hennepin County Park Reserve, MN, to Kansas in December 1984, lost her mate during the southward migration. She returned with her cygnets to Minnesota by 26 February 1985. The female then paired with a new male during the spring and nested successfully (Gillette 1985). In Wyoming, pair bond initiation occurred in mid to late winter and through the spring. Pairing followed soon after the late winter dissolution of sibling group bonds, during a swan's second or third winter. Soon after the marked sibling groups disassociated, marked individuals were consistently observed in pairs with unmarked individuals, presumably in early bond formation. A peak of courtship activity was observed in April and early May (Lockman *et al.* 1987).

If temporal or behavioral mechanisms exist which prevent pair formation (and thus cross matings) between the Interior Canada and Tri-state subpopulations, then these two groups should be recognized and managed as separate populations. If Rocky Mountain trumpeters form pair bonds in a manner favoring selection of a mate from a swan's own natal area, the possibility that the ICSP and TSP are genetically isolated is increased. Lack of genetic interchange increases the probability of either inbreeding or loss of heterozygosity due to genetic drift occurring in isolated breeding flocks (see Chap. 11).

Marked pairs of trumpeters remained together yearlong and rarely separated. Pairs normally returned to the same territory each year (Lockman *et al.* 1987). Of seven marked breeding swans observed in Grande Prairie between 1978-80, all returned to the lakes where they nested the previous year, presumably with the same mates (Holton 1985).

Instances of aberrant pair bonding have occasionally been reported. Two females cooperated in the defense of a territory in YNP for at least three consecutive summers and incubated two clutches of infertile eggs each year (Shea 1979, 1980). At Red Rock Lakes NWR, three swans were reported tending one nest during the incubation period and all assisted in tending the cygnets. The three adults raised six cygnets from a clutch of seven eggs (RRLNWR Ann. Narr. Rept. 1959). Banko (1960:98) reported a similar situation involving three adults in Idaho in 1956 and 1957.

Age at First Breeding

Wild trumpeters usually begin nesting at between four and six year of age although some individuals are able to nest successfully as three-year-olds. Delayed nesting has been suggested to be a response to competition for suitable nesting habitat, however cygnets transferred from Red Rock Lakes NWR to the National Elk Refuge in Jackson Hole in 1938 did not attempt to nest until they were six years old, even though no other swans occupied the habitat (Banko 1960:95). Cygnets transplanted from Red Rock Lakes NWR to Lacreek, Malheur, and Ruby Lakes National Wildlife Refuges first nested and raised cygnets as three-year-olds (Hansen 1973, Fjetland 1974).

Among marked female trumpeters at Grande Prairie, two first nested at age 3, one nested at 4, and one nested at 5. The two females that nested at 3 years of age were siblings and were known to overwinter at Red Rock Lakes NWR where they fed supplemental grain. This enriched winter diet may have contributed to their precocial reproductive behavior (Turner 1987). One marked male was also observed to nest successfully at 3 years of age (Turner and Mackay 1981).

Seven trumpeters, marked as cygnets in Grande Prairie, were located during consecutive winters in the Tri-state Area. The age at which these swans were first observed arriving on the winter range accompanied by cygnets was four-years-old for two males, five-years-old for three males and one female, and seven-years-old for one male (CWS Grande Prairie neck-band records).

K. Kalenak of Saginaw, MI reported that Trumpeter Swan cygnets hatched from eggs that were taken from Red Rock Lakes NWR in 1977, pair bonded at 9 months and first laid eggs at 21 months (Kalenak 1982).

Establishment and Defense of Territories

Trumpeters often visited territories at Red Rock Lakes NWR in early February, long before open water was available, spending much of the day sitting on the ice miles from food or open water (Banko 1960:99; T. McEneaney pers. comm.). In YNP, trumpeters occupied their territories any time from mid-March to late May, depending upon elevation and the rate of thaw

(Shea 1979). Grande Prairie trumpeters moved on to their nesting lakes as soon as a portion was ice-free, usually about 20-25 April (Holton 1985, Turner 1987). Establishment of the territory weeks prior to incubation may insure an undiminished food resource and reduce the time spent later in defense recesses by the incubating female. Territorial defense by both sexes is a major component of the trumpeter's breeding strategy and may have resulted from the critical food requirements of the cygnets and the limited nesting habitat (Cooper 1979).

Banko (1960:109) suggested that territory size in the continuous marshes at Red Rock Lakes NWR was related to the quantity of food available at a given distance from the nest site, as well as the arrangement of various terrain features in an area and the number of potential nesting pairs. In 1971-73, Refuge territories averaged about 13 ha in size, with a range from 2 ha to slightly over 40 ha. The average territory on Lower Lake was almost twice as large as territories on River Marsh and Swan Lake (Page 1976:48). Nesting pairs were most numerous and had the smallest territory size in areas that had the most irregular shorelines (Page 1976:112).

Banko (1960:107) observed that in years with higher numbers of nesting pairs at Red Rock Lakes NWR nesting swans did not compress territory size but rather moved into peripheral, and possibly less desirable, areas. In 1971-73, however, studies showed that nests in the Refuge's major nesting areas were dispersed uniformly. Compression of territory size was documented in two instances when new pairs nested in close proximity to established pairs and the territory size of the established pairs decreased. Such compression of territory size could increase the total number of territories available on the Refuge as the numbers of nesting pairs expands (Page 1976:108).

In most of the other RMP nesting areas, territory size approximates the size of the selected nesting lake, because most breeding pairs will not tolerate use of their lake by other swans. In YNP, median territory size was 10.7 ha and the range was from 1.2 ha to 111 ha (Shea 1979). In the Targhee National Forest, Idaho, territory size averaged 15.2 ha (Maj 1983). Territory size was much larger in Grande Prairie, although there too it approximated nesting lake size with a mean of 127 ha (Holton 1982).

Rarely, nesting swans will tolerate other swans on their nesting lake. Indian Lake on the Idaho/Wyoming border and Silver Lake at Harriman State Park, ID, were each usually occupied by at least one nesting pair and several nonbreeders. In 1985, Silver Lake was occupied by two successful nesting pairs and at least 20 nonbreeders (G. Eyraud, pers. comm.). Two pairs have also nested simultaneously at Hermit Lake in Grande Prairie (G. Holton, pers. comm.).

At Red Rock Lakes NWR, swans rarely left their territories during nesting except on short forays or in defensive flights (Banko 1960:103). Page (1976:47) observed that on five occasions, usually after a territorial defense, the cob (presumed) left the territory. Twice he was accompanied by the pen. They loafed and fed in neutral areas for short periods. Two of 18 pairs in YNP were observed to leave their territories during incubation in order to feed at other ponds up to 2.5 km distant. During incubation, simultaneous absences of both adults from the territory lasted up to 70 minutes (Shea 1979).

In the continuous nesting habitat at Red Rock Lakes NWR, frequent intra-specific encounters are likely. In 1971-73, Refuge territories did not have a vertical component; nesting pairs ignored other swans flying over their territories. Trespass and territorial defense were rare, with only six occurrences observed during three years (Page 1976:47). Banko (1960:101), in contrast, described territorial defense directed at swans flying nearby and noted that aggression was most frequent during nesting and early brood rearing, and then tapered off. While never common, greater aggression was directed at strange swans that at swans from adjacent territories.

At Red Rock Lakes NWR a buffer zone, not used by either pair, separated most territories. Pairs nested closer together when visual barriers, such as emergent vegetation, existed. In two cases neighboring pairs shared a common territorial boundary without a buffer zone. In at least five instances the neighboring cobs swam toward each other, met and then swam side by side, each on its own side of the boundary, for the length of the common boundary (Page 1976:47).

Although most intra-specific confrontations do not involve physical injury (Banko 1960:92), a near-fatal battle was observed on the Madison River in YNP (J. Foott, pers. comm.). During the previous nesting season one pair had been flooded out during incubation in early May and a second pair subsequently moved in 0.5 km upstream and nested successfully. When two pairs returned to the territory during the following April, a fierce struggle occurred. The victor had almost killed its battered opponent by holding its head underwater when a park visitor intervened (Shea 1979).

Pre-laying Period

The energy stored by the female in the weeks preceding egg-laying and incubation is a critical factor in determining subsequent reproductive success (King 1973, Scott 1973, Ankney and MacInnes 1978, Krapu 1981). Breeding mute swans spent most of the daylight hours feeding during the pre-laying and laying periods, and body condition appeared to be a determinate of both the timing of breeding and clutch size (Reynolds 1972). de Vos (1964) noted that both prior to incubation and after the hatch, the female trumpeter spent more time feeding than did the male, and conjectured that her increased feeding was somehow tied to the physiological demands of incubation. Captive Minnesota trumpeters went through a lengthy pre-laying phase of from 15 to 35 days that was devoted primarily to feeding, and that enabled the female to attain the energy reserves needed for egg laying and incubation (Cooper 1979).

Egg-laying and Incubation

Trumpeters usually lay one egg every other day. de Vos (1964), Hansen et al. (1971) and Tillery (1969, in Cooper 1979) measured egg-laying intervals at 48-50 hours. Cooper (1979) reported minimum and maximum intervals to be 39 and 48 hours.

Several workers have studied the incubation behavior of captive and wild trumpeters. de Vos (1964) described daily activity patterns, reproductive behaviors, comfort movements, vocalizations, and aggressive behaviors of one pair of captive trumpeters at Delta Waterfowl Research Station, Manitoba. Cooper (1979) observed four successful nest attempts by two pair of captive trumpeters in Minnesota, and described and quantified their behavior. Incubation behavior of wild trumpeters has been studied in YNP (Shea 1979), Idaho (Hampton 1981), and Grande Prairie (Holton 1982).

In contrast to the shared incubation of the Tundra and Bewick's Swans, incubation by captive trumpeters is normally accomplished entirely by the female (de Vos 1964, Cooper 1979). Detailed observations of the incubation behavior of wild marked trumpeters of known sex, however, have not been conducted. Holton (1985) described one instance of apparent incubation by a neck-banded, wild male in Grande Prairie. During the 22 minutes in which he sat on the eggs, the male's actions included sitting, rocking, rearrangement of nest materials and frequent reorientations. These actions were indistinguishable from those of the female.

The onset of incubation in captive swans was progressive, with a rise in attentiveness during the latter half of the laying phase, reaching a plateau two days after the last egg was laid (Cooper 1979). Studies of trumpeter incubation have measured certain components that provide an indication of the attentiveness of the incubating female. These components are constancy (the percent of the day spent sitting on the eggs), recesses (periods of time off the nest), and sessions (periods of time on the nest).

Constancy

After building their energy reserves during the pre-laying period, captive trumpeters basically fasted throughout incubation. They showed an average constancy of approximately 95% which did not vary significantly in relation to day of incubation, air temperature, rainfall, or solar radiation , nor did they leave the nest during the night (Cooper 1979). Wild Grande Prairie trumpeters demonstrated a similar fasting strategy during incubation, with an average daily constancy for six females ranging from 93.5% to 96.6% (Holton 1982).

Lower constancies were measured among wild trumpeters in the Tri-state Area. In YNP, Shea (1979) reported an average constancy of 85%, with a range for six females of 78% to 89%. On a daily basis, constancy ranged from 58% to 100%, with the highest rates occurring on cold, snowy days. In Idaho, average daytime incubation constancies for five females ranged from 66.5% to 81.9%. Daily constancies ranged from 19% to 100%. Unlike captive trumpeters, Idaho swans took nighttime recesses, resulting in average nighttime constancies that ranged from 68.1% to 86.5% (Hampton 1981).

Recesses

Cooper (1979) classified recesses as either normal recesses, prior to which the pen covered the eggs with nest materials; defense recesses where the pen left hastily without covering the eggs; and brief material gathering recesses during the first two days of incubation. Normal recesses comprised 60% of the total and were taken most often between 1500 and 1900 hours. They averaged 21 minutes in length and on the average occurred three to four times per day. Defense recesses made up 40% of the total and were common both morning and evening when Canada geese were most active. Defense recesses averaged 18 minutes in length and were more frequent early in incubation. Material gathering recesses made up less than 1% of the total recesses. Normal recess activities included 41% comfort movements, 38% feeding, 17% locomotion, and 5% displays. Defense recess activities were 64% displays, 16% comfort movements, 10% feeding, and 9% locomotion.

In YNP, recesses (all types combined) occurred an average of 3.9 times per day and averaged 37 minutes in length. Sixty-seven percent of all recesses lasted from 10 to 40 minutes, however, one undisturbed recess lasted 314 minutes during the second week of incubation (Shea 1979).

Hampton (1981) reported that average recess length for five Idaho swans was 33.9 minutes. Average recess lengths ranged from 23.8-50.7 minutes for different individuals, but due to high variance none of the means were significantly different. In contrast to captive trumpeters (Cooper 1979), Idaho trumpeters took numerous nighttime recesses throughout the incubation period. Hampton found no significant difference in the average recess length on any lake during the light and dark hours. The total number of recesses per twenty-four hour period averaged 12.1.

During recesses, Idaho swans spent 70-74% of the time feeding, 24-30% in locomotion, and less than 2% in comfort and alert postures. Compared to Cooper's (1979) captive swans, Idaho trumpeters spent much more time in feeding and locomotion (Hampton 1981). Smith (1985) reported that one incubating trumpeter in Grand Teton National Park spent 10% of the daylight hours in recesses. Of these hours, 60% were spent feeding, 30% preening, and 10% swimming.

In Grande Prairie, nesting trumpeters had a mean recess length of 44.3 minutes. The average number of recesses per day was not determined (Holton 1982).

Sessions

Cooper (1979) found that the distribution of session lengths was bimodal with peaks occurring at 2 hour and 15 hour intervals, the latter reflecting the overnight steadfastness of the incubating females. No significant relationships were detected between session lengths and environmental factors or the day of the incubation period.

Mean length of daytime sessions was 133 minutes in YNP. No nighttime observations were made (Shea 1979). The mean session length of 192.5 minutes in Grande Prairie was significantly longer than that reported from YNP. The difference may even have been greater than indicated because Holton made no observations on days with continuous rain when the swans may have taken even fewer or no recesses, as was the case in YNP (Holton 1982).

Relationship of Incubation Behaviors to Cygnet Production

In YNP, all six pairs observed by Shea (1979) hatched at least one egg, however only three pairs fledged any cygnets. The pairs which fledged cygnets laid larger clutches, had higher incubation constancies, took fewer recesses per day, and had higher hatching success than did the pairs whose entire broods died prior to fledging. The female that laid the largest clutch (6 eggs) and incubated with the highest constancy (89%), occupied a territory in the geothermally-warmed Madison River. This territory never froze and abundant food was available year-round. This was the only territory in Yellowstone which remained totally unfrozen in winter and early spring, thus providing this pair with unique food resources during the pre-laying period. Not surprisingly, in the last 20 years, this territory has fledged more cygnets than any other territory in YNP (USFWS Tri-state Trumpeter Swan Survey data). Shea concluded that YNP trumpeters that had energy reserves sufficient to provision a clutch of five or six eggs also possessed adequate reserves to allow them to incubate with relatively high constancy. Even the most attentive incubators in YNP, however, could not achieve the high constancy of Cooper's (1979) captive trumpeters, which were able to incubate with a fasting strategy after laying clutches containing from 7-9 eggs.

Holton (1985) showed that Grande Prairie trumpeters incubated with higher average constancy than the YNP trumpeters. They also laid larger clutches, experienced higher hatching success, higher nest success, and higher cygnet survival. Holton suggested that differences in physiological condition and incubation behavior between these two flocks influenced their relative reproductive success.

Length of Incubation and Hatching Periods

Among captive trumpeters that incubated with 95% constancy, cygnet hatching spanned approximately 48 hours on days 31 and 32 of incubation. In all four nest attempts, cygnets exited the nest in the morning of day 33. Factors which cause the female to reduce incubation constancy would lengthen the incubation period. If the female basically employs a fasting strategy and has limited internal reserves, then extension of the incubation period much beyond the norm would not be possible (Cooper 1979). Estimates for the length of the incubation period of wild trumpeters ranged from 33-37 days (Banko 1960:115).

Response to Human Disturbance

Occasional nest abandonment due to disturbance by fishermen and photographers has been observed at RRLNWR (Page 1976:67) and human disturbance in Refuge nesting areas has increased substantially since the 1930s and 1940s. Early Refuge managers permitted no one to venture into the Refuge breeding grounds between 1 May and 1 August (Hull 1939). Since the late 1940s, however, research studies and management activities have increased in the marsh, and public fishing and boating access was permitted in later years. Several Refuge workers noted that the airboat, which has been used on the Refuge since 1952, caused the Refuge swans to react strongly. Within one year of the airboat's acquisition, adults had become very wary of the craft and hid themselves and their young whenever it was used. Swans up to three miles away were observed to go into hiding when the engine was started (RRLNWR Ann. Narr. Repts. 1953, 1954). Airboat use during the early brood rearing period caused pairs to abandon their broods temporarily and the family groups scattered widely (RRLNWR Ann. Narr. Rept. 1960).

In the 1950s, 1960s, and 1970s, the airboat was used to capture flightless swans during the banding and transplant operations. Since the early 1950s, many of the nesting swans on the Refuge have been captured with the airboat at least once during their subadult years. In addition to its use in capturing swans, in some years the airboat was also used extensively in the nesting areas for clutch size checks, egg removals, hatching success checks, vegetation studies, fishery studies, duck surveys, graduate student research and Refuge tours. Page (1976:58) observed that airboat use in 1971-73 caused incubating females to leave their nest while the boat was still several miles away. Many left without covering their eggs. Page also noted that during his study swans were locating their nests in areas of the Refuge which were less accessible to the airboat. Although we could not reconstruct and quantify airboat use on the Refuge in detail by year, we concluded that its regular use for routine research and management purposes, interspersed with use of the craft to chase and capture flightless swans, was a significant disruption to the swans during nesting and early brood rearing, and likely contributed to nest failures and early cygnet mortality.

Because of the very real possibility that the airboat use has contributed to nest failure and loss of small cygnets, the current Refuge staff has drastically curtailed its use, particularly during incubation and early brood rearing (T. McEneaney, pers. comm.).

Shea (1979) observed that when undisturbed, incubating females in YNP spent three to five minutes covering their eggs before leaving the nest. When disturbed, however, most females quickly slipped off the nest and left the eggs exposed. On two occasions a female left her nest to aid her mate in territorial defense, then returned and covered the eggs before continuing the recess. In contrast, swans which left the nest due to human disturbance gave no attention to their exposed eggs until the human left the territory. Exceptions to these patterns were few, but some trumpeters were extremely tolerant of human presence. Three pairs in YNP had habituated to high levels of human activity and defended their nests vigorously. One pair frequented occupied lakeshore campsites in search of handouts.

Although nests were occasionally destroyed by human vandals (Hampton 1981), overall the pairs which nested in disturbed sites in YNP did not exhibit decreased reproductive success per nest attempt. However, the 14 territories classified as disturbed were occupied by a nesting pair significantly less often than were undisturbed territories, and three historic territories were no longer suitable for nesting due to high levels of disturbance. If a pair was able to habituate to the level of human activity at a territory, it would be on the average as productive as swans at undisturbed territories. However, after the death of the breeding pair, disturbed territories were less likely to attract a new pair and experienced a higher overall vacancy rate than undisturbed territories (Shea 1979).

Disturbance at lakes in the Grande Prairie region was less than at lakes in YNP (Holton 1985). Holton compared incubation behaviors of swans at disturbed agricultural lakes and undisturbed forested lakes. There were no significant differences in recesses, session lengths, nor incubation constancies, but swans did spend significantly less time sleeping and more time in head high alert postures on agricultural lakes. Reproductive success was not detectably different at disturbed and isolated lakes. However, as was noted in YNP, trumpeters nested less frequently on lakes which were exposed to agricultural and other human activity, than on isolated lakes (Holton 1982).

Cygnet Behavior and Parental Care

de Vos (1964) found that parent-cygnet relationships were very close immediately after hatching but gradually relaxed thereafter. During the first few weeks the cygnet was closely guarded by both parents. Normally the female was the leading parent on land and water, followed by the cygnet, with the male bringing up the rear. Although much more closely associated with the female prior to 24 July, the cygnet followed the male whenever he took the lead. The cygnet first took the lead at the age of 30 days and was first left to swim alone at the age of 50 days.

Leach (1977) observed that during the first four weeks cygnets fed on materials which came to the surface as a result of the parents' feeding activities. Cygnets were never observed to submerge their heads until they were four weeks old.

Holton (1985) noted that the Grande Prairie adults frequently fed using a "puddling" motion when accompanied by cygnets less than two weeks old, but used this same feeding technique infrequently during incubation. Smith (1985) concluded that the adults, particularly the male, actively assisted the cygnets to feed in the first two weeks of life by using a puddling or foot-treading technique to bring food items to the surface where they were picked up by the cygnets. The adults rarely used this method of feeding when not accompanied by the cygnets. During the first week after hatch, the adults spent 58% (female) and 65% (male) of recorded intervals assisting the cygnets to feed. This dropped to 10% and 20% respectively by 6 July, and then fluctuated below 20% for the next month. Feeding assistance was near zero by mid-August.

Brood Movements

Page (1976:61) observed six cases of brood relocation at Red Rock Lakes NWR, including one overland move of 0.9 km. Two broods moved when muskrat activity disrupted their nesting/loafing site; causes of the other moves were not determined. Families in YNP walked up to three km when the cygnets were less than two months old. Moves appeared to be precipitated by human disturbance, the drying up of the nesting pond, and the death of a cygnet from predation (Shea 1979). Holton (1982) also observed the movement of Grande Prairie broods along water courses and overland from their nesting lakes to other nearby waters.

Brood movements may also be induced by nutritional need. One Wyoming pair annually moved its 3 to 5-day-old cygnets about 2 km. Studies showed that feeding sites at the nesting pond contained few invertebrates and a monotypic vegetation community. The parents moved the brood to a site which contained a more diverse vegetative community and about ten times greater abundance of invertebrates (Lockman et al. 1986).

Brood Amalgamation

Banko (1960:119) reported the amalgamation of several broods at RRLNWR into a group of ten cygnets accompanied by two adults. Page (1976:62) also reported two instances of prefledging brood amalgamation at the Refuge. In one case the combined brood later disbanded. The willingness of trumpeters to care for unrelated cygnets has also been shown by the success of attempts to cross-foster wild cygnets. In two instances, wild parents with their own broods have adopted introduced cygnets, and in one case a pair without cygnets adopted an introduced cygnet and maintained a bond with it at least through the following autumn (RRLNWR Ann. Trumpeter Swan Rept. 1983, Lockman et al. 1987).

Sibling Associations

Trumpeters maintain long-lasting family associations. Observations of marked Grande Prairie swans showed that the bond between parents and cygnets remained intact through fall migration and throughout most of the first winter. This bond began to dissolve during February and by the time the swans reached spring staging areas in southwestern Alberta, cygnets were not usually attended by adults. Yearlings later were tolerated on their parents' territory and were observed to associate with their parents during the following winter (Turner 1987).

The bond between siblings may endure for several years. During banding operations in the Centennial Valley, yearlings have frequently been captured with their broodmates (Page 1976:62). Bonds among Grande Prairie siblings remained intact among nine groups of yearlings, five groups of 2-year-olds, and one group of 3-year-olds. The incidence of sibling association at age three is probably low because some birds begin to pair at age two (Turner 1987).

Lockman et al. (1986) observed that brood separation occurred at the end of the second winter in two groups of three, and at the end of the third winter in one group of three. Courtship interactions appeared to function in the disruption the sibling bonds.

Summary

Few details regarding the timing and mechanisms involved in the pairing of RMP trumpeters have been studied. Although the subpopulations winter together, to date marking studies have provided no evidence that bonding occurs between individuals from different subpopulations. If the ICSP and TSP do not pair bond and cross mate, they should be regarded and managed as separate populations rather than subpopulations.

Wild trumpeters may delay first breeding attempts until at least six years of age, although some individuals of both sexes are capable of breeding successfully at age three. Captive trumpeters have been observed to pair as early as nine months and lay eggs at 21 months.

Territories at Red Rock Lakes NWR were less than 40 ha in size and averaged about 13 ha. Elsewhere in the Tri-state Area, territory size approximated the size of the nesting lake and averaged 10-15 ha. In Grande Prairie, territory size averaged 127 ha. While most breeding trumpeters exclude other swans from their territory during nesting and brood rearing, others nest successfully within sight of neighboring pairs and nonbreeders. Pairs often tolerate their own subadult offspring in their territory after incubation ends.

Trumpeters go through a lengthy pre-laying period in which feeding is the main activity. During this critical period the female acquires the energy reserves needed for egg-laying and incubation.

Captive trumpeters and wild Grande Prairie trumpeters laid larger clutches and incubated with higher constancies than Idaho or YNP swans. Grande Prairie swans also had higher nest success, hatching success, and cygnet survival than YNP swans. In YNP, pairs that laid larger than average clutches also incubated with higher constancy and with fewer recesses, and had higher hatching success and prefledging cygnet survival than pairs that laid smaller clutches.

Human disturbance occasionally caused direct nest destruction or abandonment. Incubating females that are disturbed by humans leave their eggs uncovered and unprotected, and will temporarily abandon small cygnets. On the average, trumpeters which could accommodate to human disturbance and chose to occupy disturbed territories showed no detectable reduction in productivity. However, disturbed territories were less likely to be occupied than undisturbed territories, and therefore total nesting effort was reduced.

Since 1952, frequent airboat use on Red Rock Lakes NWR during the incubation and brood rearing periods caused obvious stress to swans and may have contributed to nest failure and early cygnet mortality. Recently, however, airboat use and other human disturbance at the Refuge have been greatly reduced.

Adult/cygnet bonds are strong and long enduring. Adult feeding activity provides a major source of cygnet food in the first week of life. Adult/cygnet bonds usually persist through the first winter and occasionally longer. Sibling associations often persist for two or three years, and subadult offspring may reassociate with their parents.

CHAPTER 5. MORTALITY

Despite the supplemental winter feeding of grain at RRLNWR and concerted efforts to halt the illegal shooting of trumpeters throughout the Tri-state Area, annual losses of up to 35% of the Tri-state Subpopulation have occurred between consecutive late summer USFWS Tri-state Surveys. Because relatively few swan carcasses have been found over the years, managers speculated that these annual losses might be due, in part, to dispersal of swans out of the Tri-state region.

To determine survival rates as well as movement and dispersal patterns, tarsal-banding studies were started at the Refuge in 1945 and continued intermittently to the present. Between 1945 and 1982, 678 adults and 372 cygnets were banded in the Centennial Valley (Anderson *et al.* 1986). Neck-banding studies were also conducted at Red Rock Lakes NWR from 1966-68 (RRLNWR-1) and from 1977 to 1985 (RRLNWR-9). While these studies detected occasional movements of a few individuals beyond the Tri-state region (Appendix I), they did not provide any evidence that dispersal was a regular or significant factor contributing to the observed annual losses.

Trumpeters from the Grande Prairie flock were marked from 1954-57 (Mackay 1957) and from 1973-78 (Turner and Mackay 1981). Other less intensive efforts to mark Canadian swans were conducted in Saskatchewan in 1972 and 1973 (Nieman and Isbister 1974), the Yukon in 1980 (McKelvey *et al.* 1983), and the Northwest Territories in 1986 (L. Shandruk, pers. comm.). Lower elevation Wyoming trumpeters have been tarsal- and neck-banded since 1982 (Lockman *et al.* 1987).

Survival of Fledged Swans

Tri-state Subpopulation

The annual survival rate of adults from the CV flock was estimated to average 80-88% based upon 1949-82 banding data. Several potential biases, such as emigration, neck-band induced mortality, and neck-band loss probably caused survival rates to be underestimated. Post-fledging survival rates for the 372 swans marked as cygnets could not be estimated because recoveries were too few for analysis (Anderson *et al.* 1986). The low resightability of marked cygnets suggested that their first winter survival was substantially lower than that of adults. Of 20 cygnets neck-banded at Red Rock Lakes NWR in 1977-1978, only four were resighted as yearlings, and only one was known to survive to age two. Neck-band loss and neck-band induced mortality were factors which confused interpretation of the resighting data. The extremely low resightability of cygnets, however, resulted in the decision to neck-band only yearling and older swans after 1979 (RRLNWR files, RRLNWR-9).

Lockman *et al.* (1987) calculated age-specific survival rates of 38 neck-banded lower elevation Wyoming trumpeters (Table 15). In contrast to other marking programs, neck-band loss in the Wyoming study was minimized by the recapture of marked individuals and the replacement of damaged neck-bands. First-winter cygnets and yearlings experienced markedly lower survival than swans in the older age classes, while survival of swans in age classes 3+ was higher than that estimated by Anderson *et al.* (1986) for adults/subadults of the CV flock.

Table 15. Age-specific survival rates from observations of neck-banded birds.

Interval	Wyoming Flock[a]	Grande Prairie Flock[b]	
Age class 1			
Entire year	0.47	nd[c]	nd
Hatch-fledge	0.78	nd	nd
Fledge-early November	0.78	nd	0.72
November-June	0.75	nd	0.47
Fledge-June	0.60	0.43	0.34
Age class 2 (yearling)	0.66	0.71	nd
Age class 3-5	0.93	0.82	nd
Age class 6+	0.93	0.82	nd

[a] Data from Lockman *et al.* 1987.

[b] First column data from Turner and Mackay (1982); second column data computed by Gale (this study) based upon resightings of Grande Prairie birds wintering in the Tri-state Area.

[c] No data

Lockman *et al.* (1987) have also shown that these survival rates, when used in a deterministic population model, result in simulated population growth that closely mimics fluctuations in the actual population.

We estimated annual survival rates of all post-fledging age classes combined for the Tri-state Subpopulation using data from the late summer USFWS Tri-state Surveys (Table 16). The annual survival rate was calculated as:

number of adults/subadults censused in year$_{x+1}$
total number of swans censused in year$_x$

The estimated annual survival rates (adults and cygnets combined) averaged by decade ranged between 79-84%, although annual estimates in the 65-75% range occurred occasionally. Overall, this analysis tends to confirm the estimates of Anderson *et al.* (1986) and Lockman *et al.* (1987) and their suspicions that survival rates based upon band recoveries are conservative. We strongly suspect that adult survival rates are actually 90+% in most years.

The lowest estimate (65% survival between September 1985 and September 1986) occurred after a near-record production season, further reinforcing the point that cygnet survival during the first winter must be very low. Of the 507 adults and cygnets censused in 1985, only 331 adults were censused in 1986. The possibility that a portion of these birds dispersed beyond the survey area as yearlings cannot be dismissed but past marking studies have not shown dispersal to be an important contributing factor to annual losses. Until significant dispersal of Tri-state subadults is documented, we assume that the bulk of the losses were due to overwinter mortality.

Interior Canada Subpopulation

Turner and Mackay (1981) estimated age-specific survival rates for the Grande Prairie trumpeters (Table 15) based upon resightings and recoveries of neck-banded swans. Like the estimates for the CV flock, these also were minimum estimates because dispersal and neck-band loss were inseparable from mortality.

We used CWS neck-banding data to estimate the survival of cygnets during their first autumn migration from Grande Prairie to the Tri-state Area (Table 17). Of 57 Grande Prairie families that were marked in September 1973-78, 23 were relocated in the Tri-state Area during their first winter. Because marked cygnets were rarely observed to arrive at the wintering grounds apart from their families, autumn survival rates were calculated from the number of cygnets in these 23 families that survived from marking in September to arrival in November in the Tri-state Area. Survival averaged 77% for 53 male cygnets and 65% for 46 female cygnets and did not differ significantly between sexes ($X^2 = 1.8$, 1 d.f.).

The reasons for the low autumn survival in 1975 and 1978 are not known. Shea made intensive searches of all known wintering sites in 1978-79 and concluded that the losses were real, and not due to observation conditions. Reduced survival during the autumn of 1978 probably included yearlings as well as cygnets: six of eighteen yearlings marked in 1978 were never resighted after marking. This loss of cygnets and yearlings was also apparent in late-summer censuses at Grande Prairie that recorded a drop from 131 cygnets and unpaired white birds in 1978 to 15 unpaired white birds in 1979 (Appendix VI).

Table 16. Mean annual survival rates of the Tri-state Subpopulation, by decade, 1940-86.

Decade	N^a	Mean	Range
1940s	4	0.81	0.74-0.89
1950s	7	0.84	0.77-0.96
1960s	8	0.83	0.75-0.98
1970s	No data		
1980s	3	0.79	0.65-0.94

a N= number of annual survival rates computed from consecutive late summer USFWS Tri-state Surveys. Annual survival rate equals the number of white birds in year $_{x+1}$ divided by the total number of white birds and cygnets in year$_x$.

[b.] Annual survival rates could not be calculated due to triennial survey schedule.

Table 17. First autumn survival of cygnets from 23 broods that were marked at Grande Prairie and resighted on Tri-state Area winter range, 1973-78[a].

Year	Cygnets marked in September		Cygnets resighted in the first winter		Percent survival both sexes
	Males	Females	Male	Female	
1973	5	3	4	2	75
1974	4	2	3	2	83
1975	7	5	4	2	50
1976	12	8	12	7	95
1977	18	19+	15	13	76
1978	7	9	3	4	44
All years	53	46	41	30	72

[a.] Data were summarized from the files of Bruce Turner, Canadian Wildlife Service, Edmonton, Alberta

Using CWS neck-banding data, we also estimated the survival rates of Grande Prairie cygnets during their first winter in the Tri-state Area and during their first spring migration (Table 15). From 1976-78, 28 male and 25 female cygnets that were neck-banded in GP were located in the Tri-state Area during their first winter. Of these 53 cygnets, 13 males (46%) and 12 females (48%) were either found dead or disappeared after their first winter and were presumed dead. Relatively low survival of cygnets during their first winter was also indicated by a positive relationship between the number of cygnets fledged in Grande Prairie and overwinter losses from the GP flock (Turner and Mackay 1981).

Relationships between Annual Survival Rates and Environmental Variables

Red Rock Lakes NWR

A variety of methods for estimating annual survival rates were examined, but the most useful proved to be a crude survival rate calculated from the total number of swans counted in late summer versus the number present the following spring. The spring numbers were inferred from the count of white birds during the following late summer, corrected for adults removed during the spring and summer. Thus, the mortality rate (1-survival rate) was really a mortality plus emigration rate. The actual rate was expressed as an instantaneous mortality rate (d) so that factors could be treated as additive:

$$d = ln \text{ (previous late summer total swans) - } ln \text{ (spring white birds)}$$

This was converted to a finite survival rate (s) as follows: $s = e^{-d}$. Net movement onto the Refuge occurred in 7 years, producing estimates of the instantaneous mortality rate which were below 0. These years were deleted from the subsequent analyses.

Eight factors were examined for potential relationships with winter mortality:
1) the proportion of the fall flock that was cygnets (PROPCYG);
2) the number of months that water releases at Island Park Dam on the Henry's Fork River were below about 250 cfs (LOWHF);
3) the number of bushels of grain fed per bird (BUFEDBIRD) during the winter at RRLNWR which was estimated from the total grain fed (BUFED) divided by mean number of wintering swans;
4) the mean number of swans counted at the Refuge feeding ponds during the winter (WINTSWAN);
5) previous late summer count of all swans at Red Rock Lakes NWR (FALLFLOCK);
6) a spring severity index (SPRSEV);
7) a winter severity index (WINTSEV); and
8) the date that Refuge lakes froze during the previous fall (LAGFREEZ).

Our reasons for selecting most of these independent variables should be reasonably obvious but one variable (LOWHF) should be explained at this point. As we will discuss in Chapter 8, the amount of water released at Island Park Dam has varied greatly over the years. Based upon our own observations, and those of observers at Harriman State Park and on aerial surveys, we judged that near total freezing of the Henry's Fork at Harriman State Park occurred at releases below 250 cfs. Marking studies (Chapter 9) have shown that Refuge swans primarily used two winter habitats, the Refuge feeding ponds and the Henry's Fork River. Wintering swans have frequently been reported moving between the two areas since the 1940s, and we suspected that their winter survival could be influenced by the availability of open water on the Henry's Fork.

Simple correlations between instantaneous mortality rates and environmental variables fell into 4 categories (Table 18):
1. Positive, highly significant ($P< 0.01$): the proportion of cygnets in the fall flock and the number of months of low water releases at Island Park Dam.
2. Negative, significant ($0.01<P<0.05$): bushels of grain fed per bird, and total bushels of grain fed per winter.
3. Negative, slightly significant ($0.05<P<0.1$): mean wintering swan numbers at Red Rock Lakes NWR.
4. Not significant ($P> 0.1$): Size of fall flock, winter severity, spring severity.

These results indicated that the rate of loss of swans from the Refuge flock increased in years when the proportion of cygnets was relatively high (as expected because their mortality rate is higher than that of older swans), when water releases at Island Park Dam were low for three or more months, when less grain was fed, or when fewer swans wintered at Red Rock Lakes NWR.

All of the significant correlations made biological sense except for the inverse density-dependent relationship with mean number of wintering swans. We suspected that this inverse relationship was due to correlations with other factors and took two approaches to testing this possibility. First we calculated partial correlations between mortality and mean wintering swan numbers while holding other important factors constant (Table 19). Partial correlations with grain feeding and low water flows in the Henry's Fork were not different from the simple correlations.

However, the partial correlations with wintering swan numbers all were substantially less than the simple correlations and were nonsignificant. This partial correlation fell to zero when the proportion of cygnets in the fall flock was held constant. Thus it seems that there is little solid evidence for density-dependent effects on winter mortality/emigration at Red Rock Lakes NWR, other than the indirect effect through grain feeding. If the amount of grain fed were held constant while the number of wintering swans increased, the declining amount of grain fed per bird would predict an increase in mortality rate.

These ideas were further supported by the regression of instantaneous mortality on low water flows, mean number of wintering swans, a winter severity index, and the bushels of grain fed per bird. The regression model predicted a slight increase in winter mortality/emigration with increasing numbers of swans but the regression coefficient could not be shown to be significantly different from 0 with the 19 years of complete data. The regression coefficients provided further support for the conclusion that longer duration of reduced water flows in the Henry's Fork, less grain fed per bird at RRLNWR, and increasing winter severity resulted in higher winter losses of swans.

Perhaps the most significant finding concerning winter losses of Refuge swans is that two of the major factors are under the control of managers. The amount of grain fed per bird and the procedure by which the grain is distributed is determined by the managers of Red Rock Lakes

Table 18. Correlations of the instantaneous winter mortality rate of Red Rock Lakes NWR Trumpeter Swans with environmental and other factors[a].

Variable	r	P	n
Proportion cygnets in late summer count (PROPCYG)	0.49	0.004	33
Number of months of low water releases at Island Park Dam (LOWHF)	0.49	0.01	28
Bushels of grain fed per bird at RRLNWR feeding ponds (BUFEDBIRD)	-0.44	0.02	29
Total bushels of grain fed per winter at RRLNWR (BUFED)	-0.39	0.04	29
Mean number of swans wintering at RRLNWR (WINTSWAN)	-0.33	0.08	30
Previous late summer flock size at RRLNWR (FALLFLOCK)	-0.14	0.43	33
Spring severity index (SPRSEV)	0.17	0.44	24
Winter severity index (WINTSEV)	0.04	0.83	25

[a.] Simple pair-wise Pearson correlation coefficients were calculated. Non-parametric correlation coefficients were also calculated but were not pursued further because they gave similar results. Note that various numbers of years were used for each calculation because of missing values.

Table 19. *S*imple partial correlations between mortality rate and selected independent variables.

Correlations with variable	Simple Correlation	Partial Correlation	Variable held constant
WINTSWAN	-0.329*	0.032^{ns}	PROPCYG
WINTSWAN	-0.329*	-0.196^{ns}	BUFEDBIRD
WINTSWAN	-0.329*	-0.187^{ns}	LOWHF
LOWHF	0.486**	0.486**	BUFEDBIRD
BUFEDBIRD	-0.443*	-0.444*	LOWHF

* $P<0.05$
** $P<0.01$
ns = not significant

NWR (see Chapter 8). Water flow volumes, and thus the timing and extent of freezing of the Henry's Fork wintering area at Harriman State Park, are determined by water releases from Island Park Dam. The regression model suggests that during a severe winter on the Henry's Fork in which water flows were low for 3-4 months, the effects on Refuge swans could be compensated for by feeding at least 5.5 bushels of grain per bird at Red Rock Lakes NWR. Such a suggestion rests on a large number of tenuous assumptions. We would feel confident in predicting that feeding increased grain in such winters would partially ameliorate the effect of low flows in the Henry's Fork on the Refuge flock.

Centennial Valley Flock and Tri-state Subpopulation

The mortality rate of the entire CV flock showed virtually the same relationships with the independent variables as did the Refuge flock. The non-Centennial Valley portion of the Tri-state Subpopulation showed no significant correlations with any of the factors, except the proportion of cygnets in late summer. Again, the Subpopulation consists of a number of flocks wintering in a variety of locations. Perhaps there are no simple overriding factors associated with mortality in these small flocks. Alternately, it would be better to examine each flock separately if the environmental and population data allowed such an approach.

Grande Prairie Flock

Marked Grande Prairie trumpeters have exhibited little tendency to utilize the supplemental grain at RRLNWR and have been observed to winter at sites throughout the Tri-state Area where they are subject to a variety of environmental conditions (see Chapter 9). The only significant correlation with winter losses that we detected was with the winter severity index ($r= 0.658$, $P<0.0003$, $n=26$). We concluded that without the benefit of supplemental grain feeding, the Canadian migrants are much more vulnerable to winter severity than are the CV birds.

Prefledging Survival Rates

Low cygnet survival has been documented at Red Rock Lakes NWR since the 1940s and at YNP since the 1920s (Banko 1960:131; Page 1976; Shea 1979, 1980). Recent studies (Maj 1983, Holton 1985, Lockman *et al.* 1987) have shown that cygnet survival is higher in Grande Prairie and in the lower elevation Wyoming and Idaho portions of the Tri-state Area (Appendix IX).

We had difficulty determining cygnet survival rates in past years at Red Rock Lakes NWR because Refuge reports routinely computed survival by dividing the number of late summer cygnets by the number of cygnets observed one to three weeks after the peak of hatch. Comments in the Annual Narrative Reports indicated that in many years, significant mortality probably occurred before the first count, and therefore Refuge calculations tended to overestimate cygnet survival. Based on the annual maximum cygnet count and final late summer counts, prefledging cygnet survival at the Refuge since 1964 has averaged less than 46%. The highest estimated survival occurred in 1985 and was somewhat less than 67% (Table 20). In a 1971-73 study which closely followed the hatching success and fate of each clutch and therefore more accurately documented the actual number of cygnets hatched, Page (1976:70) found that annual cygnet survival at Red Rock Lakes NWR ranged from 17% to 37%.

Using Refuge data for maximum and final cygnet counts, we also estimated maximum cygnet survival rates by nesting area (Table 21). The accuracy of these estimates was reduced due to deaths of cygnets before the maximum counts, and movement of broods from one lake to another between counts. Although some movement undoubtedly occurred, Refuge Annual Narrative Reports generally indicated that most broods remained on their natal body of water until they fledged. These data indicate that cygnet survival on the Refuge as a whole declined in the 1980s despite better than average survival in 1985 and 1986. Particularly large reductions in survival rates occurred on Lower Lake, Swan Lake, and the ponds. Most of the ponds were newly constructed in the mid-1960s and initially showed very high cygnet survival rates. In 1971-73, cygnet survival was 56% at Swan Lake, 50% at the ponds, 35% at Upper Lake, 11% at Lower Lake, and 8% at River Marsh (Page 1976:74).

Table 21 also shows that in addition to low prefledging survival, the number of cygnets hatching on Lower Lake, Upper Lake, and River Marsh has declined. As discussed in Chapter 3, the declining number of nesting pairs and low hatching success, in addition to low prefledging survival, have combined to reduce cygnet production on the Refuge.

Based upon late summer USFWS Tri-state Survey results (Appendix III), low cygnet production also has been chronic in YNP. Prefledging survival in Yellowstone was only 25% for 131 cygnets hatched between 1977 and 1979. Sixty-five percent of the cygnets died within six weeks of hatching. In 1978, 55% died within 10 days of hatching (Figure 13). Mortality usually involved entire broods: in 1979, eight of the seventeen broods suffered 100% mortality within three weeks of hatching. Brood mortality was strongly site and/or pair specific (Shea 1979, 1980). Survival of cygnets in Idaho was 33% and was also strongly site and/or pair specific, with most deaths occurring between two to six weeks of age (Maj 1983). In Grande Prairie, cygnet losses were greatest during the first two weeks after hatching and then decreased as the cygnets aged. Survival rates were 77% in 1970-72 (Burgess 1972) and 56% in 1978-80 (Holton 1985).

Cygnet Mortality Studies

From 1962-66, special efforts were made at Red Rock Lakes NWR to collect cygnets immediately after death for detailed necropsies by Drs. W. T. Jensen and M. McDonald, at the USFWS Bear River Research Station in Utah. Mortality was found to be most frequent in the first four weeks after hatching although dead cygnets were found throughout the summer. These studies were hampered by the difficulty of locating cygnets soon after death; many carcasses were unsuitable for detailed examination due to postmortem tissue changes and bacterial contamination (RRLNWR Annual Narrative Reports 1962-66).

These necropsies revealed that, in general, waterfowl at the Refuge showed severe parasite infections. Compared to Bear River Refuge, birds at Red Rock Lakes had two to five times as many helminth species and ten to twenty times as many individual worms. Examination of 14 dead trumpeters of various ages found that three suffered from heavy acanthocephalan infections (7,000, 7,000, 2,300 parasites, respectively) and two were infested with approximately 70 nasal leeches each. Birds apparently died of suffocation due to blockage of the nasal passages and trachea (Summary of findings, B. Tuggle letter to RRLNWR files, 1 November 1983).

Leech infestations and associated mortality were also observed among prefledging Alaskan

Table 20. Estimated maximum annual prefledging cygnet survival at Red Rock Lakes NWR[a].

Year	Maximum Cygnet Count	Final Cygnet Count	Maximum % Survival
1964	60	24	40
1965	53	16	30
1966	133	65	49
1967	38	20	53
1968	142	90	63
1972	33	20	61
1974	82	20	24
1976	48	25	52
1977	82	39	47
1978	76	38	50
1979	117	52	44
1980	29	5	17
1981	94	37	39
1982	17	4	24
1983	48	19	40
1984	30	4	13
1985	63	42	67
1986	26	15	58
All Years	1,171	535	46

[a] All data are from RRLNWR files. Maximum pre-fledging survival = final cygnet count/maximum cygnet count.

Table 21. Estimated maximum prefledging cygnet survival and number of cygnets hatched in each major nesting area at Red Rock Lakes NWR, for three periods between 1964-86[a]. Data are presented as maximum percent survival (number of cygnets hatched).

Years (n)	Lower Lake	Upper Lake	Swan Lake	River Marsh	Ponds	Entire Refuge
1964-68 (5)	59% (95)	64% (64)	49% (93)	37%(161)	92% (13)	50% (426)
1970-79 (6)	43% (60)	47% (73)	42% (119)	49%(189)	71% (38)	48% (479)
1980-86 (7)	36% (31)	61% (46)	27% (96)	37% (98)	53% (40)	39% (311)
1964-86(18)	50%(186)	56% (183)	40% (308)	42%(448)	66%(91)	46%(1,216)

[a] All data are from RRLNWR files. Maximum survival rate = Final cygnet count/ Maximum cygnet count.

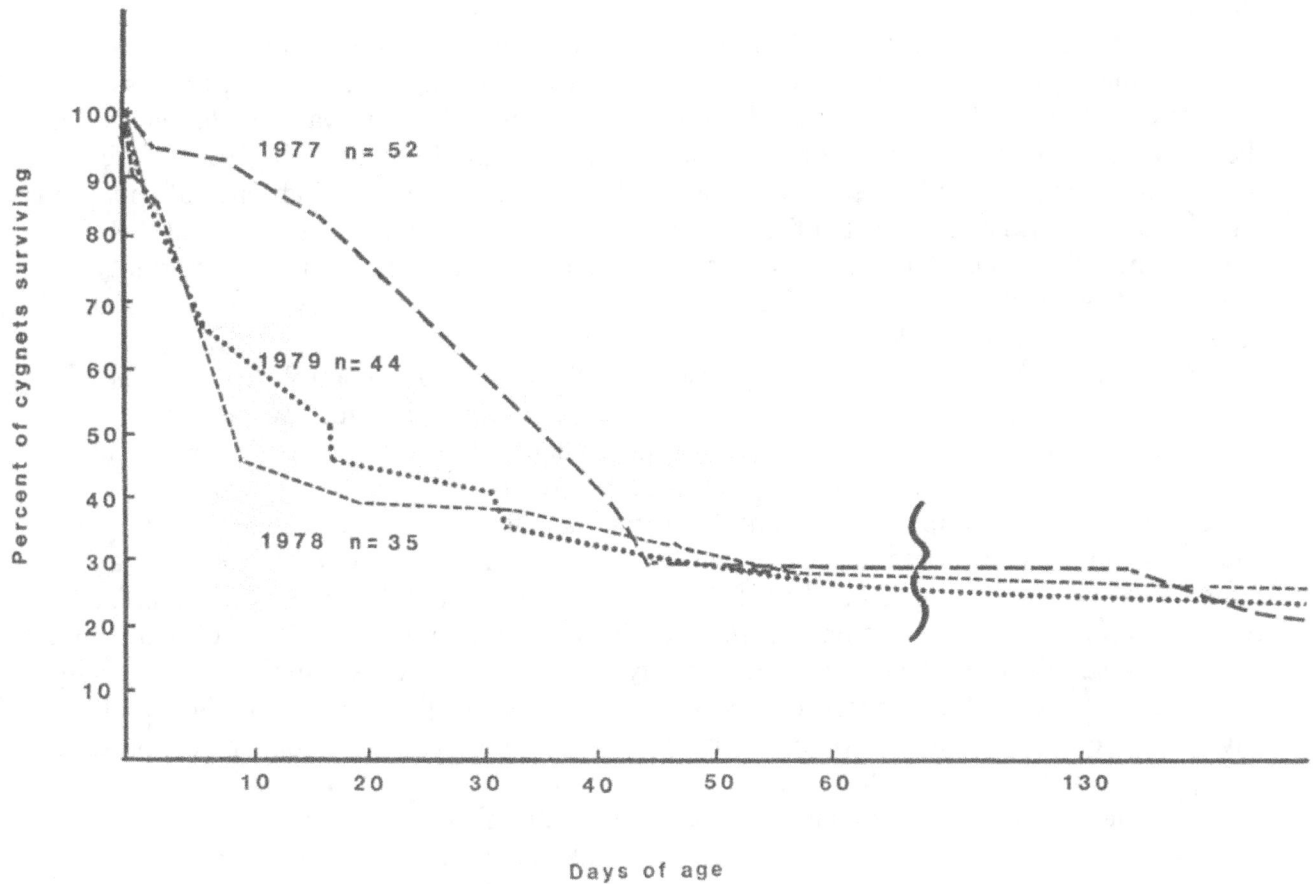

Figure 13. Survival of cygnets in Yellowstone National Park, 1977-79 (from Shea 1980).

cygnets (Bangs *et al.* 1981). The nasal chamber is the most prevalent site of infection in waterfowl. Leeches may indirectly cause mortality through impaired vision, retarded growth and development, obstruction of the respiratory tract, and the transmission of blood parasites. Sick and disturbed birds are particularly prone to leech infestations (Trauger and Bartonek 1977).

The Refuge trumpeters showed a great many infections with forms not normally expected in swans, and apparently were susceptible to almost anything available. The researchers speculated that this vulnerability might be a physiologic response to a swan population level at the carrying capacity of its range, or that Refuge conditions might favor high parasitic infections due to year-round contact between swans and the host invertebrates or the absence of conditions which would cause periodic death of the invertebrate hosts (B. Tuggle letter to RRLNWR files, 1 November 1983). Dr. Jensen summed up the investigations by stating that the "mortality characteristics are obscure and multifaceted" and that "parasitism is usually the result not the cause of a weakened physiological condition" (R. Sjostrom memo to RRLNWR files, 10 February 1978). Thorne *et al.* (1985) suggested that the lower water control structure has maintained relatively stable water levels and eliminated fluctuations which would have interrupted parasite life cycles, thus allowing unusual concentrations of parasites to develop.

In three years of cygnet mortality studies in Yellowstone National Park, necropsies of 14 carcasses were often hampered by postmortem contamination and tissue changes. No specific cause of death was diagnosed for any cygnet, even specimens that were retrieved within a few hours of dying. Dr. L. Siegfried, USFWS National Wildlife Health Laboratory, summarized the results as follows: "A combination of certain findings in common in several of the small cygnets are of interest, this being the presence of a variety of mixed parasites within the gastrointestinal tract, the presence of blood-engorged *Theromyzon rude* [leeches] within the nasal and oral cavities, the observation of generalized muscle atrophy and tissue emaciation, and the presence of apparent anemia" (Shea 1980).

Despite the difficulty of diagnosing specific mortality factors, the pattern of mortality was evident in Yellowstone. Cygnets hatching from small clutches had a significantly lower survival rate than cygnets that hatched from large clutches ($P=0.02$, $X^2=5.44$, 1 d.f.). Cygnets from clutches of four or fewer eggs ($n=12$) had a survival rate of only 10%, and only 25% of these broods fledged even one cygnet. In contrast, cygnets from clutches of five or six eggs ($n=12$) had a 31% survival rate, and 58% of the broods fledged at least one cygnet. Higher survival of cygnets from larger broods was noted at Lacreek NWR in 1971. Records showed that these larger broods were from the older and more established parents (Leach 1977). Parental age was not the likely determinant of clutch size and cygnet survival differences between pairs in Yellowstone, however, because small clutch size and poor survival were characteristic of particular pairs and/or sites year after year.

We believe that clutch size was influenced by the availability of spring food resources, and the ability of adults to accumulate adequate energy reserves prior to nesting. In other avian species, the adult female's nutritional state has been shown to affect the survival of her young in the critical period immediately after hatching. Scott (1973) summarized the situation:

"...the nutritional requirements for reproduction in birds are very similar to the requirements of the young birds for survival and optimum growth for the period from hatching through the critical starting period of approximately 3-6 weeks. The dam must consume sufficient amounts of all of the essential amino acids, vitamins, and minerals to

produce an egg and to supply that egg all of the nutrients needed by the embryo throughout its entire incubation period. For optimum survival and growth of the young, the dam also must provide sufficient reserves in the spare yolk to supply nutrients to the newly hatched bird".

Lockman *et al.* (1987) suggested that brood survival on some Wyoming territories was strongly linked to the availability and abundance of aquatic invertebrates and macrophytes during the first four to six weeks following hatching. Several pieces of evidence led to this hypothesis. On sites with a consistent history of low cygnet survival beyond four weeks of age, water levels were relatively stable and invertebrate and vegetation production was sparse and low in diversity. In contrast, on good cygnet production sites water levels fluctuated more dramatically between years and within years. Such fluctuations exposed soils and allowed oxidation, and upon reflooding promoted nutrient releases which increased invertebrate production and vegetation diversity. Successful brood rearing sites had up to 10 times the density of invertebrates found in unsuccessful sites.

The abundance and availability of cygnet food items in June is likely related to environmental conditions during the preceding months. Adverse weather that prolongs cold water temperatures and ice cover in April and May will slow plant development and thus the availability of food supplies both to prenesting adults and subsequently to newly hatched cygnets.

Page (1976:120) concluded that cygnet mortality related to food availability was minimal at Red Rock Lakes NWR, but his conclusion was based upon incidental and unquantified observations of invertebrates and vegetation. To date, no studies of cygnet food habits or food availability during the first days and weeks following the hatch have occurred at the Refuge.

Most cygnets at Red Rock Lakes NWR died in the first few weeks after hatching in 1971 and 1972, and deaths were concentrated in periods of rain, hail, snow, and cold. Cygnets appeared to have low resistance to environmental stress (Page 1976:70). Over the years, numerous observations in the Refuge Annual Narrative Reports have associated smaller broods, periods of high cygnet mortality, and low hatchability with cold wet springs and summer storms. Refuge Manager O. Vivion noted that cygnet losses of up to 70% occurred during the first two weeks after hatching, in what is usually a rainy period (RRLNWR files, letter to W. Banko, 30 October 1968). In 1969, 90% of the Refuge cygnets died during a late June storm within two weeks of hatching (RRLNWR Ann. Narr. Rept. 1969).

As discussed in Chapter 3, our analysis of cygnet survival showed that it is negatively correlated with increasing July rainfall and water levels, as well as a late peak of hatching. Peak of hatch is later in years when spring weather is unusually cold and/or wet. It has frequently been noted that late hatching cygnets die in the fall when they fail to fledge prior to freeze-up.

Page (1976:104) noted that captive trumpeters which were fed special diets had extremely high production rates. He hypothesized that the increased cygnet survival at Red Rock Lakes NWR in 1973 and increased resistance to early summer storm-associated mortality was linked to the doubling of the winter feeding in 1972-73. Page (1976:103) suggested that reduced winter feeding caused swans to go into the breeding season in poor physiological condition, resulting in weakened cygnets that were especially vulnerable to various mortality factors, mainly weather, and reduced parental care to brood and nest. As can be seen in Table 20, cygnet survival increased on the Refuge in 1985 and 1986. These years were characterized by unusually dry July weather, as well as increased amounts of grain fed during the preceding winter.

Other Mortality Factors

Our review of necropsy reports from 73 trumpeters found dead in the Tri-state Area between 1960 and 1986 showed the leading factors involved in death to be emaciation of undetermined origin (26%), parasites (24%), accidents (15%), lead poisoning (15%), and disease (15%). Deaths due to shooting and accidents were probably under-represented. Agency files contained references to swan mortality due to shooting or powerline collisions, but few of these birds were necropsied because the cause of death seemed obvious. Likewise, lead poisoning was also under-reported because few carcasses were tested for tissue lead levels prior to the mid-1980s.

In Wyoming, accidents, primarily collisions with powerlines, caused 62% of the recorded mortality (n=13) during 1981-86. No illegal shooting was documented. Swans from the younger age classes were more frequently involved in accidents, and older birds seemed more susceptible to accidents when human disturbance or food shortages forced them to pioneer new areas (Lockman *et al*. 1987).

About 21% of the mortality documented by band returns of Grande Prairie trumpeters was due to illegal shooting (Brechtel 1982). McKelvey (1981) summarized mortality of 71 wild British Columbian trumpeters and found parasites (51%), infections (20%), accidents (16%) and shooting (13%) to be the factors involved in death.

Bacteriology

During the 1960-66 mortality studies at RRLNWR, no known pathogenic bacteria were found with the exception of *Pasturella multocida* in an adult swan that died in November 1964 at the winter feeding pond. Although the spread of fowl cholera was feared, this carcass was picked up soon after death and no other cases were found (Gritman and Jensen 1965). Few cases of disease have been diagnosed among Rocky Mountain trumpeters. However, in addition to fowl cholera, avian tuberculosis, aspergillosis, generalized peritonitis-pericarditis, and systemic infections of *Pseudomonas* and *Escherichia coli* have been identified in necropsies.

Dr. Jensen suggested that the lowered resistance due to heavy parasitism or other causes might increase the trumpeters' susceptibility to organisms that would not otherwise be pathogenic (Letter to RRLNWR files, 1 October 1969). Subsequent to high mortality among wintering swans on the Henry's Fork during late winter 1985, two emaciated and louse-infested trumpeters from the Henry's Fork were treated in April 1985 by Idaho Falls veterinarian Dr. T. Moe. Both swans died within a few days of capture. *E. coli* was cultured from the heart of one swan and from the liver of the second. Dr. Moe diagnosed the cause of death as *E. coli* septicemia, possibly precipitated by the swans' debilitated condition. A systemic *E. coli* infection was also diagnosed as the cause of death of a Tundra Swan cygnet found at Harriman State Park in November 1986 (Dr. T. Moe, pers. comm.). Excessive levels of coliform bacteria have been documented to enter the Henry's Fork from the septic systems of the many stream bank summer homes. Concern over this contamination led to the installation of a centralized sewage collection and treatment facility in 1982-83 in the Mack's Inn area, and the planned installation of a similar system at Last Chance in 1987 (B. Dixon, State of Idaho Health Department, pers. comm.).

During July 1982-84, venous blood samples were taken from 65 flightless nonbreeders in the Centennial Valley. Samples were tested for pullorum, typhoid, *Mycoplasma gallisepticum*,

M. synoviae, M. meleagridis, avian influenza, and hemoparasites. Twenty-one throat and fifty-four vent swabs were tested for *Salmonella* and other pathogenic bacteria in 1982. *E. coli, Enterobacter, Klebsiella*, and *Pseudomonas* were recovered from vent swabs. *P. multocida* was recovered from one throat swab. The presence of *P. multocida* indicated the potential for avian cholera to occur and emphasized the importance of minimizing stress, especially during the winter and early spring. Field tests for *Salmonella* antibodies suggested exposure of a few birds, but lab tests conducted on the sera did not support the presence of *Salmonella* (Thorne et al. 1985).

About 33% of the swans captured in 1982 had pox-like lesions on their feet although pox virus was not demonstrated. No lesions were observed in 1985. None of the lesions was life threatening and pox is usually self-limiting in mature birds (Thorne *et al*. 1985). Similar lesions were noted at Red Rock Lakes NWR in 1970 (RRLNWR files 1970).

Twenty swabs from 11 nests and 19 eggs from Wyoming and Red Rock Lakes NWR were cultured in 1982 and all were negative for pathogenic bacteria. All fertile eggs were bacteriologically sterile (Thorne *et al*. 1985). Twenty-four unhatched eggs from Yellowstone and Idaho were examined in 1979 and 11 genera of bacteria were isolated by the USFWS National Wildlife Health Lab. *Salmonella* was specifically looked for but not found and most of the bacteria were probably postmortem invaders (Shea 1980).

Blood Parasites

During September 1978 and 1979, 75 Grande Prairie trumpeters were captured and examined for hematozoa. Twenty eight (37.3%) were infected with blood parasites, primarily *Haemoproteus nettionis*. One cygnet was infected with *Leucocytozoan simondi* and two cygnets harbored *Plasmodium circumflexum*. The prevalence of *H. nettionis* was 72.7% in yearling and older males, almost twice the infection rate noted in females and male cygnets. No serious pathogenic effects have been attributed to *H. nettionis*, and blood parasites apparently were not limiting factors to population growth or expansion in the Grande Prairie area (Bennett *et al.* 1981).

Emaciation/Non-specific Winter Mortality

Most dead trumpeters found at Tri-state wintering sites, particularly along the Henry's Fork, were moderately to severely emaciated. Eight of ten dead swans found on the Henry's Fork by Hampton (1981) showed emaciation of undetermined origin. Post-mortem contamination and tissue changes have often made determination of bacterial or viral diseases very difficult, and emaciation is thus the most frequent diagnosed mortality factor at wintering sites. In many cases the emaciation has been accompanied by heavy parasite loads and elevated tissue lead levels.

The first documented incident of numerous deaths among wintering trumpeters in the Tri-state Area occurred during the winter of 1984-85. Twenty-three dead trumpeters were found in the Mack's Inn section of the Henry's Fork River, and seven were found at Harriman State Park and the Buffalo River. A few weeks later, a grizzly bear was reported to be feeding on swan carcasses in the canyon several miles downstream from Harriman SP. With the relative lack of islands or other obstacles, currents at Harriman SP carried an undetermined number of carcasses to inaccessible areas downstream. Carcasses were more likely to remain and be found in the slow shallow meanders in the Mack's Inn area. A search of the Teton River in April revealed only two carcasses, including a banded swan from Grande Prairie. No other marked swans were found dead in any of the wintering areas. At least 10 carcasses (species undetermined) were seen during the winter along the Yellowstone River in YNP (R. Landis, pers. comm.). Known carcasses in Idaho and YNP totaled about 50. Most of the mortality in the Mack's Inn area probably occurred in March, as few of the carcasses had been scavenged when collected on 10 April. Eighteen necropsies performed at the USFWS National Wildlife Health Laboratory found that most of the trumpeters showed various degrees of emaciation coupled with heavy parasitism by a variety of parasites including *Eimeria christenseni* (coccidia), *E. brontae* or *E. hermanni*, *Notocotylus sp.*, *Echinaria sp.*, *Sarconema sp.*, *Diorchis sp.* and *Theromyzon sp*. Lead poisoning was diagnosed in five cases and was the suspected cause of death of a sixth trumpeter (Table 22).

As ranked by our winter severity index, the winter of 1984-85 was the third coldest on record at the Red Rock Lakes NWR weather station. It followed a very severe autumn freeze in Canada which abruptly forced migrants to move southward (B. Turner, pers. comm.). Unusually high numbers of Tundra Swans (700+ at Harriman SP) were noted in the Tri-state Area in November and December 1984 (J. Snyder, pers. comm.). We suspect that the extremely cold winter, possibly in conjunction with reduced autumn food resources, caused the swans to deplete their energy reserves by March, and the emaciated birds were unable to depart the River in early March as they normally do. Individuals carrying heavy parasite loads or contaminated by ingested lead were particularly stressed. No specific diseases were identified.

Because of the high mortality during winter 1984-85, the Idaho Department of Fish and Game made a concerted effort to search for dead swans during the winter of 1985-86. Trumpeters left the Henry's Fork by 1 March 1986, and a search of the river in mid-March found eight carcasses. Three of the five dead adults had been marked in the Centennial Valley, in contrast to the previous year when no marked Centennial Valley swans were found among the more numerous carcasses.

Lead Poisoning

The loss of trumpeters due to lead poisoning has been a concern in the Tri-state Area since the winter of 1936-37 when Refuge Manager Hull found four dead trumpeters at the Culver Pond feeding site. The gizzards of these dead swans contained 3, 11, 17, and 19 lead shot pellets. Waterfowl hunting was frequent at the pond prior to the establishment of the Refuge in 1935 and lead deposition no doubt occurred (Banko 1960:137). Waterfowl hunting has continued to the present in portions of Lower Red Rock Lake. In 1986, Refuge hunting regulations were changed to require the use of non-toxic shot (B. Reiswig, pers. comm.). The susceptibility of the Refuge swans to lead poisoning may have been increased by the winter feeding of grain. Waterfowl that feed on whole or part-grain diets are more susceptible to lead poisoning than waterfowl on grainless diets (USFWS 1986a).

Table 22.   Necropsy results from 18 trumpeters found dead on the Henry's Fork River in April 1985[a].

Diagnosis	N
Lead poisoning (7.7-21.0 ppm)	5
Emaciation/ elevated lead (3.6 ppm)	1
Emaciation/esophagitis/coccidiosus/visceral gout	1
Emaciation/esophagitis/Impaction	1
Emaciation/myocarditis	1
Emaciation/coccidiosus	1
Emaciation/cause open	1
Emaciation/sinusitis	1
Coccidiosis	1
Pericarditis/valvular endocarditis	1
Unsuitable for examination	1

[a]   Summary of final results from USFWS National Wildlife Health Lab report to RRLNWR, 10 June 1985.

Compared to other species of waterfowl, swans are even more likely to ingest lead shot pellets because they often dig deeper than other waterfowl as they seek out the tubers of aquatic plants. Their extensive digging and feeding on tubers exposes them to lead pellets that were deposited years earlier and had sunk into the substrate beyond the reach of most other waterfowl (USFWS 1986a).

Since about 1982, wildlife managers in the Tri-state Area have increased their efforts to retrieve and necropsy trumpeter carcasses. Of 34 carcasses from the CV and Henry's Fork areas which were necropsied in 1980-86, 11 swans died from lead poisoning. Of the 11 lead poisoned swans, 5 contained lead shot pellets, 2 contained lead fishing sinkers, and the remainder had passed the ingested lead objects prior to death (Necropsy reports in RRLNWR files). The known cases of lead poisoning in trumpeters in the Tri-state Area are summarized in Table 23.

Concern about the sublethal effects of lead on trumpeters has recently increased. Because of their very large body size, individuals may ingest and absorb lead, and carry sublethal concentrations in their tissues for some time until an additional dose of ingested lead or unusually stressful conditions cause their death. Sublethal concentrations of lead may suppress the immune system and impair normal physiological functions (Franson 1986, Wobeser 1986), thus reducing an individual's resistance to diseases, parasites, or severe environmental stresses.

Two recent investigations have examined apparently healthy Tri-state trumpeters in order to determine their exposure to ingested lead. Lockman et al. (1987) obtained blood samples from 10 trumpeters in 1985 and found that 3 birds had elevated lead levels of 0.220, 0.525 and 0.675 ppm wet weight, respectively. The lead levels of the other seven swans ranged from 0.070 to 0.165 ppm wet weight. Blood samples taken from 22 CV trumpeters in 1984 were analyzed by L. Blus, USFWS Pacific Northwest Field Station. Of the 22 samples, 9 had non-detectable levels of lead, 8 had levels ranging from 0.05-0.20 ppm wet weight, and five had elevated levels ranging from 0.21-0.71 ppm wet weight. Blus concluded that about 60% of the Red Rock Lakes NWR flock contained detectable lead, and that the most likely origin of the lead was from ingestion of shot (Letter from L. Blus to RRLNWR, 6 August 1986).

Gizzard shot surveys conducted at Red Rock Lakes NWR during five hunting seasons between 1976-80 showed 3.5% of 744 hunter-killed diving ducks had ingested lead shot. Canvasbacks (*Aythya valisneria*) ($n=51$) had the highest prevalence (11.8%). During the same period ingested shot was found in 4.7% of 235 Mallards (*Anas platyrhynchos*). During the 1984-85 hunting season, ingested lead shot was found in four (4%) of 99 hunter-killed diving ducks. Five (5%) of 97 ducks had liver lead concentrations above 8 ppm wet weight. Of 97 diving ducks, 17 had elevated liver lead concentrations greater than 2 ppm wet weight (NWHL Final Report, 1984-85 lead poisoning monitoring program at RRLNWR).

Other Contaminants

No significant pesticide residues were found in examinations of eggs from Red Rock Lakes NWR (Page 1976:119) and YNP (Shea 1979). To determine whether Yellowstone trumpeters were ingesting selenium and radium during their association with geothermal areas, feathers and eggs were examined. One feather and one egg tested by the Wyoming Game and Fish Department showed detectable selenium, but surface contamination was suspected and no deleterious effect was suggested (W. Hepworth, pers. comm.). Initial analysis of egg shells and feathers at the University of Wyoming revealed consistently elevated levels of gross alpha activity in the egg shells (Kennington et al. 1980). Subsequently it was found that the materials

Table 23. Known cases of lead poisoning in the Tri-state Subpopulation[a].

Date	Location	Comments
1937	RRLNWR	4 swans contained 3, 11, 17, and 19 shot in their gizzards (Banko 1960:137)
1960s	RRLNWR	3-week-old cygnet had one shot in gizzard
10/13/63	RRLNWR	Cygnet had 115 ingested shot; found near lower structure
Spring/83	RRLNWR	Swan contained 32 lead wafers
1/24/84	RRLNWR	Swan with five pellets in gizzard, 14 ppm liver lead wet weight
2/28/85	RRLNWR	Swan diagnosed with impaction/enteritis sublethal lead toxicosis; 4.9 ppm liver lead wet weight
3/85	Henry's Fork	5 swans with toxic lead levels ranging from 7.7 to 21 ppm liver lead wet weight.
3/85	Henry's Fork	1 swan diagnosed emaciation/elevated lead with 3.6 ppm liver lead wet weight
4/85	Salt River	Lockman (1985) reported 33 ppm lead from dried wing bone.
12/5/86	RRLNWR	Cygnet found dead at feeding ponds, 59 lead shot pellets in gizzard, 11.3 ppm liver lead wet weight.

[a] All reports are on file at RRLNWR.

had been contaminated in the lab and that the earlier results were invalid. Uncontaminated samples were rerun and alpha levels were within expected limits (G. Kennington, pers. comm.).

Analysis of blood samples collected during 1984 indicated that some trumpeters may be accumulating other contaminants such as copper and selenium, in addition to lead. Due to the high potential of this long-lived, heavy-bodied species to accumulate contaminants, the USFWS began efforts in 1987 to identify potential sources of heavy metal contaminants at Red Rock Lakes NWR (Memo to RRLNWR from USFWS, Billings, MT, 17 February 1987).

Retarded Development and Abnormalities

During incubation studies at Red Rock Lakes NWR in 1963, the Refuge staff noted that most of the newly hatched cygnets were weak, had poor coordination, and seemed to lack basic instincts. Some had difficulty breaking through the shell membrane; others could not adapt to cool temperatures comfortably. The staff suggested that these traits would contribute to the low survival of wild cygnets (RRLNWR Ann. Narr. Rept. 1963).

Poorly developed cygnets, some still flightless in late October and early November, have been noted at Red Rock Lakes NWR on numerous occasions (RRLNWR Ann. Narr. Repts. 1941, 1943, 1945, 1946, 1948, 1953, 1958). Low body-weight cygnets were also common. Twenty-one cygnets captured at the Refuge on 5-7 September 1950 ranged in weight from 12-22 lbs, with a mean weight of 15.0 lb. The average weight of thirteen male cygnets was 16.0 lb (range 12-22 lbs) and the average weight of eight female cygnets was 13.3 lb (range 12-15 lbs). The average weights of 9 female and 14 male cygnets captured on 28 August 1979 were 11.1 and 11.5 lbs respectively, with a range of from 6-17 lbs (RRLNWR banding files). Although some of the variation in weights may be attributable to differences in hatching dates, the observers often noted that the "runts" had apparently normal siblings.

Seven (18%) of the 39 cygnets that survived to September in the Yellowstone and Targhee NF study area in 1977-79 were retarded in development. In an attempt to diagnose the cause of this stunting, one abnormal cygnet from YNP was sacrificed at 41 days of age. It weighed only 475 gm (1.05 lb). Approximately 60 leeches clung to its breast, legs, and feet. Eight leeches were found in its nasal cavities. The pectoral muscles were very underdeveloped but all organs appeared normal. Bacterial cultures of liver and heart yielded a heavy growth of *Pseudomonas*. In his diagnosis, Dr. L. Locke, USFWS National Wildlife Health Laboratory, stated "One could postulate that these leeches caused a marked anemia, weakening the bird and allowing *Pseudomonas* to become septicemic." In conjunction with the sacrifice of the stunted cygnet, an apparently healthy six-week-old cygnet from Red Rock Lakes NWR was also sacrificed as a control. Surprisingly, this control bird weighed only 1,100 gm (2.42 lb) and was found to be moderately emaciated with very underdeveloped pectorals, and no subcutaneous, coronary, nor abdominal fat, although no disease was detected. Thirteen leeches were found in the nasal passages (Shea 1980). In contrast, captive descendants of Red Rock Lakes NWR trumpeters raised in captivity at Hennepin County, MN, weighed between 3,640-5,450 gms (8-12 lbs) at 42 days of age (The Trumpeter Swan Society 1977).

Two live eggs, one from a very late hatching clutch and the other egg left in a nest after the rest of the clutch had hatched, were collected in Yellowstone in 1978 and hatched artificially. Both cygnets were very weak and had leg and foot deformities. Similar deformities were seen in two of five cygnets hatched in 1978 by a private propagator from eggs received from Red Rock Lakes NWR (J. DeSarro, pers. comm.). Cygnets showed a backward curling of the toes, neck asymmetry, and rapid head shaking (Shea 1979).

Two Idaho cygnets, observed in the wild by Shea (1979), swam adequately when 2 days old but on land they lunged forward with a strenuous hopping motion. Foot deformities were suspected, however the cygnets disappeared within four days of hatching before they could be captured and examined. Foot deformities have also been noted at Red Rock Lakes NWR (RRLNWR Ann. Narr. Repts. 1946, 1959; Banko 1960:140).

Predation

Isolated observations of predation indicate that it is a regular source of mortality, but does not appear to be a major factor impacting swan numbers. Banko (1960:131-135) provided a detailed summary of incidents of predation by a variety of avian and mammalian predators.

Summary

The combined annual survival rate of yearling and older trumpeters was estimated from band recoveries to be at least 80-88% in the CV flock. Survival of trumpeters in age classes 3+ was estimated to be 82% in the Grande Prairie flock, and 93% in the Wyoming flock. Due to neck-band loss, neck-band induced mortality, and emigration, these are minimal estimates, particularly for the CV and Grande Prairie flocks. Total post-fledging annual survival in the Tri-state Subpopulation has averaged 79-84% in the decades since the 1940s. During the worst years, the Subpopulation has experienced survival of 65-75%. Although some portion of the annual losses may be due to dispersal of trumpeters beyond the Tri-state range, resightings and/or band returns derived from the marking of over 1,050 CV trumpeters gave evidence of dispersal by only a few individuals.

Survival of first-winter cygnets and yearlings is lower than that of older birds. Cygnet survival from fledging to the following June is about 60% in the Wyoming flock and 43% in the Grande Prairie flock, and probably similar or lower in the CV flock. Yearling survival is about 66%-71%.

Factors influencing the loss of swans from Red Rock Lakes NWR were analyzed using a mortality rate estimate (includes dispersal) based upon the loss of swans from the Refuge flock between consecutive annual late summer surveys. This mortality rate was positively correlated with the duration of low water releases at Island Park Dam (P=0.01), and the proportion of cygnets in the fall count (P= 0.004). The mortality rate was negatively correlated with the bushels of grain fed per swan (P= 0.02) and total bushels of grain fed per winter (P= 0.04) at Red Rock Lakes NWR.

The detrimental effect of low flows in the Henry's Fork on Red Rock Lakes NWR trumpeters can be partially mitigated by increased feeding of grain at the Refuge. There was no evidence of density-dependent effects on winter mortality at the Refuge, other than if the amount of grain was held constant as the number of wintering swans increased, making less grain available per bird. A regression model predicted that increasing duration of low water flows in the Henry's Fork, less grain fed per bird at Red Rock Lakes, and increasing winter severity result in higher winter mortality in the Refuge flock. Analysis of the mortality rate of the CV flock showed similar relationships.

The mortality rate of the non-Centennial Valley portion of the Tri-state Subpopulation showed no significant correlations with any factors except the proportion of cygnets in the fall population. These swans winter in a wide variety of locations that are not adequately described by the available data. The mortality rate of the Grande Prairie flock was positively correlated with Tri-state winter severity ($P = 0.0003$).

Low prefledging cygnet survival has been chronic in the Tri-state Area, particularly at Red Rock Lakes NWR and in Yellowstone, and to a lesser extent in Idaho and lower elevation Wyoming habitats. Since 1964, cygnet survival at the Refuge has averaged somewhat less than 46%. Despite above-average cygnet survival in 1985 and 1986, cygnet survival in the 1980s at Red Rock Lakes NWR has averaged somewhat below 38%. Declines in survival rates have occurred in all parts of the Refuge except Upper Lake and the ponds. These declines in cygnet survival have been compounded by reductions in the total number of nesting pairs, and low hatching success.

In Yellowstone and the Targhee NF, early cygnet mortality averaged 75%, typically involved entire broods, and was strongly specific to sites and/or pairs. Cygnet mortality was significantly higher among cygnets from clutches containing four or fewer eggs, and was probably related to factors that limited the food supplies available both to prenesting adults and to newly hatched cygnets.

Heavy parasitism has been repeatedly documented at Red Rock Lakes NWR, however it may be the result, rather than the cause, of the cygnets' weakened condition. Emaciation/parasitism is commonly involved in the deaths of prefledging cygnets and wintering birds of all ages.

The most frequently diagnosed mortality factors in the Tri-state Area were emaciation, parasites, lead poisoning, disease, and accidents. Approximately 50 trumpeters were known to have died in Idaho and Yellowstone during the winter of 1984-85. Of 18 trumpeters necropsied, most were severely emaciated and suffered from high parasite loads. Toxic levels of lead were found in five carcasses and sublethal lead toxicosis was involved in the death of a sixth bird. Lead poisoning has been diagnosed as the cause of death of 11 of 34 trumpeters found dead in the Tri-state Area and necropsied since 1980. Elevated blood lead levels were detected in 3 of 10 live Wyoming trumpeters and 5 of 22 live CV trumpeters.

Attempts to diagnose diseases have been hampered by postmortem contamination of specimens. Only isolated instances of disease related mortality have been detected, including cases of fowl cholera, avian tuberculosis, aspergillosis, generalized peritonitis/pericarditis, and systemic infections by *Pseudomonas sp.* and *E. coli*.

The retarded development of cygnets at Red Rock Lakes NWR and Yellowstone, with cygnets still flightless in late October and November, has been reported on several occasions. Eighteen percent of the cygnets that survived to September in Yellowstone in 1977-79 were stunted. Weakness at hatching and leg abnormalities have also been observed among Red Rock Lakes NWR and Yellowstone cygnets several times since the 1940s.

CHAPTER 6. SUMMER HABITAT
Ruth E. Shea, Leonard J. Shandruk, Kevin J. McCormick

Tri-state Area

Nesting habitat in the Tri-state Area includes two quite different types: the interconnected shallow marshes and lakes at Red Rock Lakes NWR and the discontinuous forest/lake and sagebrush/lake habitat typical of Idaho and Wyoming. Somewhat intermediate between these two types is the oxbow/pothole habitat along the Red Rock River in the Centennial Valley, downstream from the Refuge. Banko (1960:38-52) described the Red Rock Lakes NWR and Yellowstone NP habitats in detail and his monograph should be consulted for an overview of these areas. We will primarily focus on information gathered in these latter areas since 1957, and will summarize information from other habitats not described by Banko.

Red Rock Lakes NWR

Since its acquisition by the U.S. Government in 1935, Red Rock Lakes NWR has provided habitat for up to 78 pairs of trumpeters (Appendix X), as well as up to 250 molting nonbreeders (Banko 1960:151). Banko (1960:40) described the most important characteristics of trumpeter breeding habitat as stable, quiet waters without marked fluctuations and with little current, and shallow waters where foraging for roots and tubers is possible. These characteristics for the most part typify Refuge habitats. The Refuge (Figure 14) contains about 5,200 ha (13,000 acres) of shallow lakes, productive marshes, and extensive sedge meadows in the Centennial Valley of southwestern Montana. At 2,012 m (6,600 ft) above sea level, the Centennial Valley is a broad east-west oriented trough, bounded to the north and south by major faults (Banko 1960:41). The Continental Divide follows the crest of the Centennial Mountains, which rise abruptly along the Valley's southern border, reaching elevations of approximately 2,895-3,048 m (9,500-10,000 ft).

Banko (1960:107) and Hansen et al. (1971) suggested that one indicator of habitat quality for nesting trumpeters was the morphological development of an area's shoreline. This important habitat attribute has been quantified by a shoreline development index (SDI), which is the ratio of the length of a lake's shoreline to the circumference of a circle with an area equal to that of the lake. Page (1976:112) also concluded that high shoreline development characterized optimum nesting habitat at Red Rock Lakes NWR. Areas with high shoreline development had larger numbers of swan nests per unit area because habitat requirements were met within a small area and nesting pairs were dispersed more evenly than in areas with low shoreline development. Average territory size was smallest on bodies of water where the SDI was around 9 or 10. Little further reduction in territory size was evident in River Marsh even though it's SDI was 34.5.

The principal nesting areas at the Refuge are Swan Lake, River Marsh, Upper, and Lower Lakes. Banko (1960:45-48) provided photographs and a detailed description of each area. Surface areas, shoreline development indices, and recent mean water depths for these waters are given in Table 24. A discussion of the aquatic vegetation of these waters is presented in Chapter 7.

In addition to the large natural bodies of water, the Refuge contains several smaller ponds which were constructed to increase waterfowl habitat. Shambow and Culver Ponds were built

Figure 14. U.S. Fish and Wildlife Service map of Red Rock Lakes NWR, Montana.

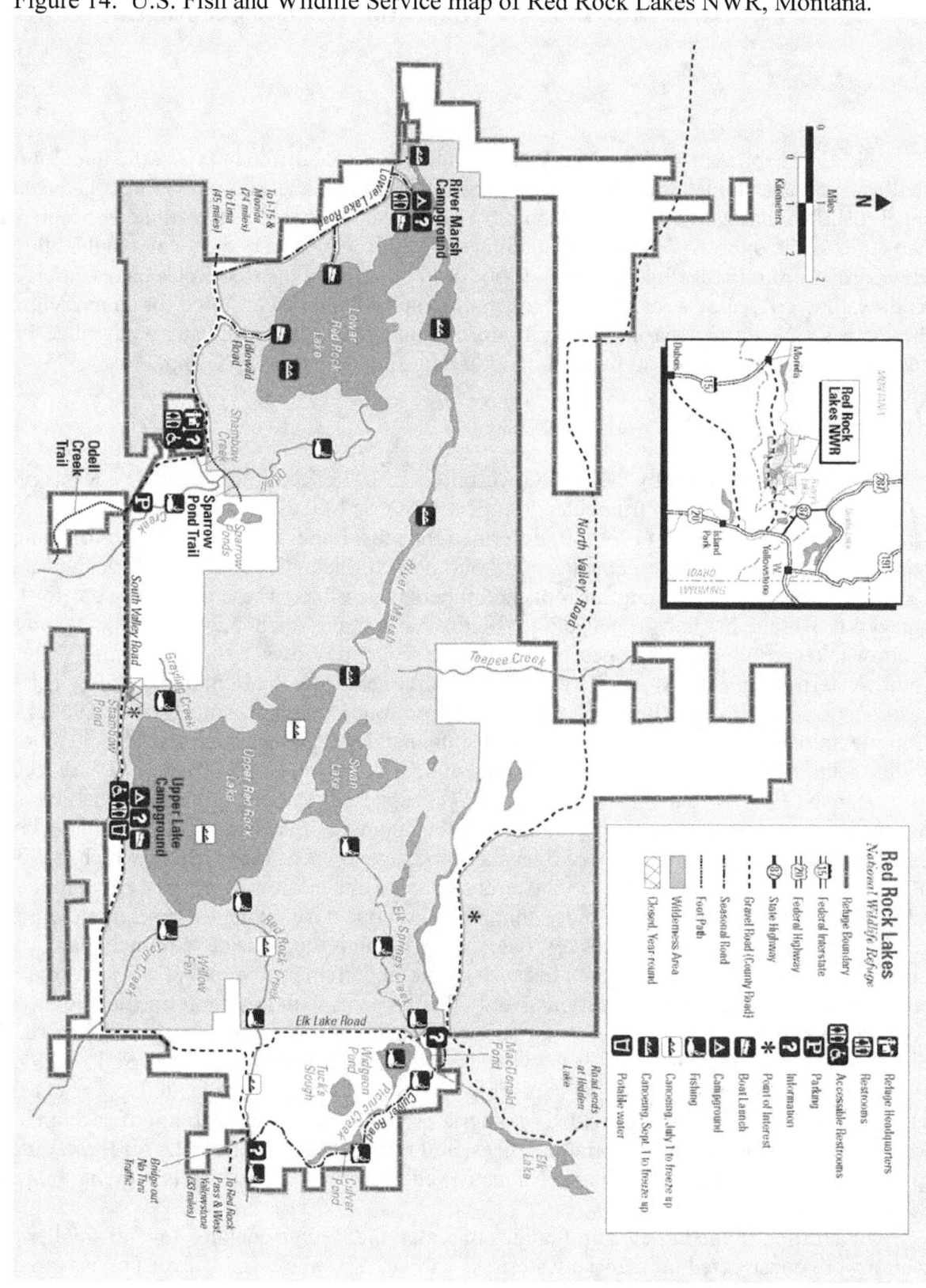

before the Refuge was established. Fed by a warm spring, Culver Pond provided the only dependable open water on the Refuge for wintering swans prior to 1953. To increase wintering habitat, MacDonald Pond was built in 1953 and Culver Pond was improved in 1959. Other ponds were constructed in the following years: Teal, Harlequin, and Shoveler in 1962, Widgeon in 1964, Pintail Ditch and related ponds in 1966, and Sparrow in 1968 (Page 1976:8).

Most trumpeter nests at Red Rock Lakes NWR were constructed on top of muskrat houses located on semi-floating sedge mats (Banko 1960:111). Between 1971-73, all Refuge nests were built on muskrat houses (Page 1976:63). When the Refuge was first established, Manager A. Hull believed that there was a shortage of adequate nest sites due to the destruction of muskrat houses by horses which grazed in the marshes during winter. Numerous artificial mounds were constructed in the 1930s to increase the supply of nest sites (Hull 1939). In order to elevate nest mounds and prevent their destruction by flooding, floating nest platforms were built and placed beneath selected high-risk nests in 1983-85 (RRLNWR Trumpeter Swan Repts. 1983-85).

Weather Precipitation and temperature data have been recorded at Refuge headquarters, Lakeview, MT, since 1939 (RRLNWR Ann. Narr. Rept. 1939) and snow course measurements have been made since 1948 (RRLNWR Ann. Narr. Rept. 1948). Annual precipitation (1965-1984) averaged 53.34 cm (20.6 in) with the greatest amounts falling in May and June. Average snowfall was 381 cm (150 in) and mean annual temperature was 1.6 °C (34.9 °F). Summers are usually short and cool with the highest temperatures reaching about 32 °C (90 °F). The average frost-free period is 51 days from mid-June to mid-August (Page 1976:18).

Refuge lakes are usually frozen from mid-November through April, although the marsh edges become ice-free several weeks earlier. Refuge files report that ice-out has been as early as 1 April in 1943, and as late as 28 May in 1975. Final freeze-up dates ranged from 21 October in 1984 to 28 November in 1954 and 1965 (Appendix III). An unusually early freeze-up occurred on 9 October 1985, but the lakes thawed after about a week, and final freeze-up occurred on 9 November (RRLNWR Trumpeter Swan Report 1985). Unusually early freezes, such as occurred in 1985, could decrease the survival of newly fledged cygnets. The early freeze in 1985 was the only environmental factor that we could identify which might have contributed to the loss of cygnets in 1985-86.

Water Levels The natural patterns of water level fluctuations in the Red Rock Lakes and marshes have been altered by man for over fifty years. Before the Refuge was established, a water control structure was built in 1931 by a private hunting club on the Red Rock River below the outlet of Lower Lake. Page (1976:6) reported that this structure raised the water level of the marsh and lakes approximately 0.6 m (2 ft), but he provided no source for this information and we found no pertinent data in Refuge files. Manager Hull (RRLNWR Ann. Narr. Rept. 1935) kept water level records, but they can no longer be found in the Refuge files. Apparently the area between Upper and Lower Lakes was fairly dry in the early years of the Refuge; specific mention was made of unusually high water flooding this otherwise dry area in 1938 and 1941 (RRLNWR Ann. Operations Repts. 1938, 1941). Early Refuge reports indicated the intent to enlarge the control structure. Records indicate that 1,000 pieces of piling, 322 large dump truckloads of rock, and 16 loads of sand were stockpiled until funds became available to work on the structure (RRLNWR Annual Operations Repts. 1938 and 1939). No further references to use

Table 24. Acreage, mean water depths, and shoreline development indices (SDI) of major nesting areas at Red Rock Lakes National Wildlife Refuge.

Area	Acreage[a]	Mean water depth (ft)[b]	SDI[c]
Upper Lake	2,800	3.6	1.6
Lower Lake	1,540	2.3	6.7
Swan Lake	400	1.4	9.1
River Marsh	8,000	1.8	34.5

[a] Banko (1960)

[b] RRLNWR Aquatic Vegetation Survey files, 1983-85

[c] Page (1976)

of these materials were found in Refuge files, and it appears that no major changes to the structure were made until 1957.

Following several years of low precipitation and drought conditions in 1953-56, plans were made to build a new water control structure at the outlet of Lower Lake, upstream from the old structure. Like its predecessor, the 1957 structure controlled water releases in the Refuge's only outlet and could potentially back up water throughout most of the Refuge. Construction occurred during the autumn of 1957, and the structure was operational in 1958 (RRLNWR Ann. Narr. Repts. 1953-58).

The immediate changes in Refuge water levels due to construction of the 1957 structure were poorly documented. Although gauges had been occasionally maintained at the old structure since as early as 1935, we could not evaluate the comparability of pre-1957 and post-1957 readings. Remarks in Refuge reports, however, described a marked rise in water levels in 1958. During the first summer of the new structure's operation in 1958, the mid-August water levels on Lower, Upper, and Swan Lakes were at least 0.3 m (1 ft) higher than in 1957 (RRLNWR Ann. Narr. Rept. 1958). This increase occurred despite little variation between 1957 and 1958 in snow course water content on 1 April, or total April to July precipitation. Refuge reports also noted that water levels were unusually high in 1957, due to above average precipitation (RRLNWR Ann. Narr. Rept. 1957). Therefore, the 1958 water levels were considerably more than 0.3 m (1 ft) higher than the levels which had been typical in the Refuge during the drought years of the mid-1950s.

The long-term effects of the 1957 control structure on water levels, marsh ecology and swan productivity are poorly understood. Water levels have been measured at a gauge at the new structure since December 1957. Monthly median readings and peak annual flows are presented in Appendix XIV and summarized in Table 25. The gauge readings show that water levels at the structure have continually increased since 1958. July water levels, which appear to be inversely related to cygnet survival (Chapter 3), averaged 0.4 m (1.3 ft) deeper in 1983-86 than in 1958-62. Refuge Narrative Reports refer very briefly to the total rebuilding of the lower structure in 1968-70. Approximately 1,272 m^3 (1,680 yd^3) of riprap and fill were hauled to improve the spillway and overflow. The net effect of this work on subsequent water levels could not be documented.

In several years, particularly in the 1980s, cygnet production at the Refuge has been reduced by the flooding of nests. The inability of the fixed-sill lower structure to pass adequate flows during high water periods has contributed to the nest flooding problems (RRLNWR Ann. Narr. Repts. 1980, 1982, 1984). Our review of the Refuge files found no mention of concerns about flooding of nests during the first 23 years of Refuge management. The first reference to nest flooding problems concerned the loss of nests on Lower Lake in the spring of 1958, the first spring that the new structure was in operation (RRLNWR Ann. Narr. Rept. 1958).

Researchers have suggested that during the more than fifty years in which the control structures have raised water levels, the drying and oxidation of the marsh soils during drought cycles has been reduced. Many aspects of the marsh's ecology, including sedimentation patterns, nutrient cycling, vegetation productivity, and parasite life cycles have likely been altered (The Trumpeter Swan Society 1984, Thorne et al. 1985). The USFWS is currently working with Ducks Unlimited to rebuild the water control structure during 1987. The new design will enable managers to release a larger volume of water during peak runoff periods in an effort to prevent nest flooding. It will also permit water levels in Lower Lake to be lowered to allow a greater

Table 25. Monthly median and annual peak gauge readings at the lower structure, Red Rock Lakes National Wildlife Refuge. Data are presented as five year averages[a].

Years	June	July	August	Peak
1958-62	8.2	7.7	7.7	9.0
1963-67	8.6	8.3	8.0	9.0
1968-72	8.9	8.6	7.9	9.4
1973-77	9.2	8.6	8.3	9.5
1978-82	8.7	8.3	8.0	9.5
1983-86	9.0	9.0	8.7	9.6

[a] Data were assembled from RRLNWR files and are presented by year in Appendix XVI. Values represent actual elevation reading minus 6,600 ft.

degree of drying and aeration of the soils (B. Reiswig, pers. comm.).

Siltation. Refuge Narrative Reports from the years 1958-67 made numerous references to significant siltation problems in both Upper and Lower Red Rock Lakes. The deposition of silt into Lower Lake dramatically accelerated following phosphate mining activities on Sheep and Taylor Mountains in 1956. A poorly designed access road was constructed in the narrow canyon carved by Odell and Spring Creeks: in places the road actually ran up the creek channels. Although the mining activity lasted for only one year, the access road eroded each spring, pouring silt into Lower Lake and Shambow Pond. Interagency investigations in 1959 and 1960 were held to identify and oversee the implementation of mitigation measures (RRLNWR Ann. Narr. Rept. 1959, 1960), but despite these efforts continuing erosion of the road bed was still evident in 1985.

Siltation problems were evident in Upper Lake by 1958, and continued for at least 10 years. Beed (RRLNWR Aquatic Vegetation Survey files 1958) reported that between 1956 and 1958, wave action had totally eroded a three foot high embankment along the north shore of Upper Lake, blocking the outlet of Swan Lake and depositing a layer of silt in Upper Lake. Also in 1958, the Refuge Narrative Report expressed grave concern over the accelerating deposition of silt into Upper Lake from Red Rock Creek. By 1959, silt bars blocked the entrance of Red Rock Creek into Upper Lake (RRLNWR Ann. Narr. Rept. 1959). An estimated 178,000 tons of sediment entered Upper Lake from Red Rock Creek in the spring of 1960, and the Creek cut a swath across grazing unit 12-G from 3-5 m (10-16 ft) deep, over several hundred surface acres (RRLNWR Ann. Narr. Repts. 1961, 1963).

The cause of the accelerated siltation in Upper Lake was also attributed, in part, to mining activity on Sheep Mountain (RRLNWR Narr. Rept. 1958). Tons of silt from the mining activities were deposited in Upper Lake by spring runoff from Snow Gulch and Lone Willow Creek (RRLNWR Ann. Narr. Rept. 1959). Paullin (1973) suggested that increased water depths due to the 1957 water control structure also resulted in increased wave action and erosion of the lakeshores. Severe overgrazing on adjacent private lands (RRLNWR Ann. Narr. Rept. 1961) and stream bank damage by cattle (RRLNWR Ann. Narr. Rept. 1963) were also identified as causes of the Red Rock Creek erosion problems. Cattle grazing on the Refuge reached a peak use level of 19,017 animal unit months (AUMs) in 1952, and commonly was between 15,000 to 18,000 AUMs from 1949 to 1965. Slight reductions occurred in 1966, and the AUMs ranged from 12,751 to 14,408 between 1966 and 1974. Substantial reductions in the mid-1970s reduced AUMs to 4,148 by 1980. Use has remained at or below 5,004 AUMs to the present (RRLNWR Ann. Narr. Repts. 1949-85). Grazing use has also been changed from season-long to rest rotation, with two years of rest in the three year cycle (B. Reiswig, pers. comm.). W. Banko (pers. comm.) has also suggested that the substantial reduction in grazing may have reduced the rate of nutrient recycling on the Refuge and the amounts of nutrients carried in runoff into the marshes, and thus could have contributed to a decrease in marsh productivity.

In an attempt to halt the erosion, Refuge staff rechanneled approximately 1.6 km (1 mi) of Red Rock Creek, upstream from the Upper Lake, in 1960 and 1961 (RRLNWR Ann. Narr. Repts. 1960, 1961). By 1967, however, it appeared that the rechannelization had actually resulted in increased erosion. Refuge staff therefore plugged the man-made channel and returned Red Rock Creek to its most recent natural channel (RRLNWR Ann. Narr. Rept. 1967). In 1969, 2,600 cubic yards of riprap were used to further stabilize the creek (RRLNWR Ann. Narr. Rept. 1969).

Another event which apparently altered water levels and may also have influenced siltation patterns in the Refuge lakes was the Hebgen Lake earthquake of 17 August 1959. This quake measured 7.8 on the Richter scale and its epicenter was approximately 27.4 km (17 miles) north of Refuge headquarters. Flows in the Refuge springs increased after the quake, however neither the extent nor the duration of changes in flow were recorded (RRLNWR Ann. Narr. Rept. 1959). Other evidence indicated that the valley floor shifted due to the quake. The 1964 Narrative Report noted that the U.S. Geological Survey came to the Centennial Valley in 1964 to resurvey its benchmarks which were altered by the 1959 quake. Our efforts to obtain information from the USGS regarding the extent of the change in the benchmark elevations were unsuccessful.

Changes in water depths within Upper Lake between the 1956 and 1966 surveys (Table 26) could also have been related to the 1959 earthquake. In 1956, Beed ran a water depth transect across Upper Lake from the north to south shores. When the transect was rerun in 1966, the observer concluded that the mean water depths had not changed (RRLNWR Aq. Veg. Survey Files 1966). When we compared the water depth changes at individual stations, however, a distinct pattern was apparent. The water depths at every station (1-14) in the north half of the lake were shallower in 1966 than in 1956, and depths at every station (15-33) in the south half were deeper. Also, with the exception of the stations nearest to the shoreline, the changes in depth were least in the center of the transect and greatest toward its ends. Some sites experienced changes in water depth of greater than 0.6 m (2 ft). We suggest that the bed of Upper Lake may have tipped during the 1959 quake, effectively lowering its southern portions. A very small shift of the valley floor could have resulted in accelerated down-cutting of valley bottom sediments by Red Rock Creek and contributed to the siltation problems of the 1960s.

Other Waterfowl Use at RRLNWR. Despite the obvious decline in swan use at the Refuge since the 1950s, trends in use of the Refuge by other waterfowl species are less obvious, due primarily to variations in survey effort. The intensity of survey effort, the observers, and methods used to estimate waterfowl production have varied considerably. Thus it was not possible to establish confidence limits on the annual waterfowl survey data nor to assess its year-to-year comparability. In reviewing the Annual Narrative Reports we found numerous instances where observers had estimated the production of hundreds of ducklings based on observation of less than ten broods. We felt that the gross nature of the annual production estimates did not allow detailed analysis. The following general summaries are based on both numerical estimates and commentary in the RRLNWR Annual Narrative Reports.

Canada Geese (*Branta canadensis*). Peak numbers of geese on the Refuge were in the range of 50 to 450 from 1943-1957. From 1958-1966, peak estimates ranged from 600 to 3,500 geese and reflected the heavy use of the Refuge, particularly Upper Lake, by large molting flocks. This period of high use by geese corresponded with the period of high Trumpeter Swan use. Since 1969, peak goose numbers have ranged from 250 to 560.

Goose production estimates have varied greatly. From 1943-49 annual estimates of production ranged from 30 to 65 goslings. Refuge staff were puzzled by the total absence of goslings in 1950 to 1952. Since 1953, estimated gosling production ranged from 20 to 75 in most years. Substantially higher production (80 to 144) was estimated in 1954, 1955, and 1963-1968, again during years when use of the Refuge by swans was at its peak. Since 1978, gosling estimates have averaged 30 per year.

Table 26. Changes in summer water depths on a transect run across Upper
Lake, Red Rock Lakes National Wildlife Refuge, in 1956 and 1966[a].

Survey Station	1956 Depth Ft.	In.	1966 Depth Ft.	In.	Amount of Change (Inches)
1	1	2	0	8	-6
2	2	0	1	6	-6
3	3	0	1	4	-20
4	3	6	2	10	-8
5	4	0	3	0	-12
6	4	1	3	2	-11
7	4	2	3	4	-10
8	4	8	3	11	-9
9	4	6	4	1	-5
10	5	0	4	3	-9
11	4	6	4	6	0
12	5	0	4	9	-3
13	5	3	5	0	-3
14	5	3	5	2	-1
15	5	0	5	2	+2
16	5	0	5	3	+3
17	5	0	5	3	+3
18	4	11	5	4	+5
19	5	0	5	6	+6
20	5	0	5	5	+5
21	5	2	5	5	+3
22	5	0	5	5	+5
23	4	8	5	3	+7
24	4	8	5	3	+7
25	4	6	5	3	+9
26	3	9	5	0	+15
27	3	5	4	9	+16
28	2	9	4	6	+21
29	1	8	4	0	+26
30	1	3	3	6	+27
31	0	7	1	10	+15
32	dry		1	0	+12
33	dry		0	8	

[a] Data were extracted from the Red Rock Lake NWR Aquatic Vegetation Survey files.

104

Ducks. In 1935, Manager Hull estimated duck production at 20,000-30,000, however there are no supporting data with which to assess the validity of these very high estimates. Later estimates of duck production ranged from 1,151 in 1951, to 18,661 in 1974. There has been a general trend toward higher estimates of duck production in recent years, however the effect of survey variation is unknown. Prior to 1965, most estimates were less than 5,000. Since 1965, estimates for average production years were 7,000-8,000 ducklings. In exceptionally good years, production estimates ranged from 11,600-12,000. Estimates of production in the decade 1976-1985 were all between 6,748-8,769, except for a high of 11,778 in 1976, and a low of 4,004 in 1980, when a late spring and flooding impacted nesting. Burgess (The Trumpeter Swan Society 1984) summarized the estimated duck production, by species, for the years 1954-83 and concluded that there had been a threefold increase in duck production from 3,450 per year in the 1954-63 period to 8,900 per year in the 1974-83 period, with the largest increase occurring among the diver species.

We were less confident of the validity of Refuge waterfowl production data. Remarks in the Narrative Reports, such as "Production of 10,000-12,000 ducks would probably be more realistic, rather than the 18,000 reported" (RRLNWR Ann. Narr. Rept. 1975) increased our skepticism regarding the accuracy of the production estimates. It was particularly disturbing when the higher estimate was retained as the official figure in Refuge reports, even though the original observers knew it to be inaccurate. We therefore concluded that duckling production has at least not shown a dramatic decline comparable to that experienced by the trumpeters, and may have increased somewhat since the 1950s. We could not determine the direction or extent of change from the 1930s to the present.

Coots (*Fulica americana*). Coot production has occasionally exceeded duck production at RRLNWR, however estimates have declined since the 1950s. From 1957-70, annual production estimates averaged 4,292 with a range of 2,200-5,500. Since 1970, estimated coot production averaged 2,389 with a range of 1,116-3,880. Peak numbers of coots exceeding 15,000 were estimated in seven years between 1956-70. Peak estimates have not been recorded in recent years. The record estimate of 25,000 in 1958 corresponded to a record high duck estimate and a very high goose estimate in that year. It is possible that these high estimates reflect habitat increases caused by the flooding of additional acreage by the newly installed lower structure. It is also possible that the complete change in both Refuge managers and biologists in 1958 contributed to the unusual estimates.

Tundra Swans. Migrating Tundra Swans usually visited RRLNWR in late March and April, and in October and November. Spring numbers usually have been less than 150 and in many years none were recorded. Fall numbers varied greatly with frequent peak estimates of 1,500-3,000, more than ten times the number of trumpeters on the Refuge. Although Tundra Swans moved through continually from about 20 October to 20 November, total use has been estimated infrequently. In the autumn of 1983, 6,000-8,000 tundras were estimated at the Refuge.

Centennial Valley, Outside Red Rock Lakes NWR

The Centennial Valley downstream from the Refuge has received relatively little management attention although it is a nesting area of increasing importance and in several years has fledged more cygnets than the Refuge (Appendix V). Thirty-two historical nest sites have been identified, of which eighteen were active in 1985 (J. Roscoe, pers. comm.). In contrast,

105

even though total swan numbers were higher in the Tri-state Subpopulation in 1949-57, no nesting occurred west of Blake Slough, which lies about 1.6 km (1 mi) downstream from the Refuge (W. Banko, pers. comm.).

No studies of the swan habitat in the lower CV have been made. The following general description of the area was written by James Roscoe, Wildlife Biologist, Bureau of Land Management, Dillon, MT.

"Trumpeter Swan nesting habitat in the CV, excluding RRLNWR, is characterized by two habitat types; extensive lateral wetlands adjacent to the Red Rock River, and small, isolated ponds surrounded by rangeland. Wetlands along the river are relatively continuous from the refuge downstream to Lima Reservoir ...[and include] numerous small, temporary ponds and oxbows isolated from the main river channel. Portions of these wetlands are seasonally inundated during spring runoff, but dry out during the summer. Several large oxbow ponds provide permanent habitat and range from about one-half acre (0.2 ha) to approximately 10 acres (4 ha).

"Typically, river bottom wetlands do not represent a well-developed vegetative community, as opposed to wetlands on the refuge. Very little tall emergent vegetation (cattails, bulrush) is available. Low Carex and Juncus types provide most nesting habitat. Nests are generally located in the more extensive wetland areas associated with permanent water and exclusively use muskrat houses for foundations. Breeding territories appear to be approximately the same size as on the refuge and are used continuously until fledging of young occurs.

"Pond habitat used by breeding Trumpeter Swans differs greatly from river bottom habitat. Ponds are generally isolated, being one to two miles (1.6 to 3.2 km) from other permanent water without any intervening wetland habitat. These ponds vary from 1 to 35 acres (0.4 to 14 ha) in size and are permanent. Typically, there is less vegetative cover on these ponds than along the river, and nests are often totally exposed on a muskrat house or hummock in open water. Again, Carex and Juncus vegetative types are predominant. Apparently in these areas where nesting sites, and taller, dense vegetation are severely limited, the denser vegetation is avoided to afford maximum visibility for swans. In a few instances, a pond territory will include one or two adjacent ponds when they are very closely associated, but more often the territory includes only the nest pond. These swans are then very dependent on a limited food supply for brooding and are quite susceptible to disturbance.

"A unique aspect of Trumpeter Swan habitat in the (off-Refuge) Centennial is that virtually all habitat is exposed to livestock grazing. This use has a definite adverse impact on availability of tall, emergent vegetation, and in some territories, the presence of any vertical cover at all. Along the river, cattle use and lack of residual cover probably are the reason that most nests are in wetter, more extensive wetland areas which are generally inaccessible to cattle early in the season. However on ponds, such buffers around nests are often reduced to only a few feet of shallow water. Despite this impact, swans on existing territories persist in using these areas and are usually successful in fledging cygnets.

"Another difference in habitat in the lower CV is that the area generally receives much less snow than the Refuge, and consequently opens up earlier in the spring.

"Aquatic vegetation in ponds and along the river is basically the same as on the Refuge, although on some ponds, production is very low."

106

Idaho

Nesting habitat in Idaho consists of scattered lakes, located mainly in the forested volcanic plateau south and west of Yellowstone NP within 64 km (38 mi) of the Henry's Fork wintering area. Elevations are generally between 1,425-2,000 m (4,674-6,560 ft). Two nest sites have recently been established on sloughs on the edge of the sagebrush covered Snake River Plain. Nesting lakes range in size from beaver ponds less than 1.8 ha (4.5 acre) to the 2,480 ha (6,200 acre) Henry's Lake. Normally only one pair of swans nests per lake. However, at least two pairs have nested on widely separate arms of Henry's Lake in some years, and two to three pairs regularly nest on Silver Lake at Harriman State Park, where they are neither vocally nor visually isolated from one another.

In 1986, 9 of Idaho's 19 trumpeter nests were located on ponds within the Ashton District of the Targhee National Forest, adjacent to the southwest corner of Yellowstone NP. Maj (1983) described the water chemistry, morphology, vegetation, and invertebrates of presently-used, historically-used, and unused lakes on the Ashton District, and determined that trumpeters selected the more eutrophic lakes for nesting. Nesting lakes had greater shoreline development, greater vegetative diversity, and greater invertebrate diversity than the unused lakes. Presently-used lakes had greater total vegetative coverage than either historically-used or unused lakes.

Since 1932, when trumpeter use was first documented on the Targhee National Forest, 29 (44%) of the 65 lakes on the Ashton District have been used for nesting. Nesting lakes were characterized by an average depth of 1.2 m (3.9 ft), shoreline development indices greater than 1.6, no less than 83% vegetative cover, greater than 26% of the total area less than 1 m (3.3 ft) deep, and diverse macrophyte and invertebrate communities. *Nuphar polysepalum*, which provided security cover by hiding a swan amidst the white reflections of sunlight off the leaves, was the predominant plant species on presently-used nesting lakes (Maj 1983).

In 1981, ice left the Targhee nesting lakes between 1 April and 9 May, with most lakes thawing after 22 April. The variation in ice-out dates between lakes was attributed to differences in elevation, amount of vegetation, and spring runoff. By the middle of April, feeding sites were available at ephemeral ponds and marshes. New growth of *Ranunculus, Sparganium, Utricularia*, and *Typha* occurred by mid-April (Maj 1983).

Trumpeters were opportunistic in their selection of nesting sites. Five of thirteen nests were built on anchored mounds of sod, vegetation, and mud, often within a stand of emergent vegetation such as *Carex, Typha,* or *Scirpus*. Other nests were located on small islands, in the root mass of an uprooted lodgepole pine, and on a beaver lodge. One man-made mound was used when the pond depth was increased by the U.S. Forest Service in order to prevent the pond from drying out by late summer. Swans preferred to use stands of *Typha* or *Scirpus* even when *Carex* was abundant. Nests were located an average of 42.9 m (141 ft) from shore, and nest locations were dependent upon the location of islands, beaver lodges, mounds, or emergent vegetation. Nests averaged 2.5 x 2 m (8.2 x 6.6 ft) and rose an average of 44.1 cm (17 in) above the water surface. Water depth around the nests ranged from 10.6-95.5 cm (0.35-3.1 ft) (Maj 1983).

Yellowstone National Park

The Park habitat is marginal compared to the marshes at Red Rock Lakes NWR. Shoreline development is low, shorelines are often more timbered, feeding areas are often peripheral due to deeper waters in the center of lakes, and lake elevations are up to 670 m (2,200 ft) higher. Most sites have changed little since the 1950s when they were described in detail by Banko (1960:49-52). In 1977 and 1978, four of 20 territories were on slow meanders of the Madison and upper Yellowstone Rivers. All other nests were on widely scattered lakes and ponds at elevations ranging from 1,830-2,515 m (6,000-8,249 ft). Nesting lakes ranged in size from 1.2-110.8 ha (3-274 a). Although volcanic rhyolite flows cover substantial portions of Yellowstone, all nesting lakes occurred either on basalt or Quaternary detrital deposits. Lakes on the rhyolite flows were typically more dilute, slightly acid, and less productive than the lakes which were used for nesting (Shea 1979).

Although YNP contains over 200 lakes greater than 0.420 ha (1 acre) in size (Condon 1941), only 37 have been used by nesting swans between 1931-78. Only 11 lakes have successfully fledged broods in five or more years since 1931, and these few lakes contributed some 78% of all cygnets fledged in the Park. Many lakes were unsuitable for nesting due to high elevation, oligotrophic conditions, fluctuating water levels, or unusual water chemistry due to geothermal influences. Three historically productive territories (Shoshone Lake, Grebe Lake, and Beach Spring Lagoon) did not support successful nesting pairs in the 1970s due to human disturbance near the traditional shoreline nest sites (Shea 1979). In 1985, the National Park Service began efforts to compensate for the loss of nesting effort at territories where traditional shoreline nest sites were vulnerable to human disturbance. Floating nest platforms, designed by T. McEneaney for use at Red Rock Lakes NWR, were placed at several traditional territories and contributed to increased cygnet production at two territories in 1986 (C. McClure, pers. comm.).

Muskrat houses were not important to Yellowstone trumpeters. Of 44 nests, 10 were built on abandoned beaver lodges, 24 were on islands, and 10 consisted of mounds of aquatic vegetation. Some of the latter mounds may have contained a muskrat house base although this was never apparent. Trumpeters nested on islands whenever one was available in the territory. Island size ranged from less than 1 m^2 (10.7 ft^2) to about 2.6 ha (6.4 acres) (Shea 1979).

Shea (1980) mapped the vegetation in seven nesting lakes and recorded over 20 species of submerged aquatics. No studies have attempted to quantify vegetative or invertebrate production, or other habitat differences between used and unused lakes in Yellowstone. From casual observations, it appears likely that in YNP, as in Idaho, trumpeters select the most eutrophic and productive of these relatively sterile, high mountain lakes.

Lower Elevations of Wyoming

Detailed studies of the nesting habitat in the lower elevations of Wyoming, outside of YNP are currently in progress (Lockman et al. 1987). Of 14 areas used by nesting pairs since 1981, only 4 have been relatively consistent producers of fledged cygnets. All four of these areas are secure from excessive human disturbance and apparently provide habitat attributes necessary for cygnet production. The most important attributes identified include:
1) feeding areas, in short flight distance from the territory, which have an early ice-off and presumably provide sufficient food for a prenesting pair;
2) more than one suitable nest site;
3) nest materials available immediately adjacent to nest sites;
4) availability and dispersion of brood-rearing habitat;
5) the juxtaposition and interspersion of emergent cover (for concealment, escape, and as a buffer to human disturbance) relative to feeding areas.

In total, within the current lower elevation Wyoming breeding range Lockman *et al.* (1987) had identified 10 occupied sites which produced cygnets without management protection and/or enhancement, 12 occupied sites which would require protection and/or enhancement to be made productive, 6 unoccupied sites requiring little or no enhancement, and 43 unoccupied sites requiring protection and/or enhancement to be made productive. Protection from excessive human disturbance in the years of territorial establishment and in the nesting and early brood rearing periods was believed to be all that is required to achieve territorial use and production on at least 14 additional sites within the current range.

Molting Areas

Flocks of molting trumpeters gathered each summer on some of the larger lakes and reservoirs in the Tri-state Area. Regular molting activity occurred on Lima Reservoir, Upper Red Rock Lake, Sheridan Reservoir, Island Park Reservoir, Silver Lake, and Jackson Lake (USFWS Tri-state Trumpeter Swan Surveys). Little is known regarding the importance of these areas to the Subpopulation, their important habitat characteristics, or long-term habitat security.

Canadian Habitats

Grande Prairie

The Grande Prairie flock nests within a 5,700 km^2 (2,200 mi^2) area of west-central Alberta and east-central British Columbia, near the city of Grande Prairie, Alberta. The flat-to-rolling glaciated topography contains numerous kettle lakes and marshes. Most lakes are shallow with maximum depths less than 3 m (9.8 ft). Interspersed aspen (*Populus tremuloides*), white spruce (*Picea glauca*), and black spruce (*P. mariana*) forests and *Agropyron, Stipa,* and *Carex* meadows comprised the original vegetation. Much of the meadow and forest has been cleared for agricultural activities (Holton 1982). Mackay (1978) observed that trumpeters seemed to prefer the aspen parkland/pothole nesting habitat but had extended their range into the boreal forest lakes in recent years. Lakes typically were free of ice from late April to late October, and were deepest during spring runoff in late April to early May (Holton 1985).

Holton (1982) compared the characteristics of present and historical nesting lakes with those of lakes which have never been used for nesting. Presently-used lakes had a greater mean width of emergent vegetation and a greater edge length of emergent vegetation zones greater than 40 m (131 ft) in width, compared to unused lakes. The predominant emergent vegetation was *Carex sp., Typha latifolia*, and *Salix sp.*

Lakes are much larger in Grande Prairie than in the Tri-state Area. The size of the average nesting lake was 140 ha (346 a) with a range of 54-394 ha (133-974 a) (Turner 1982). With only three known exceptions, only one pair of trumpeters nested per lake (Holton 1982). Thus, territory size in Grande Prairie was about four times greater than at Red Rock Lakes NWR and ten times greater than in YNP.

Several lake characteristics did not seem to influence swan use. These included total surface area, surface area of littoral zone, lake length, length in direction of prevailing wind, shoreline development, mean depth, maximum depth, volume development, area with depth less than 1 m (3.3 ft), and total biomass of macrophytes. Adult swans with cygnets did, however, use portions of lakes where the biomass of macrophytes was higher than average (Holton 1982).

Although most trumpeters built their nests within stands of emergent vegetation, its presence was not essential. Of 60 nests, 87% were located in emergent or flooded vegetation. Nests were located most frequently in *Carex* (29) and less frequently in flooded *Salix* (13) or *Typha* (10), on beaver lodges (5), on small islands (2), and on a muskrat house (1). Nest sites were located an average of 13 m (42.6 ft.) from the offshore edge of the emergent vegetation and 108 m (354 ft) from shore. The chosen sites appeared to have low vulnerability to predation and were sheltered from wind and wave action. The minimum observed distance from nest to shore was 40 m (131 ft) (Holton 1982).

British Columbia

Northeastern British Columbia nesting habitat lies adjacent to the core nesting areas of both the Grande Prairie and Yukon flocks. If these flocks continue to expand, they could occupy a vast amount of vacant, potentially suitable habitat which forms a continuous arc from Dawson Creek, to Fort St. John, and north to Fort Nelson. Most sightings of trumpeters have been east of the Continental Divide, in the lakes of the bench lands adjacent to the Rocky Mountain foothills. Unoccupied, suitable habitat also exists in the slightly lower elevation, shallow eutrophic lakes of the Alberta Plateau. Recent inventories found that potential trumpeter habitat in northeastern B.C. includes about 1,800 lakes ranging from 5-50 ha (12.4-124 a) in size. In addition, at least 40 suitable lakes greater than 50 ha (124 a) in size, and 71 streams may provide suitable nesting habitat (Churchill 1987).

Yukon

Trumpeters occur in the Boreal forest from the southeastern corner of the Yukon to the Alaska border near Dawson City. Swans breeding in the southeastern Yukon probably belong to the RMP, and are found primarily in the Liard and Beaver River drainages with a concentration of nesting in the Toobally Lakes vicinity. This glaciated terrain is characterized by rolling hills and plateaus from about 600-1,350 m (1,968-4,428 ft) in elevation with numerous eskers, kettles, and glacial terraces. The mean elevation of 17 lakes is 869 m (2,850 ft). Precipitation varies from 400-500 mm (15.6-19.5 in) and the mean annual temperature is -3 °C (26.6 °F). Below the subalpine zone, the climax terrestrial vegetation consists of white and black spruce with a moss or moss/shrub understory. Lodgepole pine (*Pinus contorta*) is prevalent on old burns and aspen is common on south facing slopes (McKelvey et al. 1983, Dennington 1987).

Water bodies are of three main types: deep elongated glacial lakes, perched basins associated with moraines and terraces, and meandering outflow streams. Beaver activity has created ponds associated with the latter. Nesting occurred both in the perched basins and beaver ponds. Ponds ranged in area from 5->250 ha (17.4->617 acres). Most of the ponds where broods have been observed had very irregular shorelines, and peripheral sedge, or sedge/willow communities. All ponds supported some emergent vegetation although the larger lakes and some of the smaller ponds were forested to their edges. The dominant emergent plants were sedges and *Equisetum*. Common submerged aquatics included *Chara, Hippuris, Nuphar, Sparganium* and various pondweeds, most commonly *P. richardsonii* (McKelvey et al. 1983, Dennington 1987).

Of 16 nest sites, 11 were on islands, 4 were on beaver lodges or dams, and one was on shore. Only one was similar to the typical Alaskan moat type nest described by Hansen et al. 1971 (McKelvey *et al.* 1983).

Northwest Territories (NWT)

Trumpeter Swans are known to occupy the southwestern portion of the NWT, which lies west of the Mackenzie River and south of approximately 63° N. Physiographically, this area is dominated by the Mackenzie Mountains and the adjacent Great Slave Plain. The interior of the Mackenzie Mountains is a sea of peaks and ridges which reach up to 2,600 m (8,528 ft), whereas the more easterly ranges are lower and divided by wide valleys and cut by deep canyons. Most of the summits in this area are less than 1,375 m (4,510 ft). The low-lying Great Slave Plain is covered by muskeg and numerous small shallow ponds and lakes. Most of the region lies below 300 m (984 ft) elevation; however, three upland areas occur north of the Liard River. The highest, the Martin Hills, attains an elevation of approximately 690 m (7,263 ft).

The montane vegetation is characterized by an altitudinal transition from the relatively dense forests of the lowlands to the alpine tundra of the mountains. In the South Nahanni River area, the treeline occurs at approximately 1,200 m (3,936 ft) on south and west facing slopes and at 1,100 m (3,600 ft) on north and east facing slopes. White spruce, willow, and balsam poplar (*P. balsamifera*) occur in valley bottoms, especially on floodplains. The shrub layer consists of alder (*Alnus incana*), squashberry (*Viburnum edule*), and wild rose (*Rosa acicularis*). On the Mackenzie Plain, recent alluvial soils support forests of white spruce, balsam poplar, and aspen in mixed or pure stands. The shrub layer consists of alder, willow, rose, and *Viburnum spp.* Older alluvial soils are characterized by black spruce and white spruce in mixed or pure stands. Labrador tea (*Ledum palustre*) and *Vaccinium spp.* comprise the shrub layer, whereas sphagnum and feather mosses cover the ground.

Major river systems, including the Liard, South Nahanni, North Nahanni, and Root, drain from the Mackenzie Mountains into the Mackenzie River. Breeding swans are closely associated with these river systems and the wetlands which occur on their adjacent floodplains. The wetlands include lakes, ponds, oxbow lakes and creeks, particularly where beavers have been active. Soils under these wetlands include sand, silt, peat and organic silt which are conducive to the growth of emergent vegetation. Emergent zone species include *Carex, Juncus, Typha, Equisetum*, and *Calamagrostis*. *Potamogeton, Nuphar, Ceratophyllum*, and *Myriophyllum* are some of the more common submergent zone species. The majority of swans occur below 300 m (984 ft) elevation) but individuals have been observed at 900 m (2,970 ft) elevation.

Two nest sites have been examined in this area. Both nests were small islands of mud and dead vegetation which had been constructed in shallow water. One nest was approximately 15 m (49 ft) from shore whereas the other nest was about 125 m (410 ft) from the nearest shoreline.

Summer Climates

Geographically, the Tri-state region is at the southern edge of the RMP's current breeding range. Climatically, however, Red Rock Lakes NWR and the higher elevation lakes in YNP are comparable to the Yukon nesting areas, and colder than either the Grande Prairie or Northwest Territory nesting habitats (Table 27). In April, all the higher latitude habitats in Canada are colder than the Tri-state Area, but once spring arrives, May temperatures in Canada rise rapidly and exceed those at Red Rock Lakes NWR and Yellowstone. Within the Tri-state Area, the Idaho and lower elevation Wyoming habitats experience temperatures which are comparable to those of Grande Prairie.

Severe spring weather retards the availability and development of spring food sources and thus reduces the swan's ability to accumulate the energy reserves essential to successful reproduction and reduces the foods available to newly hatched cygnets. Comparison of the reproductive performance of the various RMP flocks (Appendix IX) with the mean May temperature in each nesting habitat shows that the flocks which experience the coldest May temperatures are those which exhibit the poorest cygnet production. These flocks, the Red Rock Lakes, Yellowstone, and Yukon flocks, have recently declined while the NWT, Grande Prairie, Idaho, and lower Wyoming flocks have increased or remained stable.

Summary

Red Rock Lakes NWR provides the most important Trumpeter Swan nesting habitat in the Tri-state Area. Although quantitative data are lacking, the productivity of the Refuge marshes has likely declined since the 1930s. Water control structures have increased the water depths on the Refuge since at least 1958, and most likely since the 1930s. In addition to increasing the water depths, the water control structures eliminated the natural water level fluctuations that would have exposed the marsh soils to air during drought cycles. It is likely that the relatively stable, high water levels have decreased the rates of nutrient recycling and vegetation production on the Refuge, and allowed parasites to multiply without interruption to their life cycles. High water levels have also caused the direct destruction of swan nests, particularly on Lower Lake. Reductions in cattle grazing in the marsh may also have reduced rates of nutrient recycling.

Erosion problems in the late 1950s and 1960s noticeably accelerated the deposition of silt into Upper and Lower Lakes. The increased erosion resulted primarily from phosphate mining activities in the Centennial Mountains and from overgrazing. The construction of the 1957 water control structure and the 1959 Hebgen Lake earthquake also probably altered patterns of erosion and deposition.

Nesting habitat in YNP is marginal compared to Red Rock Lakes NWR due to its higher elevations, smaller territories, and deeper waters. Nesting lakes in Yellowstone were located on the Park's most productive soil types. At least three of Yellowstone's historically-used nesting lakes were no longer occupied by nesting swans in the late 1970s due to excessive human disturbance. In 1985, the Park Service began to mitigate for the detrimental effects of human disturbance by providing artificial offshore nest sites in historical nesting lakes where human disturbance prevented the use of traditional shoreline nest sites.

In Idaho, nesting trumpeters selected the more eutrophic lakes which had greater shoreline development and contained high plant and invertebrate diversity. In Grande Prairie, trumpeters used nesting lakes which had above average amounts of emergent vegetation to provide potential nest sites. Lake size, and territory size is much greater in Grande Prairie than elsewhere in the RMP range. Studies of the key features of breeding habitat in the lower elevations of Wyoming are currently in progress; preliminary results have indicated the importance of adequate food resources for prenesting adults and for cygnets in the first weeks after hatching, and emphasized the need for protection from human disturbance.

Table 27. Elevation, latitude, and mean daily temperatures of RMP nesting habitats. Locations are arranged in order of increasing May temperature.

| Location | Elevation (m) | Latitude (°N) | Mean Daily Temperature °C | | | | | |
			April	May	June	July	August	September
Yellowstone Lake, WY	2,358	44	-1.6	3.9	8.3	13.0	12.4	7.7
Ross River, YK	705	62	-2.2	5.4	11.0	12.8	10.5	5.1
Red Rock Lakes NWR, MT	2,012	44	0.6	6.7	10.8	14.9	14.1	9.4
Watson Lake, YK	689	60	-0.6	6.9	12.7	14.9	13.1	7.6
Ft. Simpson, NWT	169	61	-2.5	7.9	14.4	16.6	14.4	7.3
Jackson, WY	1,900	43	2.9	8.1	12.2	16.0	14.8	10.4
Ft. Liard, NWT	213	60	-0.2	8.7	14.3	16.7	15.0	9.2
Grande Prairie, AB	655	55	2.7	10.0	13.7	15.9	14.8	9.8
Ashton, ID	1,608	44	4.5	10.3	14.3	17.9	16.6	12.2

Although the Tri-state Area is at the southern edge of the RMP's current breeding range, parts of the region are colder than the Canadian habitats. The coldest RMP nesting habitat is in YNP, followed by Red Rock Lakes NWR and the Yukon, which experience comparable temperatures. Milder May temperatures occur in the lower elevations of the Tri-state Area, the Northwest Territories, and Grande Prairie. Flocks which nest in the milder habitats have increased or remained stable in recent years, while the flocks which nest in the colder habitats have declined.

CHAPTER 7. AQUATIC VEGETATION AT RED ROCK LAKES NWR
David G. Paullin, Edward O. Garton, and Ruth E. Shea

The first quantitative survey of aquatic plants at RRLNWR was conducted in 1955 and 1956 by W. L. Beed, USFWS Biologist. Beed returned to the Refuge in 1958 and 1960 to record changes in plant species composition and abundance. In 1966, annual production surveys were initiated by Refuge personnel. Each year between 1966-85, Refuge staff surveyed half of the lakes and ponds. In 1971 and 1972, all Refuge lakes and ponds were intensively surveyed by Paullin (1973), in addition to the annual production surveys conducted by Refuge personnel. No surveys were conducted in 1986, while the Refuge staff evaluated their information needs and possible improvements in survey methods.

Aquatic plant surveys conducted between 1966-75 were summarized and discussed by Roscoe (RRLNWR Aquatic Vegetation Survey files 1976). We will discuss the aquatic plant surveys for the entire period of record (1955-85) and the factors which influence the distribution and growth of aquatic plants on the Refuge. All survey data referred to in this section can be found in the RRLNWR Aquatic Vegetation Survey files and are summarized in Appendix XV.

Survey Methods

In the initial 1955 survey, single transects were run across Upper, Lower and Swan Lakes. River Marsh was surveyed in the main channel throughout its length. Growth characteristics and abundance (frequency of occurrence) of each species were recorded at each sample site. In addition to the transects, 1,000 point samples were taken in each lake. This survey measured species diversity, but no attempt was made to estimate total biomass. This survey method has not been repeated since 1956.

Biologist Beed resurveyed the marsh in 1958 and 1960. These surveys were limited to a boat trip across each lake, from which Beed wrote a narrative description of aquatic plant conditions and changes since the 1956 survey. Neither of these later visits included any sampling.

Surveys were discontinued until 1966, when Refuge staff began using the Clark-Webster method of aquatic plant sampling. This method involved sampling a .09 m^2 (1 ft^2) area with metal tongs, separating the retrieved vegetation by species, and determining the wet weight volume for each species by simple water displacement. Results were expressed as milliliters of vegetation per square foot. These data were then converted to lbs/acre (assuming a standard volumetric weight for each species), in order to reach an overall estimate of total plant biomass for each lake. Sample sites were selected using a random sampling scheme and recorded on a map in an attempt to sample the same sites each year. This method was generally followed from 1966 to 1972, when the use of wet weight, rather than volumetric measurement began. In 1971 and 1972 all major lakes and ponds were surveyed by an ocular estimation method described by Paullin (1973). This survey recorded plant distribution and abundance (frequency of occurrence, percent species composition, and total acres of vegetation by species), but did not estimate total biomass. This method has not been repeated since 1971.

With the exception of the surveys conducted by Beed and Paullin, the Refuge has used only the Clark-Webster method (Table 28). Sample sizes have been increased throughout the survey period in an attempt to reduce variance and obtain estimates with 95% confidence intervals within 25% of the mean sample weight. Water depth was also recorded at each sample point and used to calculate the average water depth for each pond or lake surveyed.

Our review of the Refuge data and historical files revealed that since the initial surveys in 1955-56, the following major changes have occurred:

1) *Elodea* has declined markedly, particularly in Upper and Lower lakes, and Widgeon Pond. Despite many difficulties in quantifying this change, it is obvious from the narrative descriptions as well as the sampling, that where *Elodea* once existed in extensive, luxuriant beds, it was virtually absent in later years.

2) Species diversity has increased Refuge-wide.

3) Wide fluctuations occurred in species abundance in the Refuge ponds.

Attempts have been made to explain these plant changes (RRLNWR Aquatic Survey files, Paullin 1973, Page 1976). Generally these explanations are tied to one of four major perturbations: siltation from phosphate mining in the Odell Creek drainage, construction of the lower structure, overgrazing by waterfowl, and erosion of Red Rock Creek.

Surveying submerged aquatic plants at RRLNWR has been difficult. The techniques employed were crude and imprecise and this was reflected by the high variance in the data. Some of the problems associated with the Refuge survey methods included:

1) Variation in survey dates significantly influenced the data gathered. Paullin (1973) determined that the standing crop of any particular species of aquatic plant at RRLNWR normally changed throughout the summer due to growth, senescence, and herbivory. Over the last 20 years, the starting dates of the surveys have ranged from 5 August to 1 September on Upper Lake, and from 29 July to 24 August on Lower Lake. The range of survey dates was similar for the other bodies of water. Paullin (1973) developed biomass production charts which showed that in any one year, between 1 August and 1 September, species experienced biomass changes of at least the following magnitude:

Potamogeton pectinatus	18%
P. praelongus	50%
Chara vulgaris	50%
Myriophyllum spicatum	75%
P. Richardsonii	78%
Sagittaria cuneata	80%
Elodea canadensis	100%
Ceratophyllum demersum	182%

In addition to the variation in survey dates, each year the timing of maximum plant growth, senescence, and herbivory varied due to annual variation in weather and dates of waterfowl migration.

2) Variance was incorrectly calculated. The sample sizes required for each lake were determined by calculating the variance of the sample weights. The variance, however, was calculated for all species combined, rather than for each species separately. Therefore, the desired confidence interval (95% C.I. equal to plus or minus 25% of the mean sample weight) pertained only to the estimate of total biomass of all species, and was not appropriate for the estimation of biomass of any particular species, although it was used in this manner.

Table 28. Summary of methods[a] used for aquatic plant surveys conducted at RRLNWR, 1955-1985.

Year	Culver Pond	MacDonald Pond	Widgeon Pond	Swan Lake	Upper Lake	River Marsh	Lower Lake
1955-6	ns	ns	ns	1	1	1	1
1966	2	ns	2	2	2	ns	ns
1967	ns	2	ns	ns	ns	2	2
1968	2	ns	ns	ns	2	ns	ns
1969	ns	2	ns	2	ns	2	2
1970	2	ns	2	ns	2	ns	ns
1971	3	3	3	2, 3	3	3	2, 3
1972	2	ns	2	ns	2	ns	ns
1973	ns	ns	ns	2	ns	2	2
1974	2	ns	2	ns	2	ns	ns
1975	ns	ns	ns	2	ns	2	2
1976	2	ns	2	ns	2	ns	ns
1977	ns	ns	ns	2	ns	2	2
1978	2	ns	2	ns	2	ns	ns
1979	ns	ns	ns	2	ns	2	2
1980	2	ns	2	ns	2	ns	ns
1981	ns	ns	ns	2	ns	2	2
1982	2	2	2	ns	2	ns	ns
1983	ns	ns	ns	2	ns	2	2
1984	2	2	2	ns	2	ns	ns
1985	ns	ns	ns	ns	ns	ns	2
1986	ns	ns	ns	ns	ns	ns	ns

[a] Methods: 1 = line transect, 2 = Clark-Webster, 3 = Ocular estimation; ns = not surveyed.

3) Random sampling was used to sample plant species that had nonrandom distributions. The distribution of aquatic macrophytes at RRLNWR is not random but rather shows marked zonation by water depth (Paullin 1973). The survey data also showed that several species typically occur in densely clumped beds.

4) The use of tongs to grab vegetation samples introduced several potential sources of error into the determination of sample weights. These errors included overestimating plant biomass/ft^2 for large, long-stemmed species when a massive tangle of stems was sampled and plants were pulled in from outside the 1 ft^2 area, and lack of standard techniques among the many different survey personnel (i.e., amount of water squeezed out of the plants, how much of the roots were grabbed, etc.). Mean sample weights were often in the range of 0.4-0.8 lb, and errors of only an ounce or two due to inconsistent sampling and weighing techniques would have resulted in large changes in the total biomass estimates when the weights per ft^2 were multiplied by factors of over 10^6 during the process of converting the data first to tons/acre, and then to total tons produced in each water body.

5) The misidentification of plant species was also a problem that was recognized in the survey write-ups, particularly in 1974 when a seasonal crew made the survey.

Frequency of occurrence data could not be statistically compared between years due to the annual shift in the location of sample points. Although sample points have been numbered and marked on a master map, locating the exact sample point in subsequent years was impossible. Workers with experience running the surveys estimated that the easiest points to relocate were in the irregular water bodies where useful landmarks existed, and these may be relocated within 6-15 m (20-50 ft). The relocation of sample points in the center of Upper Lake was much more imprecise (B. Kurtenbaugh, pers. comm.).

Using the annual survey data, Refuge reports have analyzed the percent change in each plant species from year to year, and attempted to associate these estimated changes with environmental conditions. Due to the problems that we have described, we concluded that such detailed analysis is not valid and that the survey results can only be used to detect very broad trends in the relative abundance of plant species.

Despite the statistical shortcomings of the survey data, their value should not be discounted entirely. This information, collected over a 30-year period, represents more and better information than found on most waterfowl management areas. Approximately 24 species of submerged aquatic macrophytes have been recorded on the Refuge. The data indicate some broad trends in plant abundance and distribution over time. Thus, we chose to review pertinent literature to provide insights into the interactions of key macrophyte species at RRLNWR.

Factors Influencing Vegetation

Light, Turbidity, and Siltation

In lakes, light penetration limits colonization by submerged plants and may also influence zonation (Spence 1970, Spence and Chrystal 1970, Spence et al. 1973). In general, total macrophyte biomass increases with increasing light (Forest 1977) and when light is optimal, plants respond to increases in water temperature (Barko et al. 1982). The most critical time for adequate light transmission is early in the growing season when plants initially send up shoots

(Anderson 1978, Schiemer 1979, Davis and Carey 1981).

In general, the waters of RRLNWR are clear and relatively shallow. Visibility to the lake bottom is the normal. Under these prevalent conditions, light is not a limiting factor. The only situation where visibility to the lake bottom is obscured is on windy days when turbidity is high, particularly on Upper Lake. Turbidity in Refuge waters was described by Paullin (1973) and found to be directly related to lake size, and caused by wind. Turbidity increased during the growing season as winds increased in duration and intensity.

The turbidity of Refuge waters apparently was noticeably greater in the late 1950s than it is today. When Beed examined the Refuge habitat in 1955 and 1956, the waters of Upper and Lower Lakes were described as "crystal clear". In 1958 and 1960 however, Beed observed that the lakes and River Marsh were so turbid that the vegetation could not be seen through the water and had to be sampled with a rake. This turbidity was attributed to the accelerated erosion which was discussed in Chapter 6. When examined in 1971-72 by Paullin (1973), Refuge waters were again quite clear.

The "northern *Potamogetons*" (Stuckey 1971, Davis and Brinson 1980) and *Elodea* are particularly sensitive to suspended sediment turbidity, especially early in the growing season (Barko *et al.* 1981, Ward and Talbot 1984) while sago pondweed (*P. pectinatus*) is relatively tolerant (Davis and Brinson 1980). Increased turbidity in the late 1950s, worsened by silt-laden spring run-off, was likely a major cause of the decline of *Elodea* in Upper Lake. Other instances of turbidity-caused changes in macrophyte communities have been documented (Steenis 1947, Peltier and Welch 1969, Stuckey 1971, Southwick and Pine 1975, Perverly and Johnson 1976).

Turbidity was a significant factor in describing differences in plant associations between Refuge ponds and lakes (Paullin 1973). Wind-driven currents cause the overall high turbidity of lake waters. The pattern of turbidity distribution depends on wind direction, and also strongly influences sedimentation rates and light conditions. Silts suspended by waves are deposited on plants, especially in sheltered bays, and deposition decreases with distance offshore. The effects of silt coating included: shading of assimilating parts and increasing the weight of buoyant fronds, pressing them deeper into the water column and thus decreasing assimilation (Schiemer 1979). In general, coatings are more marked on finely dissected leaves, such as *Myriophyllum*. Turbidity and the associated siltation is probably a limiting factor only on the larger Refuge lakes. Its effect is variable between plant species, being more significant for some (*P. praelongus* and *Elodea*), than for others (*P. pectinatus*).

Wave Action

Wave action has been identified by several authors as a limiting factor in macrophyte production (Brown 1975, 1979; Jupp and Spence 1977, Anderson 1978, Schiemer 1979, Richmond 1981, Ward and Talbot 1984). Waves limit macrophytes directly by hydromechanical disturbance and indirectly by affecting particle size in the substrate, turbidity in the water, and deposition of silt on macrophytes.

Lake bottoms exposed to wave action generally have coarse, sandy, nutrient-poor substrates, whereas sheltered areas have finer clay sediments that can retain more nutrients (Jupp and Spence 1977). Substrate difference associated with prevailing winds and waves have been demonstrated to be a limiting factor for milfoil and *Elodea*, among others.

In Manitoba, sago pondweed attained its best growth in sheltered shallow bays with peat and sandy loam soils. No sago was found on lakes where fetch exceeded 2,300 m (7,590 ft)

(Anderson and Low 1976). In Scotland, 79% of the total biomass removal of sago pondweed could be explained by wave action (Jupp and Spence 1977).

In the 1956 survey, *Elodea* was observed to grow in pure stands over much of the area of Upper Lake, especially in the deeper waters. The shoreward limit of *Elodea* growth appears to be determined by inshore wave action as noted by Brown (1975, 1979) and Richmond (1981). Other authors (Ward and Talbot 1984) have also noted that the lack of suitable substrate caused by prevailing winds affected the distribution pattern of *Elodea*.

The mechanical effect of wave action is considered to cause a sharp reduction in *Myriophyllum* density toward lake centers where turbulence is greatest (Schiemer 1979). Steenis (1947) noted major reduction in macrophyte abundance caused by increased fetch due to a rise in water level. This effect was greater on *Myriophyllum* than on *P. pectinatus*. The increased water depths in Refuge lakes since the 1950s probably led to decreased macrophyte abundance due to increased wave action, particularly in Upper Lake (Paullin 1973).

All of the wave related factors discussed above (substrate, hydromechanical, turbidity/light, and sedimentation) act on the lakes and ponds of RRLNWR to varying degrees. The morphology of Refuge lakes has been described by Paullin (1973). Wave-related impacts on macrophytes are greatest where fetch is great and shoreline development low (Upper Lake). High shoreline development (River Marsh), low fetch (Refuge ponds) and the presence of patchy emergent vegetation (Lower Lake) are all factors which minimize wave-related limitations.

The 1946 and 1947 Narrative Reports stated that high winds in the summer of 1946 kept *Potamogeton* torn out of the center of Upper Lake. As a result, feed for waterfowl was poorer than in previous years. Severe churning wind action was again noted in 1947 and 1948. Aquatics were observed to be scarce and did not fare well in the center and leeward side of Upper Lake (RRLNWR Ann. Narr. Repts. 1947, 1948).

Water Temperature

Low water temperatures effectively diminish the capacity of macrophytes to utilize available light in photosynthesis, thus reducing the growing season. This may partially account for the inability of macrophytes in some systems to colonize substrata to depths consistent with their maximum photosynthetic potential (Barko *et al.* 1982).

Experiments conducted on *Elodea* and other macrophytes showed that total biomass for all species increased with increasing water temperature up to 28 °C (82 °F) (Barko *et al.* 1982). These studies showed that when water temperature was optimal (28-32 °C), plants responded to light. However, light was probably less important than temperature in affecting the geographical distribution of submerged macrophytes. In addition, temperature may be as important as light in modifying competitive interactions among coexisting species.

Studies of water temperature at RRLNWR relative to macrophyte production have not been conducted. Water temperatures in July to September 1976 ranged from 15.1-17.0 °C (59.2-62.6 °F) in Upper Lake, and 17.2-18.6 °C (62.9-65.0 °F) in Lower Lake (Randall *et al.* 1978). Maximum summer water temperatures are suboptimal for plant growth, the growing season is short, and the climate is severe. Thus, we conclude that Refuge water temperatures, particularly in May and June, are suboptimal for most of the macrophyte species on the Refuge. Any factor that could increase average water temperatures (decreased water depths, mild sunny weather, low spring run-off) should increase overall macrophyte production.

Water Depth

Anderson and Low (1976) showed that pondweed growth increased with increasing water depth up to approximately 60 cm (2 ft), followed by a steep linear decline at depths greater than 60 cm. Similar results were found by Anderson (1978) and Robel (1961, 1962). Schiemer (1979) observed that sago pondweed development was associated with prolonged periods of low water levels and that when lake levels increased 40 cm (1.3 ft), plant abundance declined. These studies generally showed that macrophyte distribution follows a depth gradient centered around a unimodal "optimum depth" that varies by species. Macrophytes will often become established at an optimum depth and later invade shallower and deeper water, but are limited to a maximum depth known (Aiken *et al.* 1979).

At RRLNWR, submerged macrophytes occurred in slightly less than 0.3 m (1 ft) of water to nearly 2.8 m (9 ft). Maximum species diversity, however, occurred in waters 0.6-0.9 m (2-3 ft) deep (Paullin 1973). Water depth is one of the most important factors affecting overall macrophyte distribution and abundance on the Refuge. Increasing water depths cause regression in macrophytes (Schiemer 1979). The most important factors influenced by water level fluctuations are light penetration and the hydromechanical effects of wave action on the lake floor. Possibly a deterioration of underwater light conditions which are most critical at the beginning of the growing season is the main cause of macrophyte regression in connection with increased water levels (Anderson 1978).

Substrate

Macrophytes take up nutrients via their underground parts (Spence 1964; Martin and Clemens 1968; Peltier and Welch 1969; Denny 1972; Bristow 1975). Fine-particled muds and clays, characteristically deposited in deeper waters and sheltered situations, generally contain more nutrients than course-particled sediments. Sheltered regions might support more macrophyte biomass than exposed ones as a result of their higher nutrient concentrations in the substrate (Jupp and Spence 1977).

Anderson (1978) found that soil texture affected both plant distribution and growth, and that the most important factor influencing plant production was soil type, with greatest standing crops on very fine sandy loam soils. Chapman *et al.* (1971) found a definite relationship between particle size and plant community (*Elodea* occurring on more sandy sites than *Nitella*). Other authors (Sculthorpe 1967; Brown 1979; Spence 1979) have also considered substrate a major factor in species distribution.

The soils of Refuge lakes and ponds have not been intensively studied. A cursory examination of lake sediments summarized by Paullin (1973) showed that marsh soils are primarily clays and clay loams. Lake soils are highly calcareous, moderately alkaline, low in sodium, and moderately saline. These soils are rich in Mg, K, Fe, and Ca; deficient in P; and normal in Mn. The texture and chemical composition of Refuge soils are conducive to optimal macrophyte growth. Since much of the lake bottom is unvegetated, it seems reasonable to conclude that some factor other than substrate limits macrophyte growth.

Waterfowl Grazing

Few vertebrates directly utilize aquatic macrophytes as a food source (Kiorboe 1980), so the majority of macrophyte production normally enters the detritus food chain (Mann 1972, Fenchel 1977). Waterfowl concentrations, however, have been documented to have significant impacts on macrophyte biomass (Berglund *et al.* 1963). Enclosure studies by Jupp and Spence (1977) showed that waterfowl removed 21% of the total biomass, while in another study (Kiorboe 1980) waterfowl removed approximately 30% of the annual macrophyte production or nearly half of the maximum biomass.

Among several waterfowl species studied, Kiorboe (1980) felt that the Mute Swan (*Cygnus olor*) was the most important grazer due to its greater body weight and primarily vegetarian diet. While a food habits study of Trumpeter Swans in the RMP has yet to be done, trumpeters are also important grazers that likely can affect macrophytes, particularly on winter feeding habitat where the swans are concentrated. Page (1976:100) suggested that overuse of *Elodea* by trumpeters in early spring contributed to its decline on the Refuge. This suggestion was based upon observations that three penned trumpeters showed a preference for *Elodea*, and that *Elodea* declined on Upper Lake during the period of years when swan numbers on the Refuge were highest. We found no evidence to support this suggestion. To the contrary, grazing pressure must be much heavier at Culver Pond than on Upper Lake, yet *Elodea* has been maintained at Culver Pond over a 20 year period.

On the spring-fed Refuge wintering ponds and the ice-free reaches of the Henry's Fork, key macrophyte species (e.g. *Elodea, Myriophyllum*) remain photosynthetically active throughout the winter period. These sites concentrate high densities of trumpeters and other waterfowl for several reasons including: the limited availability of ice-free feeding areas elsewhere, artificial feeding (at RRLNWR), and protection from disturbance. Refuge Narrative Reports frequently noted that the concentrations of wintering trumpeters totally removed the available vegetation in the limited ice-free areas at the wintering ponds. The extent to which wintering trumpeters have altered the overall distribution and abundance of macrophytes on other river feeding areas is unknown.

Plant Succession, Fluctuation, and Competition

Submerged macrophytes commonly exhibit wide annual variation in total production (Forest 1977, Verhoeven 1980). Often these changes are detected in annual aquatic plant surveys and described as plant successional changes. These short term changes in plant abundance, though sometimes great, are more likely fluctuations than succession. Fluctuations are reversible changes in the quantitative ratio of the community components, mainly as a result of climatic and environmental variability, whereas successions are sequences of irreversible changes in the community composition due to long-term changes in microclimate (Rabotnov 1974).

Evaporation, precipitation, salinity, turbidity, and water levels change from year to year, and thus certain aquatic plant species are in turn at an advantage or disadvantage. Irregular changes in abiotic factors over a long period of time can affect coverage, species composition, and dominance (Verhoeven 1980).

Macrophytes compete for all their primary demands, e.g. space, light, and nutrients. Three primary factors determine the ultimate success of one species or the other: 1) the number of hibernating organs (rhizomes, tubers, and seeds) of each species that bud in the spring; 2) the growth rate and growth pattern of each species under the prevailing conditions, and 3) the ability of each species to adapt to and survive temporarily unfavorable conditions (Verhoeven 1980).

Species that can overwinter in an active state (Boylen and Sheldon 1976) or are capable of rapid growth with increasing light at low temperatures in the spring may have a competitive advantage in northern localities over species that overwinter in a dormant state and/or are incapable of growth at low temperature (Barko *et al.* 1981). Paullin (1973) listed *Elodea canadensis*, *Lemna trisulca*, and *Ceratophyllum demersum* as three species that overwinter in an active state at RRLNWR. Barko *et al.* (1981) noted that *Myriophyllum* was capable of photosynthesizing at low temperatures and may have a competitive advantage over other species under cold water conditions.

Whitaker (1975) pointed out that continuous shifts in the relative importance of coexisting species occurred in response to shifts in the environmental gradient that formed the basis of their coexistence. Harper and White (1971) noted that these shifts or fluctuations involved a complex of factors and processes both biotic and abiotic. At RRLNWR some of the factors that segregate species are turbidity, and water and soil chemistry (Paullin 1973).

Vegetation data at RRLNWR (Appendix XV) provide some insights into plant community changes over time. The best example is provided by Widgeon Pond. Widgeon Pond covers 53 ha (132 a) and was created in 1964 by damming Culver Spring. When the pond was first surveyed in 1966, *Elodea* accounted for 99% of the total aquatic plant production, and only two species were recorded. By 1971, the number of species present had increased to 12, with *Elodea* accounting for 58% of the plants sampled. By 1984, *Elodea* accounted for only 4% of the plants sampled.

Widgeon Pond gives evidence that *Elodea* is an early successional pioneer species that rapidly invades and dominates a newly flooded site given the right water and soil conditions. Over time, species diversity increases, and *Elodea's* dominance diminishes.

Species diversity is a factor in stabilizing aquatic ecosystems. Lakes with a highly diverse flora are more likely to maintain their character despite disturbances (Forest 1977). It was noted by Singhal and Singh (1978) that pure stands of any macrophyte species had greater biomass compared to stands of mixed species. This phenomenon has been casually observed at RRLNWR, and Widgeon Pond data suggest that this has occurred in recent years. Data collected at RRLNWR over the past 30 years suggest that plant succession operates at least in part in the following way: *Elodea canadensis* is a pioneer species in some sites that is replaced over time by an increase in total plant diversity. As plant diversity increases, overall biomass production may decrease. Although no surveys were made of the Upper Lake vegetation in the 1940s, Manager Sharp noted the destruction of *Potamogeton* in the center of Upper Lake by high winds during 1946-48 (RRLNWR Ann. Narr. Repts. 1946-48). This destruction of the established plant community would have increased the likelihood of pioneering by *Elodea* and its subsequent luxuriant growth observed in the mid-1950s.

On Culver Pond, *Elodea* has maintained its dominance for over 20 years. In theory, if *Elodea* is a pioneer species, it should have diminished in importance, similar to the situation at Widgeon Pond. We suggest that the heavy grazing by trumpeters and other waterfowl associated with the winter feeding program at Culver Pond is a disturbance significant enough to keep the *Elodea* in a seral stage.

Verhoeven (1980) commented on some temporal aspects of plant community change that are relevant to RRLNWR: "*Smaller and shallower habitats in particular, are very dynamic. Their existence is of limited duration. Further, new habitats originate continually. As the habitats are often isolated from each other, it takes time before all species that are capable of colonizing them have actually reached them. Hence, the plant communities can have an incomplete species composition for several years."* Many of the lakes at RRLNWR fall into this category. Species composition, abundance and distribution are constantly changing and evolving in response to temporal, biotic, and abiotic factors that influence the entire system.

Responses of Submersed Macrophytes to Environmental Change

Potamogeton praelongus

P. praelongus (whitestem pondweed) is a natural indicator of deep water (Spence and Chrystal 1970). At RRLNWR it occupies the deepest zone of any submerged macrophyte (Paullin 1973:79). *P. praelongus* never predominates and rarely occurs in water less than 1 m deep (Spence 1967). Because it grows in deeper water, it is associated with finer particle muds which typically settle out in the deeper areas of lakes. The prevalent or preferred depth for *P. praelongus* is 2-5 m with a depth maxima of 7 m (Sheldon and Boylen 1977).

P. praelongus has a high degree of shade tolerance (Spence and Chrystal 1970). It actually has higher photosynthetic rates in a reduced light environment than in bright sunlight. *P. praelongus* has developed shade-adapted leaves that have a large surface area relative to the overall biomass of the plant (Spence *et al.* 1973). This is an intrinsic characteristic that allows *P. praelongus* and other broad-leaved *Potamogetons* to grow in deep water where large leaves are more effective at capturing the limited light.

P. praelongus is most abundant in hard (20-135 ppm $CaCo^3$) waters with high conductivity (107-415 mhos), but it can range into both soft and alkali waters. It usually occurs infrequently and makes poor growth in water with total alkalinity less than 20 ppm, or in waters with sulfate ion concentrations greater than 300 ppm (Moyle 1945).

P. praelongus is an indicator of a healthy and undisturbed environment and is sensitive to changes in water quality (Dale and Miller 1978). It is characteristic of clear, cool, well oxygenated waters. Its distribution is primarily northern (north of Lake Erie), and its ecological tolerances are apparently very narrow (Stuckey 1971). Because of its northern distribution, Stuckey classified *P. praelongus* as a "northern *Potamogeton*". Other species in this group that are found at Red Rock Lakes include *P. richardsonii*, *P. friesii*, *P. foliosus*, *P. pusillus*, and *P. zosteriformis*. *Najas flexilis* and *Elodea canadensis* also have tolerances similar to the northern *Potamogetons*. The northern *Potamogetons* in general and *P. praelongus* especially are intolerant of decreased oxygen or increases in water temperature, turbidity, siltation, nutrients due to domestic sewage or other pollution. *P. praelongus* was the least tolerant to turbidity of 11 species examined by Davis and Brinson (1980) and it does not survive the long term changes that lead to eutrophication.

The sustained presence of *P. praelongus* in Upper Lake and its intolerance to turbidity seem to present a paradox due to the frequent turbidity of Upper Lake caused by wind storms. Davis and Brinson (1980) point out that where light is only occasionally limiting due to shading by high turbidity waters (e.g. Upper Lake), *P. praelongus* may be at a competitive advantage because of its high photosynthetic efficiency in reduced light conditions. This may help explain its occurrence in waters that are only periodically turbid.

Finally, *P. praelongus* is unique among the *Potamogetons* in that it can remain photosynthetically active under the ice (Boylen and Sheldon 1976). Although this phenomenon does not always occur (Rich *et al.* 1971), overwintering in the active state is most likely to occur in the deeper waters.

In summary, P. *praelongus* does well in deep water at low light intensities, but it is restricted to narrow conditions which do not include highly turbid waters for sustained periods. At Red Rock Lakes it historically has been restricted to Upper and Lower Lakes and recently has been increasing in MacDonald Pond. It is a good indicator of deep, high quality waters. If the average depth of the marsh were to decrease or if turbidity suddenly increased for sustained periods, P. *praelongus* would be expected to be one of the very first species to disappear.

Ceratophyllum demersum

C. *demersum* (coontail) is at the opposite end of the turbidity tolerance continuum from P. *praelongus*. Coontail has a very high tolerance to turbidity and is normally dominant or subdominant in disturbed systems (Stuckey 1971, Davis and Brinson 1980)

In terms of water chemistry, coontail has broad ecological tolerances. It grows best in hard water but will range into both soft and alkaline waters (Moyle 1945). Coontail is generally confined to eutrophic waters (Spence 1967); however, its extended tolerance for nutrients in dilute solution allows it to be more properly classified as a mesotrophic species (Seddon 1972). Data presented by Goulder and Boatman (1971) suggest that coontail is a nitrophilous plant with a high requirement for inorganic nitrogen. Coontail increases with eutrophication (Lind and Cottam 1969).

The depth maxima for coontail are high but difficult to ascertain because coontail is commonly a free-floating species. At RRLNWR, coontail is one of the deeper water species (Paullin 1973:79). Coontail anchors poorly in sediments and its presence is generally restricted to sheltered areas with low turbulence (Lind and Cottam 1969)

Coontail is very drought intolerant and slow to recolonize after a drawdown (van der Valk and Davis 1976). As such, coontail is a good indicator of more permanent waters that are not subject to periodic drawdown. This could explain its absence in Swan Lake, the shallowest and least permanent of the Refuge lakes, and its invasion of lower lake.

Coontail is a shade tolerant species that has photosynthetic capacities even in low light (Davis and Brinson 1980). Submerged macrophytes that form canopies (e.g. dense stands of sago pondweed or algae) generally depress plant growth under the canopy except for coontail and a few other shade tolerant species (Engel 1985).

The interspecific competition for light and space among submerged macrophytes can be intense, and coontail competes in a unique manner. While many macrophytes reach maximum biomass in June or July, C. *demersum* is a late season plant that does not reach maximum biomass until late August (Engel 1985). This delayed growth spurt was clearly demonstrated by Paullin (1973). Coontail elongates rapidly to the surface in late summer as many of the *Potamogetons* die back, thus occupying areas where other species were growing earlier.

Coontail can remain metabolically active even under the snow and ice (Rich *et al.* 1971, Paullin 1973, Boylen and Sheldon 1976). Because very few macrophytes at RRLNWR overwinter intact, coontail may be a very important food plant for waterfowl, particularly in the early spring immediately after ice break-up.

The large fluctuations in production estimates on the Refuge (e.g. Widgeon Pond in 1971 and 1972) might be due in part to the wide variation in survey dates. Surveys done in late July would coincide with peak growth for many Refuge macrophytes but grossly underestimate total coontail production which would not peak until later in the growing season. Coontail abundance and distribution on the Refuge should be monitored. Sustained increases in either abundance or distribution could be an early indicator of major ecological disturbances such as sedimentation, pollution, etc. Coontail is most prevalent in Lower Lake, an indication of eutrophic conditions. An indicator of an already disturbed ecosystem may be provided by those species which increase their abundance with disturbance (Dale and Miller 1978). One obvious factor that could have influenced the trophic condition of Lower Lake was the mining activity in the Odell Creek drainage.

Lemna trisulca

L. trisulca (star duckweed) is a minor component of the overall submerged macrophyte flora at RRLNWR (Appendix XV). It is most commonly found in Lower Lake, in varying amounts (up to 34% of total estimated production), and in Swan Lake and River Marsh in trace amounts. It has never been recorded in the other lakes or ponds. Typically *L. trisulca* occurs in sheltered areas with fertile lake bottoms. Moyle (1945) classified *L. trisulca* as belonging to the hard-water flora subgroup 2. Hard water flora occur in waters characterized by total alkalinity of 90-150 ppm, sulfate ion of 5-40 ppm, and summer pH of surface waters of 8.0-8.8.

Seddon (1972) expanded on Moyle's classification and used an elaborate ordination procedure to show that *L. trisulca* was found exclusively in eutrophic waters. Such waters were high in combined calcium and magnesium (hard), pH, total dissolved solids, and conductivity (>170 mhos). Consequently, Seddon classified *L. trisulca* as a moderately eutrophic species. Among the macrophyte flora at RRLNWR, only *P. pectinatus* and *Myriophyllum spicatum* had a more exacting requirement for eutrophic water than *L. trisulca*.

L. trisulca is a good indicator of eutrophic waters with fertile bottoms that are typically sheltered from winds. The sudden, but short-lived increase in *Lemna* in 1971 could have been caused by the sudden influx of nutrients in the water from increased erosion, or as Paullin (1973) suggested, short term increases in *Lemna* may be stimulated by declines in other species.

If future plant surveys document substantial sustained increases in *L. trisulca*, particularly in lakes or ponds where it has not previously been documented, then this increase can safely be interpreted as an indication that Refuge waters are becoming more eutrophic and Refuge personnel should be alert to identify the perturbations (grazing, upstream channelization, sedimentation from mining, etc.).

Chara vulgaris

C. vulgaris (muskgrass) is common in lime-rich regions such as RRLNWR where it typically grows in dense marginal zones. It belongs to the hard water flora previously described for *L. trisulca* (Moyle 1945). *Chara* is an indicator of oligotrophic conditions (Singhal and Singh 1978) and is highly sensitive to phosphorous. Additions of even "remarkably small" amounts (30 g/l) of phosphorus are enough to stop growth (Forsberg 1964). The sensitivity to phosphorous is apparently valid for all charophytes, thus explaining their absence in very eutrophic waters and disappearance from polluted localities (Jaag 1949). The mechanism for phosphorous sensitivity is unknown.

Charophytes (e.g. *C. vulgaris* and *Nitella flexilis*) are deep water species capable of growth at greater depths than submersed angiosperms (Hutchinson 1975). Although *Chara* is capable of growth in deep water, interspecific comparisons done by Paullin (1973) showed that it was intermediate among RRLNWR macrophytes in depth preference.

At RRLNWR, muskgrass is widely distributed; however, in the main marsh lakes it is least prevalent in Lower Lake. Lower Lake is fed primarily by the Odell Creek drainage, an area that was mined for phosphorous in the 1950s. Whether the paucity of muskgrass in Lower Lake is due to phosphorus inhibition, eutrophication, or some other factor is unknown. Lower Lake is eutrophic relative to other Refuge lakes, so its scarcity there is expected. However, since it the only Refuge Lake fed by Odell Creek it is possible that phosphorous inhibition could also be a factor in limiting the overall abundance of *Chara vulgaris* in Lower Lake.

Potamogeton pectinatus

Sago pondweed is probably the most widespread and abundant of all the North American submersed species (Davis and Brinson 1980). The importance and cosmopolitan nature of this species is consistent with its wide tolerance for several environmental factors.

The occurrence of sago pondweed is not substrate dependent. Whereas some species (e.g. *Myriophyllum spicatum*) are limited to substrates of fine sediments, sago can grow in fine, coarse, and firm substrates (Schiemer and Prosser 1976). However, Anderson (1978) found that soil conductivity, phosphorus and potassium were important substrate factors that affected standing crop biomass of sago.

Sago grows in calcareous and non-calcareous waters of high pH that can be described as rich to moderately rich in nutrients and are often saline (Spence 1967). In Minnesota, Moyle (1945) found sago in "rich waters" and Seddon (1972) went so far as to classify sago as an obligate eutrophic species, requiring very nutrient-rich waters. The only other obligate eutrophic species was *M. spicatum*. Sago belongs to the hard water flora described by Moyle (1945) which typically has alkalinities of 90-150 ppm, sulfate 5-40 ppm, and summer pH of surface waters of 8.0-8.8.

Extensive development of sago coincides with prolonged periods of low water (Schiemer 1979). Jackson and Starrett (1959) noted that sago grows best where water levels remained low and drawdowns were common. The prevalent or preferred depth of sago is only 1 m with a depth maxima of 3 m (Sheldon and Boylen 1977). The shallow water preference of sago was confirmed by Schiemer (1979) who showed that *P. pectinatus* biomass was generally highest close to shore and declined with distance from shore. Sago is intolerant of increases in water depth. Schiemer and Prosser (1976) documented "considerable declines" in *P. pectinatus* on a lake where a water control structure increased lake levels by 40 cm.

P. pectinatus is a highly tolerant species that is normally dominant or subdominant in disturbed systems (Davis and Brinson 1980). It is tolerant of highly turbid waters (Stuckey 1971), water induced turbulence (Davis and Carey 1981), municipal and industrial effluents (Dale and Miller 1978, Stuckey 1971), and siltation (Singhal and Singh 1978). Sculthorpe (1967) suggested that the linear leaves of sago remained free of settling particles of sediment, thus allowing the species to colonize areas unsuited for submersed plants with leaf forms more amenable to silt accumulations (e.g. *P. praelongus*).

The wide ecological tolerances of sago to turbidity, siltation, and pollution place it at the opposite end of the continuum from the intolerant "northern *Potamogetons*" described by Stuckey (1971). Sago is in the "tolerant" group which includes *Ceratophyllum demersum*, *Zannichellia palustris*, *Vallisneria americana*, *Najas guadalupensis*, and *Myriophyllum exalbescens* (Davis and Brinson 1980). These species have wider geographical distributions and occur in warmer, more turbid, and more poorly oxygenated waters than the northern *Potamogetons* (Stuckey 1971).

Sago is a typical canopy species that reaches maximum biomass in June and July (Engel 1985). Such canopies reduce light penetration, prevent wind from stirring the water column, and distribute heat and absorb solar radiation. The canopy reduces plant growth underneath it for all but the most shade tolerant species (e.g. *Ceratophyllum demersum*). Sago senesces in winter and does not grow under the ice (Boylen and Sheldon 1976).

Anderson (1978) found sago in sites that were lower in clay, higher in organic matter, had less wave action, shallower water, and less turbidity. Ordination models developed by Paullin (1973) also showed that sago sites were low in clay and high in organic matter. River Marsh typifies this type of site at RRLNWR. Available potassium and phosphorus, water depth, maximum exposure, and soil conductivity seem to be important factors affecting the standing crop biomass of sago pondweed (Anderson 1978). Sago pondweed is widespread at RRLNWR (Paullin 1973:147), a clear reflection of its broad ecological tolerances. While its overall abundance on the Refuge is relatively low, sago pondweed appears to have been relatively stable throughout the period of record at RRLNWR.

Myriophyllum spicatum

Davis and Bronson (1980) pointed out the confusion in the literature in distinguishing *M. spicatum* from *M. exalbescens*. Aiken *et al.* (1979) stated that this confusion also includes *M. verticillatum*. Jepson (1925) considered the different milfoils to be varieties, while Nichols (1975) considered them to be subspecies. Ogden (1974) made no ecological distinction between *M. spicatum* and *M. exalbescens*. A review of the literature shows much overlap in the discussions of *M. spicatum* and *M. exalbescens* and for purposes of this discussion they are considered ecologically similar with reference to the findings in the literature.

Milfoil is substrate limited, with maximum plant densities occurring in sites with fine inorganic silt or organic ooze substrate (Patten 1956, Schiemer 1979). It does not grow well in coarse or sandy substrate or on sloped bottoms.

Milfoil is highly sensitive to wave turbulence, particularly in the shallower water zones (Schiemer and Prosser 1976, Schiemer 1979). Wave action prevents the establishment of young plants due to the constant cycle of erosion and redeposition of sediment (Schiemer and Prosser 1976). Milfoil is readily fragmented due to wave turbulence (Patten 1956). Wave action combined with freezing and ice scouring has been known to eliminate milfoil growth in shallow water generally less than 1 m (Adams and McCracken 1974, Forest 1977). At RRLNWR, this limitation appears to be only slightly greater than 0.3 m in depth. Schiemer (1979) found that peak biomass and plant density was reached 400-500 m from shore at depths 1-2 m with patches of vegetation running parallel to the shore. This pattern is seen on Upper Lake with plant abundance greatest on the sheltered west side. Because of its sensitivity to turbulence and depth, milfoil has a lesser habitat range than *P. pectinatus* in open water areas (Schiemer 1979).

Milfoil can grow in depths up to 10 m (Aiken *et al.* 1979). At RRLNWR it is intermediate in its depth preference (Paullin 1973). Davis and Brinson (1980) concluded that the depth maxima for milfoil were controlled by hydrostatic pressure rather than light. Schiemer and Prosser (1976) documented that milfoil abundance declined in a lake where water levels were increased 40 cm by a water control structure. This could have been due to the combined effects of hydrostatic pressure and increased wave turbulence. It has an intermediate tolerance for shade (Spence and Chrystal 1970).

While milfoil is sensitive to suspended turbidity (Davis and Carey 1981), it had an intermediate turbidity tolerance among those species examined by Davis and Brinson (1980). It has been independently demonstrated (Schiemer and Prosser 1976, Schiemer 1979) that sediment coating is more prevalent on the finely divided and feathery leaves of *M. spicatum* than *P. pectinatus*. The impact of the silt was to shade assimilating parts and to increase the plant's weight, thus forcing the plant to deeper, less favorable waters. Silt coating on milfoil is greatest in sheltered bays.

M. spicatum is found exclusively in eutrophic lakes and together with sago pondweed has been classified by Seddon (1972) as obligate eutrophic species. Milfoil can grow in either calcareous or non-calcareous waters but it has a definite affinity for highly alkaline waters with a high pH and high conductivity (Patten 1956, Spence 1967, Adams and McCracken 1976). Milfoil is one of the few submersed species favored by pollution (Forest 1977).

Milfoil initiates growth early after ice out. Because it is capable of photosynthesis at low temperature it may have a competitive advantage in northern localities over species incapable of such growth (Barko *et al.* 1982). Milfoil has a unique growth pattern typified by two growth spurts (Kaul and Vass 1972, Adams and McCracken 1974). Typically, milfoil will undergo a rapid growth period in early May after ice out, followed by a summer plateau. Another sustained growth spurt occurs in late August, September, and October. Maximum standing crop biomass is usually attained in late August (Adams and McCracken 1974, Forest 1977). Milfoil largely disintegrates in the fall, overwinters as turions, and does not grow under the ice (Boylen and Sheldon 1976, Schiemer 1979, Engel 1985).

Milfoil is a cosmopolitan species occurring in a wide variety of habitats, being both eurytropic and circumpolar in distribution (Sculthorpe 1971, Hutchinson 1975). In disturbed sites where catastrophic change has occurred, milfoil acts as a "weed" or pioneer plant on soil stripped of vegetation or freshly deposited (Forest 1977). Such "invasions" are most likely where existing species diversity is already low. Milfoil will increase with increased turbidity (within limits) as native species (particularly the northern *Potamogetons*) decrease (Davis and Brinson 1980). It is capable of displacing most other species within 2 to 3 years once it becomes established (Aiken *et al.* 1979). "Invasions" of *M. spicatum* have been well documented (Lind and Cottam 1969, Nichols and Mori 1971, Bumby 1977).

At RRLNWR, milfoil is one of the more common and widely distributed submersed macrophytes. For unknown reasons it was absent from MacDonald Pond in 1971. The species distribution maps in Paullin (1973) show milfoil conspicuously absent from the center portions of Upper Lake. This absence is consistent with the literature, particularly Schiemer and Prosser (1976), who found milfoil absent in open water areas even if suitable sediment was present. The absence of milfoil on Upper Lake is apparently due to the combined effects of wave turbulence, suspended silt sensitivity, and the hydrostatic pressure of deeper waters, all of which exceed the optimum for the species.

Forest (1977) noted that large annual fluctuations in species composition and abundance are common where milfoil occurs. Appendix XV bears this out. Milfoil has been most prevalent in Swan Lake which is consistent with the sheltered, eutrophic conditions that exist there. On the other Refuge lakes and ponds the estimated abundance of milfoil has been erratic. The bimodal growth pattern of milfoil, characterized by a late season growth spurt can explain some of this irregularity. Paullin (1973:94) documented at least one area of the Refuge where milfoil was still actively growing in late August. Early aquatic plant surveys (late July and early August) will underestimate milfoil abundance at RRLNWR. Exactly when milfoil growth plateaus in the fall at RRLNWR is unknown. However, it must be pointed out that in many cases, the fluctuation in survey dates is at least partly responsible for fluctuations seen in Appendix XV.

Elodea canadensis

Elodea is indicative of clear, cool, well oxygenated waters. Its distributions are primarily northern and ecological tolerances are apparently narrow (Stuckey 1971). *Elodea* is intolerant of habitat change such as decreased oxygen and light and increased siltation, turbidity, and nutrient loads from domestic sewage. It is particularly sensitive to turbidity where suspended particles decrease light (Davis and Brinson 1980). Its intolerance is only slightly less than the northern *Potamogetons* (Stuckey 1971).

Elodea is tolerant of waters low in ionic concentrations of essential elements thus making them characteristic of oligotrophic waters (Seddon 1972). *Elodea* is shade tolerant and capable of growing underneath canopies (Engel 1985). The optimum water temperature for *Elodea* is 28 $^{\circ}$C (Barko *et al.* 1982). Shoot biomass decreases at temperatures greater than 28 $^{\circ}$C. Root biomass increases as water temperature decreases down to 16 $^{\circ}$C. *Elodea* is resistant to hydrostatic pressure as evidenced by the fact that it holds the depth maxima record (12 m) among the submersed angiosperms with the preferred or prevalent depth being 1-7 m (Sheldon and Boylen 1977).

The quantity and quality of light is considered to be the primary factor determining the lower depth limit of *Elodea* (Ward and Talbot 1984). Shoot biomass, root biomass, and shoot density will increase with increasing light at the low and mid-light levels. *Elodea* is photophobic, growing dense and prostrate in high light regimes, and vertical in low to medium light regimes (Barko *et al.* 1982). *Elodea* has a relatively limited ability to elongate to the water surface under low light conditions and thus may be retarded in aquatic systems characterized by low water clarity.

The upper depth limit of *Elodea* appears to be determined by ice scouring (Haag 1979) and wave action (Brown 1975, 1979, Richmond 1981). Growth in shallower waters is limited to sheltered sites where silts have accumulated (Ward and Talbot 1984). Chapman *et al.* (1971) found that *Elodea* occurred most frequently on sandy soils so apparently it is not substrate limited.

In disturbed sites, *Elodea* biomass is usually low (Davis and Brinson 1980). *Elodea* has a flexible strategy of vegetative reproduction ranging from an annual to a perennial habit (Haag 1979). Sustained shoot growth depends on successful establishment of roots, a process which may limit its survival. In late summer, *Elodea* puts most of its growth into roots, a strategy well suited to establishment of propagules at the end of the growing season, i.e., a winter annual habit.

Elodea lacks a dormancy mechanism and therefore is capable of growth at any time of year (Haag 1979). It had the highest growth rate under the ice of any species examined by Boylen and Sheldon (1976). *Elodea* has a high potential for early resumption of growth in the spring, being one of the earliest species to initiate growth in cold water (10°C) immediately after ice out (Engel 1985). The winter survival of shoots, even without growing, is conducive to an early growth spurt because delays in seed or turion germination are avoided. Its early growth habit may be particularly important at RRLNWR as an early spring food source for Trumpeter Swans.

Elodea winter kills easily. The lack of dormancy makes winter survival strongly dependent on the winter irradiance regime. Reduced light due to heavy winter ice cover is a prominent cause of shoot death (Haag 1979). Haag described an area where *Elodea* was dominant for six ice-free years and then nearly disappeared two years after winter ice cover was reestablished.

This would explain the sustained abundance of *Elodea* in places like Culver Pond where warm spring waters keep much of the pond either open or the ice very thin in the winter. This would not, however, explain the marked decline of *Elodea* in Upper and Lower Lake following the initial surveys in the late 1950s.

In describing *Elodea* growth at Lake Wabamun, Alberta, Haag (1979) provided some interesting insights that probably apply to the lakes at RRLNWR:

> *"Elodea populations may establish in unheated areas of the lake during summers of rapid growth and abundant reproduction, but the survival probability and longevity of these populations is low. The life cycle of Elodea may be viewed as an attempt at earlier establishment, growth and reproduction, a strategy which is a general failure in this climate. The North American centre of distribution of Elodea ranges from north-eastern to north-central U.S.A. (Fernald 1950; Adams, Cole and Massie 1973; Ogden 1974) where winters are milder, ice cover is thinner and net assimilation may occur under ice cover (Boylen and Sheldon 1976)."*

In short, in the northern portions of its range, *Elodea* can be a "boom or bust" species. This probably more than any other factor controls the distribution and abundance of *Elodea* on RRLNWR. Long harsh winters, coupled with thick ice overlain with deep snows in most years prevent *Elodea* from surviving the winter. Occasionally conditions are such that *Elodea* can survive the winters and dominate a lake, as was documented in 1956. However, for the most part, the harsh winters of the Centennial Valley probably make most of RRLNWR marginal for *Elodea* growth at best.

A possible mechanism for the rapid establishment of *Elodea* was also observed by Haag. In this instance, ice scouring and wind turbulence immediately following spring breakup uprooted a mass of mature *Elodea* plants. Wind generated currents transported these "immigrants" en masse to a previously unvegetated site, where most of them became established.

Summary and Conclusions

Although quantitative surveys of the aquatic vegetation at RRLNWR have been conducted since 1956, the data do not allow determination of annual changes in species biomass or relative abundance. Use of the survey data is limited by the very large variance of the annual measurements, which resulted from inadequacies of the sampling techniques. However, after examining the long- and short-term changes in macrophyte communities documented in the literature and comparing them with the data in Appendix XV, generalizations can be made that allow some predictive capability. Much of the earlier discussions on individual species has been summarized in Table 29.

Davis and Brinson (1980:47) provide an excellent conceptual model that summarizes how environmental factors affect submersed macrophytes. The three main forces are light attenuation (suspended sediments, eutrophication), toxicity (herbicides), and biomass removal (waves, currents, grazing, and sedimentation). The authors summarized various types of environmental alterations and how these perturbations affect submersed macrophytes. We have excerpted those alterations most likely to be found in the Centennial Valley and have summarized them in Table 30.

The distribution and abundance of aquatic macrophytes in nature reflect the totality of the environmental factors acting upon them. Numerous studies have shown dramatic changes in total biomass and species composition from one year to the next. The situation at Red Rock Lakes is no exception. Annual variations can be caused by cloud cover, nutrient runoff, water temperatures and water depth among others. There are many examples in the literature of waters that once supported luxuriant growths of macrophytes that are now devoid of these plants. The causes for decreases in abundance vary, but most are associated with increased turbidity (Davis and Brinson 1980).

Because turbidity is such an important limiting factor, Davis and Brinson developed a turbidity tolerance index for several species, many of which occur at RRLNWR. The index has been included here in its entirety (Table 31) because it provides much insight into the sensitivities of many species found on the Refuge. Simply stated, species with higher turbidity tolerance indices are better adapted for survival under conditions of low light transmission.

Also, to summarize the long-term changes in submersed macrophyte communities, Davis and Brinson (1980) developed a species survival index (Table 31) which is simply the ratio of the number of lakes in which a species was reported in earlier surveys to the number of lakes in which the species was present when last studied.

The capacity to predict changes in macrophyte communities resulting from natural or human-induced environmental alterations is low. However, Table 29 and Table 30 can be used in combination to make some predictions relative to disturbances and changes in plant abundance or distribution. For example, Table 30 predicts that stream channelization (e.g. Upper Red Rock Creek) increases suspended sediments and eutrophication. Table 29 predicts that *Ceratophyllum demersum, P. pectinatus, P. richardsonii,* and *Zannichellia palustris* would be the most tolerant of such an alteration whereas *P. praelongus* and the other "northern pondweeds" would be least tolerant. These tables can be used in various combinations to predict species responses to environmental alterations.

Table 29. Summary of environmental tolerances and preferences of selected submersed macrophytes found on Red Rock Lakes National Wildlife Refuge.

Species	Turbidity Tolerance	Water Chemistry Preference	Ecological Tolerances	Depth Tolerance	Trophic Requirements
Elodea canadensis	Intolerant		Narrow	Very high	Oligotrophic
Myriophyllum spicatum	Intermediate	Hard water	Very broad	Intermediate	Obligate eutrophic
Chara vulqaris		Hard water		High	Oligotrophic
Potamogeton pectinatus	Tolerant	Hard and alkaline water	Very broad	Low	Obligate eutrophic
Lemna trisulca		Hard water subgroup 2			Moderately eutrophic
Ceratophyllum demersum	Very tolerant	Broad: soft, hard. Alkaline preferred	Very broad	High	Mesotrophic
P. praelongus	Very intolerant	Hard water subgroup 3	Very narrow	High	Oligotrophic
Najas flexilis	Intermediate	Hard water subgroup 3	Intermediate	High	Oligotrophic
Zannichellia palustris	Tolerant	Hard and alkaline water	Broad		Eutrophic
P. foliosus *P. friesii* *P. pusillus* *P. zosteriformis*	Intolerant	Hard water subgroup 1, 2, and 3	Narrow	High	Oligotrophic
P. richardsonii	Intermediate hard-alkaline	Intermediate	Narrow	High	Oligotrophic
Ranunculus aquatilis					Oligotrophic
Isoetes sp.					Highly oligotraphic
Nitella flexilis				High	
Callitriche hermaphroditica		Soft water group 1, low pH, low alkalinity, low S0$_4$			

[a] Classification follows Moyle (1945).
[b] The ecological tolerances follow Stuckey (1971) and are dissolved oxygen, temperature, nutrient loads, turbidity, light, and siltation.

Table 29, cont.

Species	Shade Tolerance	Wave Tolerance	Substrate Preference	Growth Initiation	Miscellaneous
Elodea canadensis	High	Intolerant	Silt	Very early sheltered	Easily winter kills under ice with low light. Actively grows under ice with adequate light.
Myriophyllum spicatum	Intermediate	Highly intolerant	Fine silts organic ooze	Bimodal: very early and very late	Highly substrate limited. Depth maxima controlled by hydrostatic pressure. Is a pioneer in disturbed sites.
Chara vulqaris					Very sensitive to phosphorus.
Potamogeton pectinatus	Low		Not substrate dependent		Favored by periodic drawdowns
Lemna trisulca		Low	Sheltered from winds		
Ceratophyllum demersum	High	Intolerant	Sheltered from winds	Very late	Highly drought intolerant. Late season growth spurt.
P. praelongus	High		Fine silts		Can grow under ice.
Najas flexilis	High			Very late	Favored by drawdowns. Depth maxima limited by hydrostatic pressure, exhibits late season growth spurt.
Zannichellia palustris					Favored in disturbed sites.
P. foliosus					
P. friesii	High			Intermediate	Together with *P. praelongus* these constitute the northern Potamogetons. Tolerances are similar. *P. zosteriformis* survives best of this group.
P. richardsonii	High				Most tolerant of the northern Potamogetons.
Isoetes			Coarse, thin silt		Grows under ice.
Nitella flexilis				Late	

Table 30. Evaluation of the effects of various types of environmental alterations on submersed plant communities (excerpts from Davis and Brinson 1980).

Environmental Alteration	Suspended Sediments	Eutrophication	Toxicity	Sedimentation	Community Resiliency
Instream mining	Varies depending on bed sediments	Varies depending on bed sediments	Low	Plants buried by coarse sediments in a downstream gradient	Limited reestablishment after burial may be possible for forms like *Podostemum ceratophyllum*
Silviculture					
Selective harvest	Low	Low	Low	Low	Low level continuous siltation may be conducive to macrophyte establishment
Clear cutting	High on short term	Low on long term	Low if herbicides not used	Variable	Impact ameliorated with reestablishment of ground cover
Logging roads	High	Low	Low	High	Erosion of ditches and roadbed contributes sediments
Agriculture	High	High	High	High	Adverse effects would be minimized with best available management techniques; substantial recovery of aquatic system would be expected from their application
Road construction	High	High if eroding soils are nutrient rich	Variable	High	Medium-term pulse of pollution with community recovery except in areas of extreme siltation. Aquatic habitat may change due to "dam" effects of roadbed
Stream	High	High	Variable	High	Decreased shading conducive to Channelization increased plant growth but spate stress increased; gradual recovery of natural system possible without channel maintenance[a].

[a] O'Rear 1975.

Table 30, cont.

Environmental Alteration	Suspended Sediments	Eutrophication	Toxicity	Sedimentation	Community Resiliency
Dams and impoundments					
Upstream (lake)	High	Variable	Low	High at mouth of rivers	Establishment of submersed plants depends on width of littoral as well as extent and periodicity of drawdown
Downstream	High	Varies; some nutrient depletion in lake likely	Low	Low	Depends on discharge procedure Extensive plant beds may develop.
Nearshore mining	Variable	Variable	Variable	Variable	Effects vary according to mining method and control procedures.

Table 31. Survival and turbidity tolerance indices for selected submersed macrophytes (From Davis and Brinson 1980).

Plant Species	Survival	Index	Turbidity Tolerance	
Ceratophyllum demersum	1.0	(strong)	2.8	(most tolerant)
Potamogeton pectinatus	1.0		2.0	
P. richardsonii (ssp.perfoliati)	1.0		1.5	
Zannichellia palustris	1.0			
Elodea canadensis	0.8		1.1	
Najas flexilis	0.7		1.7	
P. zosteriformis	0.6			
P. foliosus	0.3			
P. praelongus	0.0	(weak)	0.0	(least tolerant)

Despite the shortcomings of the data in Appendix XV from a statistical standpoint, some major trends are evident, including the decline of *Elodea* on both Upper and Lower lakes, and a refuge-wide increase in species diversity.

The increased turbidity of Upper Lake in the late 1950s and 1960s was detrimental to the luxuriant beds of *Elodea* that dominated the lake in the mid-1950s. *Elodea* is an indicator of low turbidity, and is especially sensitive to suspended sediment early in the growing season.

Water temperatures at RRLNWR are suboptimal for the growth of most species of aquatic plants, particularly during the period of spring runoff. As we discussed in Chapter 6 water depths in Lower Lake were increased by at least 0.3 m (1 ft) by the reconstruction of the lower water control structure in 1957. The increased water depths would have decreased plant productivity by further reducing water temperatures, and also reducing the amount of available light during episodes of high turbidity. Water depths affect the distribution, zonation, and abundance of aquatic plants at RRLNWR, with the maximum species diversity occurring at depths of 0.6-0.9 m (2-3 ft).

Elodea appears to be a pioneer species at RRLNWR; it will invade newly flooded or disturbed sites given the right water and soil conditions. Over time, species diversity increases and overall biomass production may decrease. Heavy waterfowl grazing, as observed at Culver Pond, may maintain *Elodea* in a seral state for over 20 years.

With the exception of Culver Pond, no evidence has been found that indicates that waterfowl grazing at RRLNWR has altered the composition of the aquatic plant communities. The changes which have occurred over the last 30 years are very poorly quantified, and are likely the result of variations in light regime, siltation, wave action, water depths, and water temperatures.

The flora of RRLNWR indicates that the marsh is still very healthy and water quality is near "pristine" conditions as evidenced by the presence of many of the "northern *Potamogetons*". Despite alterations to the watershed caused by mining, grazing, channelization, and installation of the lower water control structure, the Refuge lakes remain healthy, diverse, and productive. Perhaps the most limiting factor on submerged macrophytes is the harsh environment of the Centennial Valley, (including long winters and short, cool, growing seasons, cold waters, ice scouring, wave turbulence, and predominately oligotrophic waters), all of which are suboptimal to maximum standing crop production.

Management Recommendations

1. Much is known about the abundance and distribution of submersed macrophytes at RRLNWR. Unfortunately the key piece of information still lacking is an adequate food habits study for Trumpeter Swans. This should be high on the Refuge's list of priorities.

2. Turbidity and lake levels are critical long-term data needs. Permanent staff gauges should be installed in each lake and pond, and water levels measured at least bi-monthly. The severe ice conditions will probably necessitate annual maintenance to keep gauges functional. Turbidity can be read at the gauging stations. Secchi disc transparencies are helpful but at RRLNWR where visibility to the lake bottom is common, such readings have limited value. Turbidity measurements using a Hach kit or similar equipment are preferable.

3. Consideration should be given to the establishment of permanent plots or transects to periodically sample macrophytes in each lake and pond. Without some known benchmark for reference, year-to-year comparisons are difficult at best, and statistically impossible.

4. Consideration should be given to testing remote sensing (e.g. low level infrared photography as a means of monitoring plant abundance and distribution. Given the clear waters of the Refuge, on a calm day this method may be helpful in determining plant distribution, coverage, zonation, etc.

5. The aquatic plant surveys as they now are conducted will consistently underestimate total biomass of those species that exhibit late season growth, e.g. *C. demersum*, *N. flexilis*, and *M. spicatum*. The accurate estimation of total biomass for all species would require two surveys. This may be worth doing at least once in order to gain insight into the amount of late season production that occurs in the marsh.

CHAPTER 8. WINTER DISTRIBUTION AND HABITATS

Trumpeter Swans once wintered across much of North America wherever suitable ice-free waters could be found. Flocks utilized estuaries along the Atlantic, Pacific, and Gulf of Mexico coasts, as well as the vast wetlands of the Mississippi Valley and the southeastern U.S. (Banko 1960:26, Rogers and Hammer 1978). Sightings of a few widely scattered individuals in the late 1800s and early 1900s provided additional clues to the location of pre-settlement wintering areas (Banko 1960:28-37). As the species neared extinction outside of Alaska, trumpeters no longer returned to most of their traditional wintering sites. Outside of Alaska, the only group of wintering swans to survive the decline were those that wintered in the inaccessible waters of the Tri-state Area (Figure 3).

The Tri-state region, with its unusual geothermal activity, provided an ice-free winter oasis in the otherwise frigid Rocky Mountain environment. Warmed by the runoff from geysers and hot springs, small portions of slow-moving rivers and streams remained free of ice throughout even the harshest winters. Virtually unexplored until the late 1860s, the Yellowstone habitat remained unknown and isolated from human settlement in the 1700s and 1800s (Haines 1977), while trumpeters were exterminated from the more accessible portions of their original range. With human activities further restricted by the establishment of Yellowstone National Park in 1872, and the extremely inhospitable winter weather, this region was probably the last winter habitat in the United States to provide both dependable open water and an almost complete lack of human activity during the winter months. This unique combination of extreme isolation and ice-free waters allowed a few trumpeters to survive into the 1920s when efforts to prevent the extinction of the species began.

Despite the almost tenfold increase of swans since the 1930s, the Tri-state wintering area remains the only known wintering grounds for the Rocky Mountain Population. The rapid increase of Canadian trumpeters during the past decade, however, has been accompanied by occasional sightings of wintering individuals and family groups in several other western states (Appendix I). Migrants from the Tri-state Subpopulation and the rapidly expanding Interior Canada Subpopulation are probably exploring other migration routes and wintering areas. The winter distribution of RMP trumpeters could expand rapidly in the near future if these pioneering birds reestablish the species' long-broken migratory traditions.

The number of trumpeters wintering in the Tri-state Area is probably higher today than it was prior to the species' overall decline for two reasons: 1) grain-feeding has enabled 200-300 trumpeters to winter in the Centennial Valley where little, if any, natural winter habitat exists; and 2) several of the current Tri-state wintering sites have been created or expanded by man-made changes, particularly dams.

Banko (1960:54-60) provided detailed descriptions of the key wintering sites that were used by trumpeters during the 1950s. Since that time, the total number of swans wintering in the Tri-state Area has almost tripled, resulting both in shifting patterns of use and the pioneering of new sites within the region.

Montana

Red Rock Lakes NWR

Two warm spring impoundments, MacDonald and Culver Ponds, provide a combined area of about 2-4 ha (5-10 a) of open water except during extreme cold spells when the ice-free area may be reduced by about half (Banko 1960:58). Grain has been fed during the winter months at Culver Pond since 1935, and at MacDonald Pond since its construction in 1953. In addition to the grain ration, limited amounts of aquatic vegetation are available in the springheads and outflow channels, depending upon the extent of the ice cover. The primary plant species found in the feeding ponds are shown in Table 32.

Refuge Annual Narrative Reports have frequently noted that the number of swans counted at the feeding ponds varied from week to week. Many swans loafed on snow covered flats, south and west of the ponds, where they were difficult to count. Swans also were frequently observed moving between the Refuge and the Island Park area in Idaho. Since 1956, the monthly high count of swans on the Refuge has usually ranged between 200-330 (Appendix XVI). Higher counts in March (up to 396) represent a movement of trumpeters onto the Refuge, as well as the occasional presence of Tundra Swans that pass through on northward migration.

Several lines of evidence indicate that only a small portion of the Canadian trumpeters winter at RRLNWR. Although neck-banded trumpeters from the Yukon, Saskatchewan, and Grande Prairie have been observed on the Refuge, only about 10 different marked individuals have been reported (McEneaney and Sjostrom 1983; Can. Wildl. Serv. banding records, Edmonton). Also, as recorded on the Midwinter Tri-state Surveys, the ratio of cygnets per 100 adult wintering on the Refuge annually has been about 13:100, just slightly lower than the 15:100 average for the late summer Centennial Valley flock. In contrast, the influx of highly productive Canadian trumpeters raises the ratio to 23 cygnets per 100 adults/subadults in the wintering population as a whole. In addition, despite the dramatic increase in the total wintering RMP since 1972, the number of swans wintering at Red Rock Lakes has not increased correspondingly (Appendix II).

Due to the very limited amount of ice-free water and available aquatic vegetation, the Centennial Valley probably wintered few, if any, swans prior to the impoundment of Culver Pond by early settlers. Most of the ice-free habitat that currently exists during normal or severe winter weather resulted from man-made impoundments of warm springs. Prior to the creation of these ponds and the initiation of supplemental feeding, most swans must have moved out of the Centennial Valley at freeze-up. Supplemental grain has been fed to the swans every winter since 1935-36, and has enabled some 250-300 trumpeters to winter on the Refuge. Even with a wintering flock that varied between 25-44 swans in 1935, Manager A. Hull believed the available natural vegetation on the Refuge to be inadequate for survival. Reporting in the 1935-36 Refuge Annual Operations Report, Hull wrote "*By the feeding of grain during the critical period we managed to pull all of the Trumpeter Swans through that remained during the winter without a loss of a single bird. Residents state that in past years several die each winter apparently due to starvation.*"

In Chapters 3 and 5 we demonstrated that winter feeding rates showed a very strong negative correlation with winter losses of swans from the Red Rock Lakes and Centennial Valley flocks ($P = 0.016$), and a positive correlation with hatching success the following spring ($P = 0.091$). We conclude that the feeding program has determined the winter carrying capacity of the Refuge, reduced the movement of swans out of the Centennial Valley at autumn freeze-up, and increased productivity. Because the Centennial flock contains over half of the swans in the

Table 32. Aquatic vegetation in Culver and MacDonald ponds, RRLNWR, presented in order of decreasing biomass[a]

Culver Pond	MacDonald Pond
Elodea canadensis	*Potamogeton pusillus*
Chara vulgaris	*P. pectinatus*
Ranunculus aquatilis	*P. foliosus*
Myriophyllum spicatum	*Chara vulgaris*
Potamogeton pectinatus	*Zannichellia palustris*
P. pusillus	*Myriophyllum spicatum*
P. foliosus	
Zannichellia palustris	

[a] Data are from RRLNWR Aquatic Vegetation Survey files.

142

Tri-state Subpopulation, winter feeding at RRLNWR has also had important impacts on the population dynamics of the Subpopulation.

Previous descriptions of the feeding program (Banko 1960:177, Page 1976:6) have briefly explained that the amount of grain fed increased as the Refuge flock increased, and implied that winter feeding was not a major variable in the dynamics of the Centennial Valley flock. However, our review of Refuge files shows that since 1936, the total amount of grain fed annually, the methods used to feed the swans, and the timing of feeding (Appendix XVI) have varied widely, and at times abruptly, as managers and management ideas changed. The management philosophy toward feeding has varied from feeding the swans early and long with all the grain they could eat, to an attempt to cut off the grain supply abruptly. In many years, the early feeding of grain was used as a tool to deliberately reduce the movement of trumpeters out of the Centennial Valley (Banko 1960:177). As we will discuss, there has been at least a 7-fold variation in the amount of grain fed per swan, per winter, and this variation has effectively resulted in a fluctuating winter carrying capacity.

History of Winter Feeding. The following account of the feeding program was compiled from RRLNWR Annual Narrative Reports 1935-86, RRLNWR Trumpeter Swan Reports 1983-86, RRLNWR winter feeding files, and personal communications as cited. Data are summarized in Appendix XVI. In addition, we estimated the amount of grain provided per swan each winter. We divided the total number of bushels fed by the average of the monthly maximum swan counts during the feeding period. We used the average of the maximum counts because low counts were often the result of poor viewing conditions or swan movement to inaccessible areas of the Refuge, and not necessarily due to the departure of swans from the Refuge. Other species of wintering waterfowl also ate undetermined amounts of the grain. Therefore, our calculations should be interpreted as a long-term index to the amount of grain available per swan, rather than as an actual calculation of the grain consumed per swan. While we recognize its limitations, this index provides a means of comparing the amount of grain available to wintering swans between years during the past five decades, and illustrates some previously unrecognized fluctuations in the winter food resources of the Centennial Valley flock (Figure 15).

Immediately after the Refuge was established in 1935, Manager Hull made two efforts to increase the trumpeters' winter food supply. In the 1936 Annual Operations Report Hull noted, "*As a winter source of food for the Trumpeter Swan 2,000 pounds of Wapato (Sagittaria latifolia) has been planted in the warmer spring heads where the Trumpeter Swan spend the winter and where the food supply is very scarce.*" In addition to this planting, Hull fed about 66 bu of wheat at Culver Pond.

During the winter of 1937 (1936-37), Hull increased the amount of wheat fed to about 166 bu, because he was convinced that the grain had significantly improved the swans' physical condition. In the 1937 Annual Operations Report he wrote, "*It has been clearly demonstrated that grain supplemental to the natural foods aids materially in keeping the swan in condition to protect themselves against predatory animals and starvation. Weak birds on being disturbed are subject to being preyed upon by coyotes since after alighting in the soft snow they are unable again to take to the air.*" It was obvious to Hull that when provided with ample food, the swans were quite active and regularly moved on and off the Refuge. Wintering numbers varied from 20 to 80, and Hull suspected that the swans flew east to open waters in Yellowstone National Park. Swan numbers increased on the Refuge in late February.

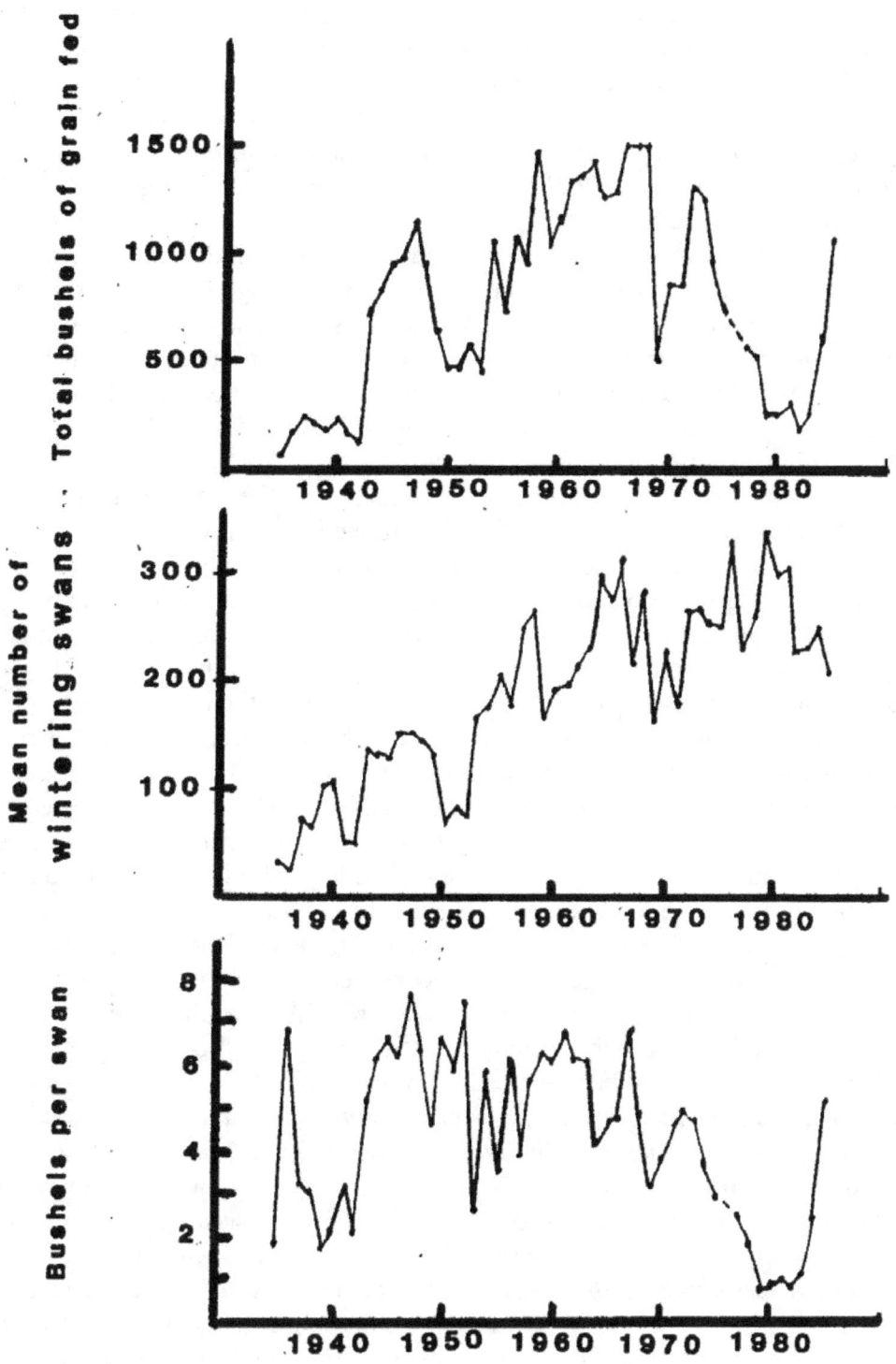

Figure. 15. Total bushels of grain fed, mean number of swans, and bushels fed per
swan at Red Rock Lakes NWR, 1935-86. Data are given in Appendix XVI.

During 1936, grain was scattered into the waters of Culver Pond from shore. In 1937 a rowboat was first used. The boat was filled with grain and rowed across the pond, while the grain was poured out in a long row. The rowboat, which was subsequently used for almost 30 years, was a very effective means of widely scattering the grain, but it was also very slow and hazardous, particularly when air temperatures were well below zero. The installation of a hopper on the rear of the boat in 1938 allowed the grain to be dispensed in piles. Hull noted, "*They [the swans] seem to pick it up better and are kept healthy by scattering it in small reefs rather than scattering it by hand. The past winter we did not lose a swan and all the birds seemed to come through in a healthy condition, which I believe speaks well for our feeding activities*" (RRLNWR files, letter to the Regional Director, 14 July 1938).

The total amount of grain fed annually remained fairly constant during the winters of 1938 to 1942, ranging from 166-233 bu. The amount fed per swan averaged 2-3 bu per winter except in 1937, when about 6.8 bu per swan were fed. Wheat was scattered from a rowboat two to three times per week, starting in mid-December or early January and continuing through March. Enough grain was provided at each feeding so that some remained uneaten at the next feeding. The start of feeding was delayed until the few swans that remained on the Refuge had eaten most of the available aquatic vegetation in the springheads.

During those first years of feeding, most swans usually left the Refuge for about six weeks after freeze-up in November. Hull presumed that they moved to open waters in the Island Park and Yellowstone areas. Swans made periodic forays to the Refuge in late December and as feeding began, approximately 44-66 centered their winter activities at the feed site by January. The number of swans on the Refuge increased rapidly each February during those early years. By late February, trumpeters grew restless and scattered over the Refuge. They fed at the spring head feeding site at night and sat on the snow on the marsh during the day.

During 1943, a change in personnel occurred and in the transition the feed was reduced by about 50% to 110 bu, with feeding occurring only once per week. The new Refuge Manager, W. Sharp, made several changes to the feeding program after 1943. He noted that in past years when feeding was delayed until late December, most of the swans left the Refuge at freeze-up. In order to prevent this movement out of the Centennial Valley, Sharp began to feed in mid-November, at or before freeze-up, and increased the amount of grain provided. Despite the unusual availability of grain at freeze-up in November 1944, most of the swans followed their established pattern of leaving the Refuge. By the second year of early feeding, however, the swans began to alter their movements and 96 remained through November 1945. By 1946, 155 remained in November and swan numbers changed little throughout the winter.

Sharp believed that earlier and heavier feeding was required to keep the swans in good condition. He wrote, "*As I see the picture the swan population will steadily increase since winter feeding starting in November has eliminated the starvation of the cygnets. Evidence obtained last year showed that the adults held the cygnets on the open water holes on the lakes until all froze over. As a result the cygnets lose weight and are too weak to leave the valley or drop down in the mountains. Last fall when the lakes froze over the swans went to Culver Pond. We had feed out and it was observed that about 70% of the cygnets had difficulty in taking off when we went to feed. Their wings were well developed and the cause was quite definitely one of weakness. After the cygnets were on the feed a short while this weakness disappeared entirely*" (RRLNWR files, letter to the Regional Director 21 June 1944).

Manager Sharp drastically increased the amount of grain fed from an average of 185 bu/winter in 1936-43, to an average of 922 bu/winter in 1944-47. The grain was scattered in water about 1 m (3-4 ft) deep to limit its availability to ducks. As a result of the earlier feeding, and the fivefold increase in the amount of grain fed, the number of swans that wintered at the Refuge increased from about 60 in 1943, to 130 in 1944, and to 170 in 1945. Despite the increased number of swans, the amount of grain provided per swan also increased substantially compared to earlier years, and ranged from 5.1-7.7 bu per swan.

Barley was added to the ration in 1943. The swans preferred wheat, however, and ate the barley only after the wheat was consumed. Problems with feeding barley were also noted in the 1960s. Refuge Manager O. Vivion (pers. comm.) observed that blood was evident in swan feces at the feeding ponds and he attributed the bleeding to the roughness of the barley hulls.

Manager Sharp noted that when the swans were provided with increased amounts of grain and kept in good physical condition, they were quite mobile and regularly flew between the Refuge and points to the southeast. Sharp was convinced that green vegetation was very important to the swans' winter diet and that the swans supplemented their grain diet with aquatic vegetation during their trips to Idaho. By feeding grain early and heavy, he thought that the swans would eat less of the available aquatic plants on the Refuge and thus the limited supply would last longer and provide a more balanced diet. The Regional Office, however, was under the impression that virtually no aquatic plants were available to the swans during the winter, and in 1944 discussed plans to truck in and store lettuce, cabbage, or celery to supplement the swans' diet. A plan was discussed to grow cabbage at Camas NWR, Idaho, and truck it to RRLNWR. Sharp discouraged this plan, pointing out that aquatic plants were in fact available at Culver Pond and Elk Springs, and with the supplemental grain the swans had a mixed diet available throughout the winter. He noted on 2 February 1944 that "*Elk Springs Creek is open and there is an abundance of aquatic vegetation including Zannichellia, Botrachium, and Potamogeton* (RRLNWR files, letter to the Regional Director 2 February 1944). When less grain was fed in 1941 and 1942, Hull had reported that the swans cleaned out the vegetation in the springs by February.

In 1948, the new manager, F. Ramelli, changed the procedures and did not begin feeding at freeze-up. Most of the swans left the Refuge by 10 November when the lakes froze, and Ramelli began to feed as the swans returned to the Refuge in late December. Confirming earlier observations, Ramelli also noted that when the swans were provided with large amounts of grain, they remained very active and mobile, regularly flying between the Refuge and points to the southeast. In the 1949 Annual Narrative Report, Ramelli observed, "*In spite of ample and regular feeding the trumpeters do not appear to depend upon this food to any extent. Even during periods of severe adverse winter weather, they come and go freely. It is interesting to speculate upon the location of other wintering waters upon which the swans are doubtless more dependent than upon our limited area*". We suggest the slightly different interpretation, that ample winter feeding maintained the swans in strong physical condition and allowed them to expend the energy necessary to fly to the Henry's Fork to supplement their diet with greens.

After resuming the pattern of movement out of the Centennial Valley at freeze-up in 1948, the swans completely left the Valley in the autumn of 1949 when grain was withheld. During the winter of 1950, Acting Manager W. Banko delayed feeding until 14 December 1949 because November and early December were relatively mild and open, and aquatic vegetation was available. Apparently the unusually mild weather in November and December 1949, coupled with the delayed feeding, resulted in the movement of trumpeters out of the Refuge to areas beyond their traditional range. An aerial survey of the Centennial Valley in early January 1950 found no swans at all; 80 returned later in the month. As discussed by Banko (1960:145) and as shown in Appendix III, a large segment of the Centennial Valley flock could not be found during the 1950 summer census. These birds apparently moved out of the region in 1950, however at least 55 returned before the 1951 census.

During the early 1950s, Manager Banko delayed the start of feeding until late December or early January and most trumpeters again left the Refuge at freeze-up and returned in late December or January. Banko noted in 1953 that only 12 swans remained on the Refuge by the end of December. Between 1950-54, the number of wintering swans on the Refuge prior to the late February influx was usually less than 100. Despite the growth of the Tri-state Subpopulation to peak numbers by 1954, during these years of delayed feeding, use of the Refuge by wintering swans was actually less than in the 1940s. A mean of 522 bu/winter was fed from 1950-54, a 44% reduction from the 1944-49 mean of 924 bu/winter. The amount of grain fed per swan varied considerably, from about 7.5 bu/swan in 1952-53 to 2.5 bu/swan in 1953-54.

Another factor closely linked to swan use of the Refuge grain during 1950-54 was the availability of alternate wintering habitat on the Henry's Fork River, 35 km (22 mi) to the southeast (Figure 3). Prior to 1951, January and February water releases at Island Park Dam were usually so low (Table 33) that much of the river was frozen for weeks at a time. Between November 1950 and December 1953, however, winter water releases were most often between 300 and 550 cfs, and these flows increased the amount of high quality, ice-free habitat available to the swans at Harriman State Park (Railroad Ranch). Releases at Island Park Dam were again shut off completely in January 1954, and probably as a result the number of swans on the Refuge increased to a record 250 by March.

MacDonald Pond was constructed in 1953 to increase the area of ice-free water on the Refuge, but little aquatic vegetation was present at first and most swans used the Culver Pond feed site. In 1955, the feeding strategy was again changed. Once again feeding began at freeze-up in November 1954 in another attempt to hold swans on the Refuge and prevent their movement to Idaho, where illegal shooting caused significant, losses. With the exception of the winters of 1956 and 1957, when weather-related travel difficulties delayed feeding until mid-December, this early feeding strategy continued until 1969.

From 1955-58, the amount of grain fed annually was higher than in the early 1950s and more swans remained on the Refuge in response to the early feeding. February swan numbers varied between 150-300. The amount of grain fed ranged from 703-1081 bu/winter and averaged 948 bu/winter. The amount fed per swan varied between about 3.5-6.0 bu. During the winter of 1956 when about 3.5 bu were fed per swan, some swans at the feed site acted weak and a few could not fly by late March. The cause of this weakness was never determined.

Banko observed that even when grain was available, the swans appeared to prefer aquatic vegetation when they could find it. In the 1955 Annual Narrative Report Banko observed *"Usually during the latter part of March or the forepart of April they desert the springheads entirely and forage tough rootstocks and other aquatic fare in the areas of open water which are beginning to collect around the edges of the lake and marsh."*

After 1957, over 200 swans were usually present at the wintering ponds, and utilization of the limited aquatic vegetation increased. On 30 January 1959, Refuge Biologist Hanson noted, *"The bulk of the green feed left in the two feeding ponds is under the ice and not available to the swan. The open water areas are comparatively bare of natural aquatic plants which indicates the over-use of these areas during the past years."*

Table 33. Mean daily water release (cfs) at Island Park Dam , Henry's Fork of the Snake River, Idaho. Data were compiled from the records of daily water releases, provided by the Bureau of Reclamation

Year	November	December	January	February	March
1938-39	232	2	6	8	9
1939-40	221	20	7	281	130
1940-41	204	5	7	8	10
1941-42	196	6	7	98	28
1942-43	208	7	8	310	354
1943-44	261	10	296	454	450
1944-45	251	7	9	24	302
1945-46	211	9	201	424	106
1946-47	260	407	8	11	198
1947-48	236	148	83	353	411
1948-49	230	78	6	170	488
1949-50	230	8	9	12	198
1950-51	251	295	359	447	451
1951-52	548	562	557	547	546
1952-53	264	123	461	491	503
1953-54	266	180	10	13	187
1954-55	236	108	10	26	342
1955-56	38	240	43	223	351
1956-57	616	148	67	374	479
1957-58	163	376	244	250	387
1958-59	50	135	10	94	171
1959-60	7	328	10	10	289
1960-61	213	30	12	13	15
1961-62	216	142	11	13	14
1962-63	217	150	15	174	403
1963-64	252	140	47	14	42
1964-65	277	117	266	276	295
1965-66	792	511	221	229	329
1966-67	421	303	8	12	14
1967-68	287	425	367	283	486
1968-69	180	517	526	601	610
1969-70	266	152	78	306	490
1970-71	318	244	449	568	335
1971-72	799	474	443	424	453
1972-73	796	504	400	496	254
1973-74	537	354	268	318	435
1974-75	579	623	616	376	535
1975-76	572	562	656	572	695
1976-77	322	459	361	235	282
1977-78	0	0	0	156	252
1978-79	277	389	384	502	496
1979-80	2	112	124	124	86
1980-81	176	209	326	316	288
1981-82	88	91	167	234	216
1982-83	572	500	504	561	569
1983-84	389	560	688	661	569
1984-85	141	328	540	575	521
1985-86	588	472	493	475	476

The Refuge staff noted in the 1960 Annual Narrative Report that the aquatic vegetation at Culver and MacDonald Ponds was almost completely consumed by the middle of January. The fluctuating availability of Henry's Fork habitat probably contributed to the increased swan use of the Refuge.

In 1958, Refuge Manager C. Markley increased the amount of wheat and barley provided, and continued to start feeding by freeze-up each year. He wrote to the Regional Director on 7 January 1959 to request that more grain be sent to the Refuge immediately. He explained "*Our swan feeding program has more than doubled and it is our intent to have the birds in a healthy and vigorous condition at the time of the spring nesting period. Increased nesting was noted on the refuge last year, and we would like to encourage this trend to the limits of available nesting territory.*" The Regional Office sent additional grain, but expressed concerns about potential overfeeding.

From 1958-66, the amount of grain provided remained relatively high and constant, with a range of 1,229-1,497 bu/winter and a mean of 1,294 bu/winter. The feed was dispensed twice a week from a rowboat. During this period the purpose of feeding was understood to be twofold: to offer supplemental nourishment during the critical winter months, and to hold the swans in the protection of the Refuge where losses to illegal shooting could be prevented. This deliberate effort to prevent the fall movement of swans out of the Centennial Valley was successful: by the mid-1960s, about 200-250 swans regularly remained on the Refuge in November.

By 1964, Refuge personnel were expressing dissatisfaction with the slow and hazardous method of scattering grain from the rowboat, and proposed to develop an automatic feeder, consisting of a 1,000 bu storage bin. "*Grain dispensing would be accomplished simply by having a 10 or 12 inch galvanized pipe going from the bottom of the bin to the water's surface. The grain would simply run into the water and would shut off as the pile built up. A slip plate would be used for cutting off the grain flow when needed*" (RRLNWR files, 26 February 1964 letter to Regional Director). This feeder design was never implemented.

During 1964 and 65, "fake feeders", consisting of lumber, old doors, and clean oil barrels were placed at both Culver and MacDonald Ponds to determine if the wintering swans would tolerate strange structures. Grain was fed in the water near the structures and the swans seemed to accept them within a few days.

In 1965 and 1966, 60 bu feeders, patterned after those in use at Lacreek NWR, were constructed and one wooden and two steel feeders were placed at MacDonald Pond. No analysis of the feeders' performance was found in Refuge files. E. McLaury, former RRLNWR Refuge Biologist, (pers. comm.) recalled that alfalfa pellets, rabbit pellets, turkey starter, and grain were fed experimentally in 1966 and that icing of the feeders frequently made the food unavailable to the swans.

For 30 years, several generations of trumpeters had fed on grain that was scattered broadly across the surface of the feeding ponds by rowboat at least twice weekly. After 1966, a variety of feeding methods was used. These changes in methods were less hazardous and required less manpower, but we conclude that they tended to constrict the area into which the grain was dispersed, and likely reduced the availability of grain to subordinate swans.

In 1966 and 1967, gravity-flow bin feeders were built at Culver and MacDonald Ponds. These 1,000 bu storage bins each had a swinging spout, 10.7 m (35 ft) long, that dispensed the grain in an arc over the water or into feeders. The amount of grain fed from 1966-69 was not well documented, but appears to have been at least 1,500 bu/winter. With the use of the new feeders, however, the frequency of feeding was reduced to once per week in 1968, and the staff experimented with feeding intervals of one to two weeks during 1969. Also in 1968, only barley was fed instead of the usually wheat, or wheat/barley blend.

Coinciding with the feeding of only barley, and the reduction to one feeding per week in the winter of 1968, the number of adults in the Centennial Valley flock abruptly declined from 327 in September 1967 to 234 in September 1968. As discussed in Chapter 3, this decline occurred within the subadult component, rather than among established nesting pairs. Although cause and effect cannot be proven, we suggest that this unprecedented loss was indeed caused by an abrupt drop in the winter carrying capacity of the Refuge as set by the availability and suitability of grain.

During the winter of 1970 an attempt was made to abruptly curtail the winter feeding program. Assistant Manager R. Papike described the results, *"We delayed feeding as long as possible, hoping that, if grain was not available at the refuge, the swans would move to the open water in the Island Park Area, Yellowstone Park, and along the Madison River where natural feed abounds. The birds refused to leave, however."* (RRLNWR files, letter to the Regional Director, 4 April 1970).

Weak swans were seen by late December and two that were too weak to fly were captured and died on 6 January. By February most of the available natural feed was consumed. Ten more dead swans were found in February and many were too weak to fly. The physical condition of the wintering swans was declining rapidly and Refuge staff feared that more would soon die; feeding began on 16 March. The total number of swans dying during this winter and following spring was not documented because Tri-state surveys were not conducted in 1969 or 1970.

Papike concluded, *"There appears to be a basic refuge wintering swan population that can be correlated with the valley nesting population and its young of the year. These birds show a strong instinct to winter in the valley even if it means starving to death. There is probably enough natural vegetation in both MacDonald and Culver Ponds to provide feed for the swans until sometime in January, depending on the weather. Supplemental feeding, therefore, should be delayed until most of the natural feed is used or is unavailable. If the winter supplemental feeding program is completely eliminated the Centennial Valley nesting population will probably be reduced by over half of what it is now."* (RRLNWR files, 4 April 1970 letter to the Regional Director).

With the benefit of hindsight, we conclude that this attempt to abruptly terminate the grain feeding program was absolutely doomed by the management strategies of the preceding fifteen years. It was not any "strong instinct" that caused the trumpeters to remain in the Centennial Valley after freeze-up. Until the mid-1950s, the Refuge swans showed strong tendencies to move out of the Centennial Valley at freeze-up, if grain was not provided. From 1955-68, however, managers began heavy feeding at or before freeze-up in November, and several generations of Refuge swans were deliberately trained to remain on the Refuge. Their earlier pattern of movement out of the Centennial Valley was intentionally, and quite successfully, disrupted. By 1969, over 250 swans had been taught to stay on the Refuge in November. Even though feeding did not begin in November 1969, 150-200 swans remained on the Refuge, gradually becoming so weak that they were physically incapable of leaving.

Feeding resumed on a once per week schedule in 1971, starting in mid-December, and water levels in Culver and MacDonald Ponds were drawn down to make more vegetation available early in the winter. About 850 bu were fed annually in 1971 and 1972.

The frequency of feeding was doubled in 1973, and the amount of grain fed was increased to 1,300 bu as part of an experiment to determine if a relationship existed between winter feeding and reproductive success. Page (1976:103) concluded "*After winter feeding was returned to a schedule of twice weekly, increases were seen in the number of nests, the percentage of eggs hatching, the total production of cygnets, and the percentage of cygnets surviving through the summer.*"

From 1974-79, the amount fed was rather inexactly estimated by timing the number of minutes that grain flowed through the spout from the gravity-flow feeders. Feeding occurred twice per week with quantities declining from about 1,250 bu in 1974 to 500 bu in 1979.

In 1978-80, floating and standing feeders were used on an experimental basis, and additional grain was poured directly into the water. Over-water feeders used in 1978 and 1979 were ineffective because of severe icing problems. A barrel feeder, set flush to the water, was tried in 1980, but water slopped into the barrel and wet the feed. Over-water feeders were abandoned because ice reduced the availability of the feed and increased labor costs.

Dry-land feeders were used in 1981 and about one ton of Ralston Purina commercial turkey finisher #8210 was mixed with the wheat in the feeders. Wheat was also delivered directly into the water. Swans seemed to accept the dry-land feeders fairly quickly; approximately 30 were observed feeding at one time. Groups of swans waited their turn to move up to the feeders (R. Sjostrom, pers. comm.).

Feeding was delayed until 8 January 1981 in order to reduce the number of ducks in the area. Duck use of the winter grain has always been a concern to Refuge managers. Banko and others noted that if feeding was delayed until late December, many ducks left the area. In some years approximately 1,500 mallards and goldeneyes were estimated to winter with the trumpeters. The total amount of grain that they consumed was undetermined.

The total amounts of grain/turkey finisher fed annually during 1980-84 were the lowest since 1943, and the amounts fed per swan were the lowest in the history of the feeding program. The amounts apparently were reduced because much of the feed was placed in dry-land feeders. Although feed was added as needed and appeared to always be available to the swans, they consumed much less than in previous years when the grain had been distributed into the water.

Several changes were made in the feeding program in 1984 and 1985 because of concern that the extremely low amounts of feed provided during the early 1980s had increased mortality rates in the Centennial Valley flock and contributed to its decline and poor productivity. In addition to increasing the amount of feed provided, the Refuge staff began experimenting with techniques to distribute the grain more widely. An electric powered spreader attached to the spout of the gravity-flow feeder at Culver Pond increased the distribution radius of the grain by about 10 m (30 ft) and a rowboat was used at MacDonald Pond to distribute grain in 1986. The suitability of several commercial feeds was also tested in the dry-land feeders (C. Young, pers. comm.). The Refuge staff are closely observing swan use of the feed and reassessing the merits of the various feeding methods.

From our review of the winter feeding program, we conclude that without the supplemental grain, few if any trumpeters could regularly winter in the Centennial Valley. The presumption that the amount of grain fed increased as the number of swans increased is not true. In reality, the amounts fed have fluctuated with changes in managers and, in effect, have caused the winter carrying capacity of the Refuge to vary. Our analysis (Chapter 5) showed a highly significant negative correlation between the rate of loss of swans from the Red Rock Lakes and Centennial Valley flocks, and both the total amount of grain fed per winter ($P = 0.036$) and the amount of grain fed per swan ($P = 0.016$). Changes in the timing of feeding have been used deliberately to reduce the autumn movement of trumpeters out of the Centennial Valley and have influenced the number of swans wintering on the Refuge. Changes in the methods used to distribute grain probably have decreased the availability of grain to subordinate swans in recent years. The extremely low amounts of grain provided in the early 1980s probably attracted swans to the feeding ponds but provided many of them with a diet that was inadequate for winter survival.

Hebgen Lake

Hebgen Lake was formed in 1915 by impoundment of the Madison River, 29 km (18 mi) northwest of West Yellowstone, Montana. The lake has become an important wintering site since 1982 (Figure 16). Between 1975-81, the Tri-state Midwinter Surveys found an average of 50 trumpeters on Hebgen Lake annually. The number of wintering swans rose to 145 in 1982, and averaged 187 from 1982-86. Use apparently dropped by about 50% in 1987, when only 92 swans were found. We could not determine whether this decline in use in 1987 was due to habitat changes at Hebgen Lake or other factors. The relatively mild winter may have allowed swans to move from Hebgen Lake to other ice-free areas which normally would have been frozen.

Although most of the lake freezes in winter, open water remains in the Madison Arm, the South Fork of the Madison Arm, and Grayling Arm. The only data on swan use of these sites are the results from the USFWS Midwinter Surveys. No information exists regarding the food resource or the influence of snowmobiling or waterfowl hunting on swan use of the area.

Madison River Valley

Small groups of swans occasionally winter on or near Ennis Lake, and on the lower reaches of the Madison River. Prior to 1980, USFWS Midwinter Surveys found less than 20 swans in these areas. However, use increased between 1980-85 when an average of 46 swans was censused, with a high of 72 swans in 1984. Since 1984, swan use has declined to 18 in 1985 and only 4 in 1987.

Idaho

Henry's Fork of the Snake River and Tributaries

With the rapid increase of the Tri-state Subpopulation by the late 1940s, the importance of the Henry's Fork and its tributaries as a wintering area became apparent to wildlife managers. Although trumpeters also winter on the Buffalo River, Sheridan Creek and Reservoir, and the Henry's Fork above Island Park Reservoir, the most heavily used habitat is a 14.4 km (9 mi) stretch of the Henry's Fork from the village of Last Chance, downstream to the Pinehaven summer home development. The core of this wintering area is Harriman State Park (HSP), previously the Railroad Ranch, which was donated to the State of Idaho by the Harriman family in 1977 (G. Eyraud, pers. comm.). Banko (1960:56) described the area in detail.

An aerial survey by the Idaho Department of Fish and Game on 3-7 February 1950 located 262 swans in the Island Park area, of which 208 were in the HSP vicinity. An estimated 200 trumpeters were also present at RRLNWR in February 1950. The combined total of swans at RRLNWR and on the Henry's Fork (462) indicated that these two areas wintered virtually the entire Tri-state Subpopulation (424 swans on 3 August 1949), plus an unknown number of migrant Canadian trumpeters.

Since the late 1940s, the Henry's Fork has been censused more frequently than any other of the Tri-state wintering sites. Both aerial and ground censuses have been conducted, with ground censuses being made almost weekly in recent years by HSP employees. These counts have shown that the number of swans at the Park varies considerably throughout the winter, as the availability of ice-free water elsewhere varies due to changing weather conditions. In order to examine long-term trends in use on the Henry's Fork, we plotted the annual peak swan counts (Figure 17). These counts show that prior to 1968, annual peak swan numbers at HSP ranged from 50-330 swans, while annual peak counts for the entire Henry's Fork drainage ranged from about 130-430. Few surveys were conducted between 1968-72. After 1968, swan use increased rapidly at HSP to some 450 by the mid-1970s. Since the mid-1970s, peak counts in the Henry's Fork drainage have varied between 400-650, with an increasing proportion of swans using sites outside of HSP.

Since the 1940s, RRLNWR managers have noted the movement of swans between the Refuge and Island Park. In 1958 the Refuge staff noted, "*Observations were made during the winter of swans flying into Red Rock Lakes Refuge from Henry's Lake and over Hell Roaring Canyon and swan counts indicate that the birds move freely between the two areas. It appears from observations and reports that during good weather the birds stay on the refuge but prefer the Island Park area in the face of a severe storm*". (RRLNWR Ann. Narr. Rept. 1958).

Manager Banko urged the development of a winter refuge on the Henry's Fork (Banko 1960:186) and Refuge staff suggested moving the winter feeding operation to the Railroad Ranch and closing RRLNWR during the winter (RRLNWR files 7 February 1963). Although these recommendations were not followed, this key wintering site most fortunately was protected as a bird sanctuary by the Harriman family, and protection from hunting and other human disturbance has continued under management by the State of Idaho.

Water Flows on the Henry's Fork. The availability of winter swan habitat at Harriman State Park is determined by the extent of ice cover on the river. Ice formation is affected by several factors, including air temperature, water temperature and volume of water flow. Historic winter flows through Harriman State Park, before the construction of Island Park Dam, would have been about 700 cubic feet per second (cfs), assuming flows similar to the present 150-200 cfs flow from the Buffalo River and 525 cfs inflow into Island Park Reservoir (memo in RRLNWR files, 10 November 1982).

Island Park Dam was completed in 1938 by the Bureau of Reclamation to provide storage and regulation of irrigation water for agricultural lands in Fremont, Madison, and Teton counties. Island Park Reservoir has a capacity of 135,000 acre feet and a surface area of over 8,000 acres (L. Busch, pers. comm.). The impoundment of the Reservoir flooded the Shotgun Creek area, which had been a preferred winter swan habitat in the 1920s and 1930s (Banko 1960:56). In 1958, the foreman of the Railroad Ranch reported that trumpeters had wintered in the water courses inundated by the Island Park Dam, and only moved to the Ranch area after the dam was built (memo in RRLNWR files, 19 February 1958). This increase in swan use below the dam also corresponded to the period of increase in the flock at Red Rock Lakes in the 1940s, when the presence of swans became more apparent to casual observers.

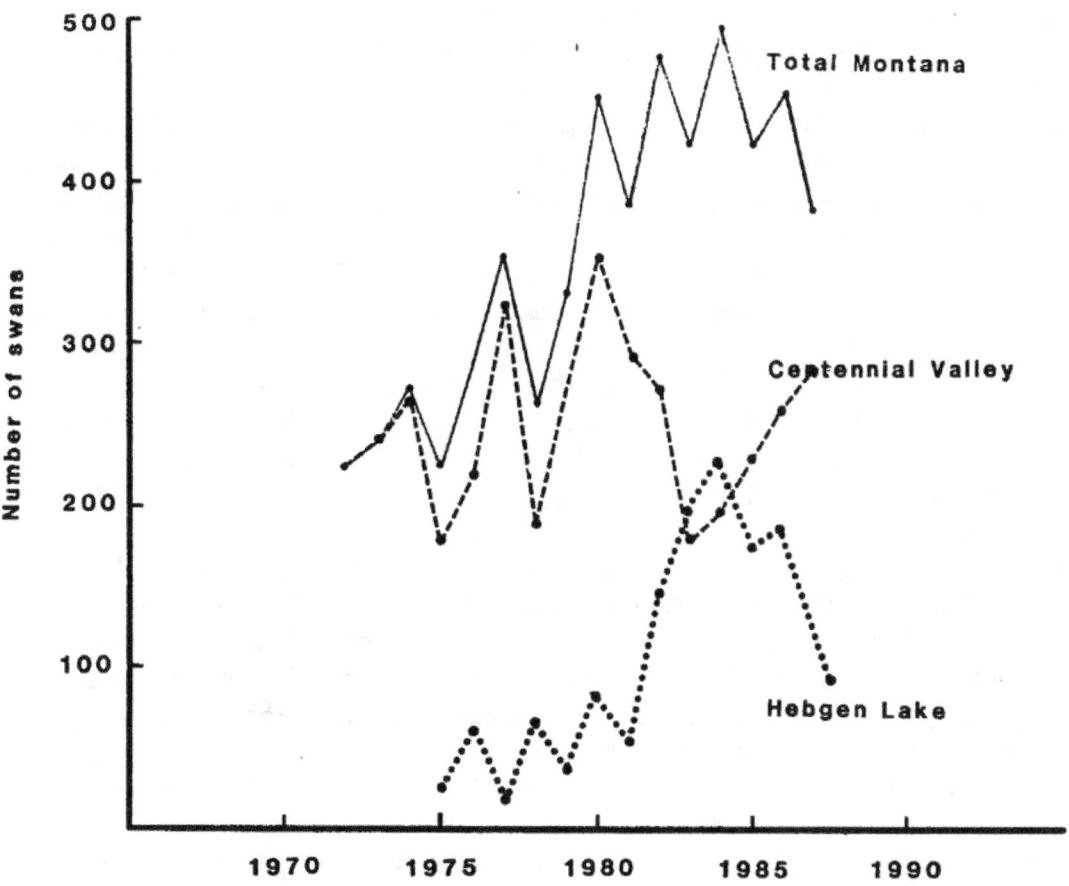

Figure 16. Trumpeter Swans wintering in Montana 1972-87. Data are from the USFWS Midwinter Trumpeter Swan Surveys.

Preliminary results from on-going studies of ice formation at Harriman State Park suggest that more ice-free habitat exists today than existed before the construction of Island Park Dam. Heat loss from water in the Reservoir is reduced by the surface layer of ice and snow cover, and in late autumn the waters become stratified, with the warmest water at the bottom of the Reservoir. Water released from Island Park Dam is drawn from the bottom of the Reservoir and in winter has a constant temperature of approximately 3 °C (37 °F) (G. Smith, pers. comm.). When the amount of water released approximates the natural flow of some 525 cfs, it prevents or reduces ice formation downstream from the dam.

Operational procedures for the dam were set forth in a 1935 contract which allowed the Bureau of Reclamation (USBR) to purchase winter hydropower water rights and store the water in the Reservoir. This purchase of hydropower water rights was necessary in order to permit the storing of winter water flows for summer irrigation by the Fremont Madison Irrigation District. The contract allowed the USBR to completely shut the gates at Island Park Dam from 15 November to 1 May of each year, until the reservoir filled and spilled (L. Busch, pers. comm.). In actual practice, the gates were rarely closed for the entire period allowed, but drastically reduced flows were common for two to four months in most years prior to 1968 (Table 30).

In 1967-68, the basic "fill and spill" procedure was informally modified to better provide for flood control, power production, and fish and wildlife habitat enhancement. Total closure of the gates became much less frequent. The modified operation provided increased water flows during most winters except 1977-78, 1979-80, and 1981-82 (L. Busch, pers. comm.).

In order to achieve a more reliable water supply for winter releases, an amended plan was negotiated in 1984. This plan called for the adverse storage of fall hydropower water rights (before 15 November), and the release of such water prior to 1 March. Releases would be made in a manner that would meet the winter energy loads of Utah Power and Light and as much as possible maintain an ice-free section of the river below Island Park Dam. Because winter power production is more valuable than fall production, this agreement benefited the hydropower users and simultaneously lessened the danger of freezing of swan habitat below the dam (L. Busch, pers. comm.).

Repairs to the dam's bathtub spillway necessitated the closure of the gates in 1979-80. In an attempt to prevent freezing of winter swan habitat at HSP, a siphon with a capacity of 125 cfs was installed to maintain partial flows into the Henry's Fork. With reduced flows in 1979-80, the river was shallower and slower moving through the Park. Much of the River was frozen throughout the winter and the number of wintering swans dropped below 70 during extreme cold weather in February. Although the River reopened by mid-February, no more than 30 swans were found in the HSP area (Hampton 1981).

In March 1987, the Idaho legislature established a legal minimum flow for the Henry's Fork, of 300 cfs measured below the confluence of the Buffalo River. Past observations have shown, however, that a minimum flow of this magnitude is insufficient to prevent the freezing of the feeding areas at HSP. Studies began during 1986-87 to determine the relationships between water releases and ice formation. This information will be used by the USBR to modify their operations, within the constraints of other water obligations, to minimize freezing at HSP. As was shown by the reduced releases in 1986-87, however, in years when water supplies are low no management guarantee exists to ensure that water releases will be adequate to maintain the availability of the HSP habitat.

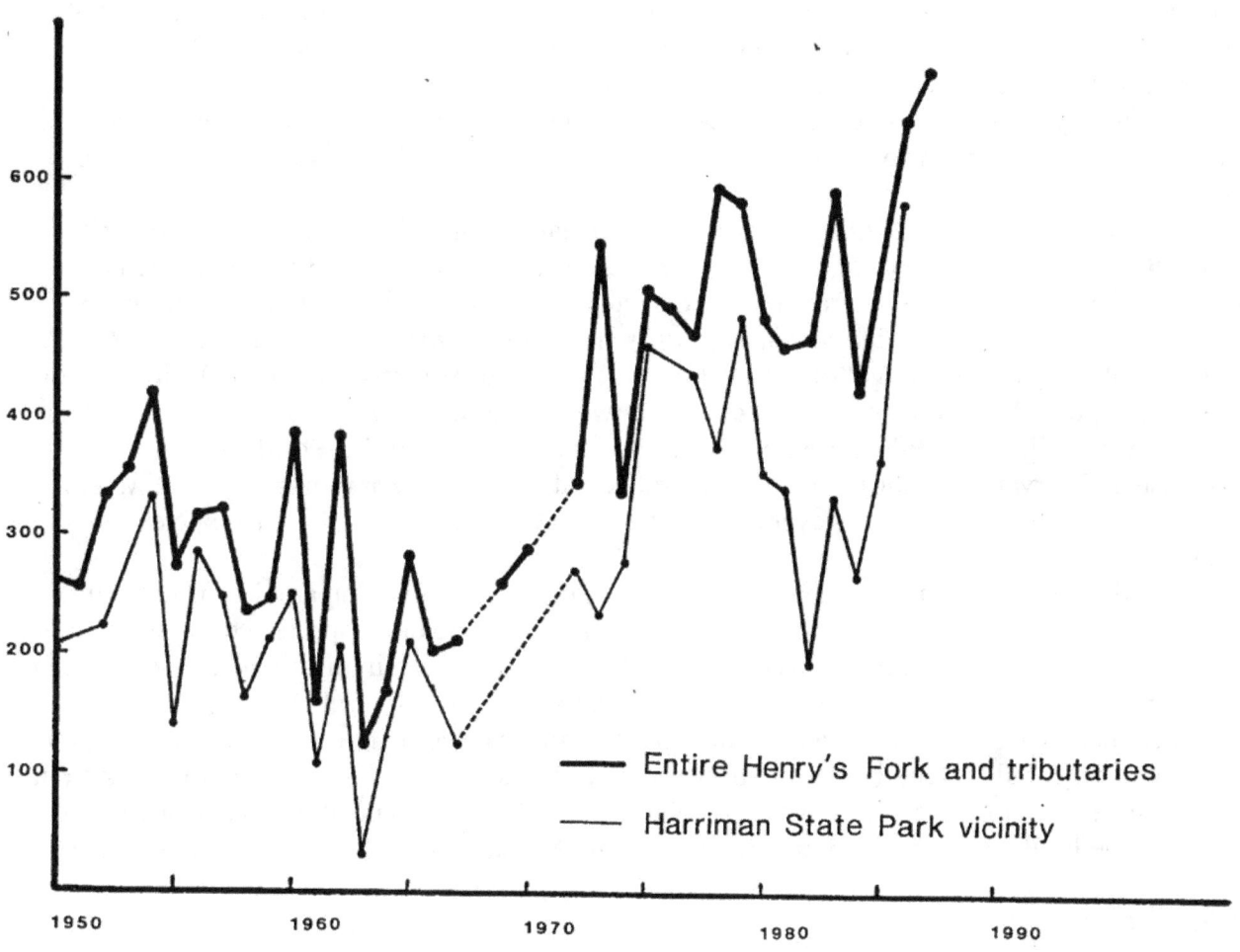

Figure 17. Maximum annual winter swan counts, 1950-87.

Aquatic Vegetation at Harriman State Park. Observations of the aquatic plants from Last Chance to Pinehaven were made in 1958 by Hansen (1959), in 1977 by Shea (1979), in 1979 and 1980 by Hampton (1981), and in 1986 by Angradi (1986). All workers determined species occurrence and percent composition by wet weight. Sampling sites were not identical, but in each study they were laid out to be representative of the wintering area. All measurements were made in October just prior to the arrival of wintering swans. Results are summarized in Table 34.

Overall, between 1958-79 species diversity appears to have decreased, while _Elodea_ and _Myriophyllum_ increased, and sago pondweed remained fairly stable. In 1986, using methods similar to Hampton (1981), Angradi (1986) detected considerable change in the community composition. Two species, _Ranunculus aquatilis_ and _Zannichellia palustris_, accounted for 36% and 10% of the total wet weight of the 1986 samples. These species were not found in 1979 or 1980. Hampton (1981) reported that _Elodea_ and two species of _Potamogeton_ made up 74% of the total wet weight. In 1986 these same species accounted for only 11% of the total weight. Our casual observations indicate that extensive beds of _Myriophyllum_ remained in the waters of HSP at winter's end in 1987, however the nutritional value of this species for trumpeters is unknown.

While the various sampling methods used are not statistically comparable, the data suggest that _E. canadensis_ and _P. pectinatus_, both of which are suspected to be preferred by trumpeters (see Chapter 10), have declined in recent years. If such a reduction in the abundance of preferred plant species has actually occurred, the carrying capacity of this wintering site for swans may have declined. Changes in the plant community could be caused by many factors (see Chapter 7) and potentially could affect the trout fishery on the Henry's Fork. Without a better understanding of ecological relationships between aquatic plants, fish, sediment deposition, water flows, and swans in the HSP area, we cannot predict whether these changes might be beneficial or adverse.

Teton River

Wintering swans have been observed on the Teton River, in the Teton Basin area between Driggs and Victor, Idaho, since at least 1949. On 11-14 January 1949, Idaho Department of Fish and Game (IDFG) Officer A. Misseldine recorded fifty swans of undetermined species in Teton Basin. In January and February 1959, IDFG officers reported 12-16 wintering trumpeters. Subsequent IDFG aerial surveys in late December and early January 1960-71 found no wintering swans (IDFG files, Region 6, Idaho Falls).   Several tributary streams in Teton Basin are intermittent because of underlying porous glacial till. Streams draining the Teton Mountains to the east and the Big Hole Mountains to the west reappear in numerous springs adjacent to the River. These springs reduce the severity of ice formation and create areas of open water except during the most severe weather. At an elevation of approximately 1,850 m (6,000 ft) the most heavily used area is a 5.6 km (3.5 mi) stretch of river from the mouth of Fox Creek to the mouth of Teton Creek.

Since the inception of the USFWS Midwinter Trumpeter Survey in 1972, these February flights have annually found trumpeters in Teton Basin. The number of wintering swans has fluctuated widely, partially at least in response to the extent and duration of ice. Severe prolonged cold can reduce the area of ice-free water to only a few acres at the mouth of Fox Creek and cause most swans to leave the area.

Table 34. Aquatic vegetation at Harriman State Park. Data represent percent of wet weight biomass. The studies used varying methods and sample site locations.

Species	1958[a]	1977[b]	1979[c]	1980[c]	1986[d]
Elodea canadensis	4	35	24	13	1
Myriophyllum exalbescens	7	23	25	25	43
Potamogeton pectinatus	40	32	33	34	9
P. richardsonii	10	3	17	27	1
Callitriche verna	1	1	1	1	0
Ranunculus aquatilis	13	1	0	0	36
Sagittaria	9	0	0	0	0
Najas flexilis	16	0	0	0	0

[a] Hansen (1959)
[b] Shea (1979)
[c] Hampton (1981)
[d] Angradi (1986)

Prior to 1978, about 30-50 trumpeters were counted annually in Teton Basin (Figure 18). The number of wintering swans rose to about 100 in 1978-80, but few swans were found in 1981 when the area was almost totally frozen. Since 1982, winter swan numbers have increased dramatically to a peak of 499 in 1986. In addition to the rapid increase total numbers, swan use of Teton Basin is characterized by the presence of a high proportion of cygnets. From 1978 to 1987, the ratio of cygnets to adults has averaged 44/100, almost twice the ratio observed among the wintering population as a whole. As we will discuss in Chapter 9, marked swans from the NWT and Grande Prairie flocks frequent this wintering site, while very few marked Tri-state trumpeters have been observed.

No quantitative studies of the aquatic vegetation have been conducted, however a preliminary reconnaissance was made in August 1985. Substrates and major beds of vegetation were mapped and the habitat was described as follows: "*Most of this section of the Teton River bottom was covered with abundant sago pondweed (Potamogeton pectinatus) with white water buttercup (Ranunculus sp.) only dominating for a short distance above Fox Creek's junction. Almost bare cobblestone bottoms were observed immediately below White Bridge and in the Big Bend downstream from Fox Creek. The east bank was eroded by excessive cattle use from Darby Creek to Teton Creek with a 20 foot wide strip of more than 6" of silt and no vegetation from Dick Creek to Teton Creek. Siltation in midstream is most detrimental to winter-feeding swans because this is often the only part with open water*" (Report to IDFG, Region 6, Idaho Falls by Harold and Ruth Burgess, 1985).

The Teton River wintering area is bordered by private lands. As of 1987, Teton County had not adopted any planning or zoning ordinances pertaining to subdivision and development of the lands adjacent to the river. Lands on the east side of the river remain as undeveloped pasture because water tables are high, but scattered residences overlook the river from the west bluff. Without protection of this wintering site, increased riverbank development and human disturbance are inevitable.

In 1987, IDFG officers observed swans wintering along the Teton River in the vicinity of the failed Teton Dam. This area was included in the USFWS Midwinter Trumpeter Swan Survey for the first time in 1987 and 51 swans were found. Due to the unusually mild weather during the winter of 1986-87, more ice-free areas existed on the Teton River than usual. The suitability of this area to support wintering swans in an average winter is unknown.

Wyoming

Yellowstone National Park

Ground counts of the accessible wintering sites in YNP have been made in most winters since 1937-38, when 51 swans were counted. Observed swan numbers were usually less than 70 although occasional higher counts (125 in 1938-39, 92 in 1970) were reported (YNP files: Park Superintendent Annual Reports). Since 1978, inclusion of Yellowstone in the USFWS Midwinter aerial survey has provided much more thorough coverage of this remote area (Figure 19).

The ice-free waters of the Madison River usually winter from 30-70 swans, although none were present during the 1987 survey. Above Madison Junction, a few scattered swans are often found on the Gibbon and Firehole rivers. Even in the harshest winters, ice-free feeding areas

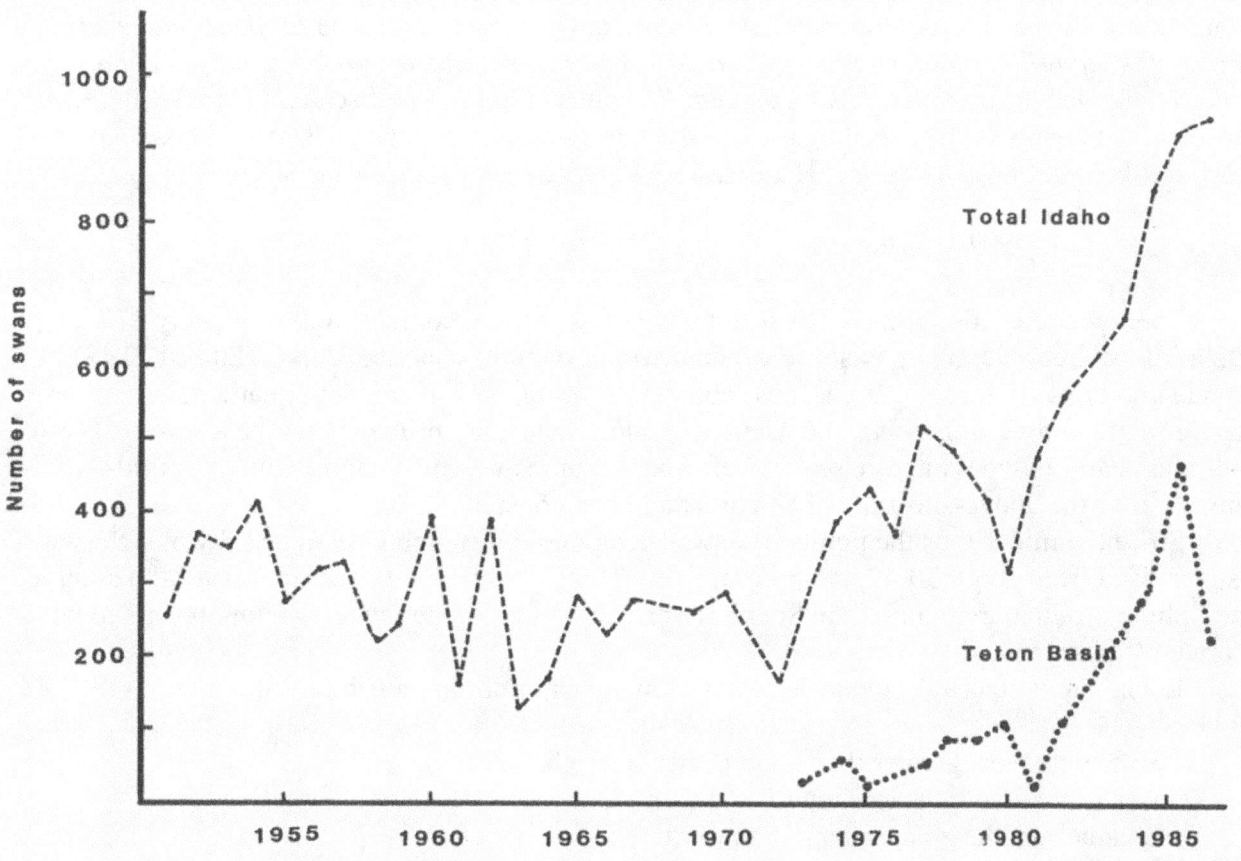

Figure 18. Trumpeter swans wintering in Idaho, 1950-87. Data are from USFWS Midwinter Trumpeter Swan Surveys and IDFG files.

160

remain widely available on the Madison and Firehole rivers due to the infusion of runoff waters from the Firehole River geyser basins.

The availability of habitat on the Yellowstone River from Yellowstone Lake outlet to the Alum Creek area varies greatly depending on ice conditions. In the very mild winter of 1980-81, much of the river was ice-free and 269 swans were censused. More commonly, extensive freezing results in less than 50 swans remaining. Shea (1978) snorkeled through the primary swan feeding sites and found the major species of aquatic vegetation to be *Elodea canadensis*, *Ranunculus aquatilis*, *Potamogeton richardsonii*, *Callitriche hermaphroditica* and *Eleocharis sp.*

Trumpeters also winter in pairs and small groups at many small scattered areas where warm springs create ice-free feeding areas. Frequently used sites include Squaw, Shoshone, Heart, Beula, and Lewis lakes, and Heart and Lewis rivers.

Lower Elevations of Wyoming

The lower elevation, Snake River drainage of Wyoming contains approximately 317 ha (792 a) of ice-free river and palustrine wetland winter habitat. About 37%, or 120 ha (298 a) provide good quality foraging areas, available winter-long. The dominant aquatic macrophytes include sago pondweed, *Elodea*, and *Myriophyllum*. Swans fed primarily on the above-substrate plant mass early in the winter. Use of tubers and rhizomes was most significant after mid-January, after the above-substrate plant parts had been consumed.

Since the mid-1970s the number of swans counted during the USFWS Midwinter Surveys has increased from about 30-42, to approximately 100 (Figure 19). Based upon 1982-86 aquatic macrophyte production and use, the Snake River, Wyoming, winter swan habitat appears to be saturated (Lockman and Brandt 1987).

Lockman *et al.* (1987) identified the following as important attributes of winter swan habitat:
1. Soft substrates, greater than 5 cm (2 in) in depth.
2. Winter water depths less than 1.3 m (4 ft).
3. Channel widths greater than 15 m (50 ft).
4. Stream or pond banks with sparse or no shrub and tree canopy to obscure vision.
5. Loafing sites consisting of water less than 10 cm (25 in) deep or gravel bars, located in close proximity to feeding areas.
6. Areas of water greater than 100 m (330 ft) in length or width.
7. Absence of physical barriers, such as fences and powerlines, crossing feeding areas or travel corridors.
8. Beds of diverse aquatic macrophytes in areas that rarely freeze for more than two or three days at a time.
9. Water velocities not exceeding 45 cm/sec (1.5 ft/sec).

Lockman and Brandt (1987) also identified about 37 ha (150 a) of habitat in the South Park area near Jackson, Wyoming, that could be developed for to produce winter forage. About 62 ha (250 a) of potential habitat exist in the Salt River drainage and opportunities to improve the suitability of this area are currently being examined by the Wyoming Department of Game and Fish.

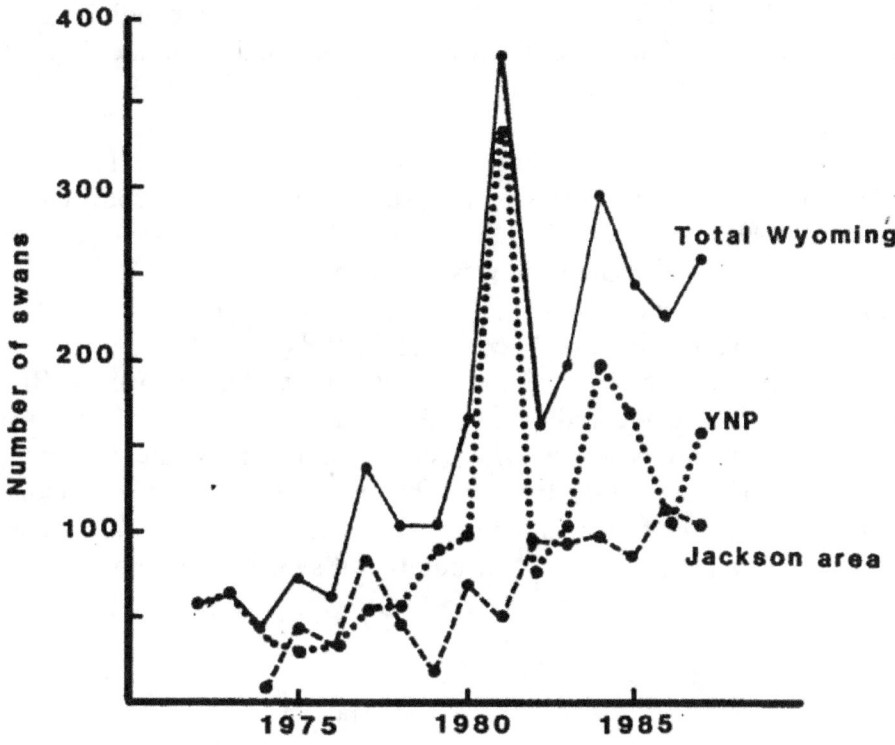

Figure 19. Trumpeter Swans wintering in Wyoming, 1972-87. Data are from
the USFWS Midwinter Trumpeter Swan Surveys.

Recent Changes in Winter Distribution

As the number of wintering swans has increased, several shifts in use have been apparent from the data gathered on the USFWS Tri-state Surveys. These are:

1) Following the increase in winter water releases from Island Park Dam in 1967-68, the number of swans wintering at HSP approximately doubled. When HSP was extensively frozen, or in unusually mild winters when other areas were ice-free, swan use increased on other stretches of the Henry's Fork. Following this increase of wintering swans on the Henry's Fork, use of other Tri-state wintering sites also began to increase by the late 1970s.

2) Since about 1979, the number of swans wintering in YNP and the Jackson Hole area has roughly doubled. Use of the Yellowstone River varied greatly depending upon ice conditions. During the mild winter of 1980-81, a record 332 swans remained in the Park. A few swans have moved south and southeast of Jackson to winter on tributaries of the Salt and Green Rivers (Lockman and Brandt 1987).

3) Prior to 1978, less than 50 swans were typically found on the Teton River. The Teton Basin area has since become a major wintering site, with maximum counts averaging 200 since 1982 and a record 499 swans counted in February 1985. In 1987, use of downstream stretches of the Teton River near the Teton Dam site was also documented.

4) Prior to 1981, Hebgen Lake wintered an average of 50 swans. Since 1981, counts have averaged 187 swans, with a high of 229 censused in 1984.

5) The only area not showing a similar increase in wintering swans is RRLNWR. Wintering numbers reached a record count of 357 in February 1980 when the Henry's Fork was extensively frozen. Overall, however, winter counts show that swan numbers have remained in the range of 150 to 350 since 1955. In years when grain is provided in abundance, an upper limit of about 350 swans is evident, suggesting that spatial competition or some other social interaction may place an upper limit on the number of swans wintering on the Refuge.

Summary

The only known wintering area for the RMP is the Tri-state region where the run-off from geothermal activity creates ice-free winter habitat. Recent sightings of marked trumpeters indicate that some swans from both the rapidly increasing Canadian flocks and the Tri-state Subpopulation are exploring new wintering areas, however no regular use of other areas has yet been documented.

Prior to the impoundment of Culver Springs, little or no dependable ice-free habitat existed in the Centennial Valley and most swans left the Valley at freeze-up. This pattern of movement, however, was deliberately disrupted by the feeding of grain at or before freeze-up. Grain feeding has allowed swans to winter successfully in the Centennial Valley, at least since the mid-1930s. Feeding rates, and the timing and methods of feeding have varied widely and abruptly, effectively varying the winter carrying capacity of the Refuge. The rate of loss of swans from the Centennial flock was highest in years when less grain was fed. In the early 1980s, the amount of grain fed per swan was the lowest in the Refuge's history and was a major factor in the recent decline of the Centennial Valley flock.

The single most important wintering area for RMP trumpeters is the Henry's Fork River below Island Park Dam, at Harriman State Park. This area provides both excellent supplies of aquatic vegetation and protection from waterfowl hunting and other human disturbance. Canadian trumpeters make only limited use of the RRLNWR feeding ponds, and are very dependent on the available open water in the Henry's Fork particularly when severe weather freezes the Teton River. Prior to 1968, winter water flows at Island Park Dam were often abruptly and almost completely curtailed, causing substantial portions of the wintering habitat to freeze for weeks at a time. A change in water management at Island Park Dam in 1967-68 provided increased winter flows and was followed by an increase in the number of wintering swans downstream at Harriman State Park. Current management of the River provides no guarantees that sufficient water will be released to prevent the freezing of the Harriman State Park wintering areas in years when water storage supplies are low.

Observations of the composition of the aquatic plant community at Harriman State Park suggest that preferred swan food species have declined in the last decade. Although extensive beds of vegetation remain after the wintering trumpeters leave the Henry's Fork, the carrying capacity of the HSP habitat may have declined. Changes in the plant community could be due to a variety of environmental factors as described in Chapter 7, as well as the effects of herbivory. A better understanding of the ecological relationships on the Henry's Fork is urgently needed, particularly to clarify relationships between sediment deposition, aquatic plant communities, herbivory by swans, and the fishery.

Trumpeters presently appear to have filled the available winter habitat in the lower elevations of Wyoming. Studies currently in progress in Wyoming have described the key characteristics of winter habitat and identified potential areas where winter habitat could be developed. Wintering trumpeters are vulnerable to human disturbance and the eventual development and loss of winter habitat outside Harriman State Park, Red Rock Lakes and Yellowstone NP.

CHAPTER 9. MOVEMENTS OF MARKED TRUMPETERS
David C. Lockman and Ruth Shea

Wyoming Studies

Since 1982, studies have documented the seasonal movements, life cycle chronology, and habitat use of 38 marked Trumpeter Swans in the Snake and Green River drainages of Wyoming (for map see Figure 3). These marked individuals included 8 adults representing 6 nesting pairs, 13 subadults representing 7 non-breeding pairs, 4 subadults with no known mate associations, one 13-year-old bachelor male, and 12 yearlings representing 5 sibling groups. Although these studies are still in progress, we present some of the preliminary findings. Additional details were described by Lockman *et al.* (1987).

Movements from Winter to Summer Range

Twelve of the thirteen marked territorial pairs wintered in the Snake River drainage near Jackson. All marked swans that summered in the upper Green and Snake River drainages, including swans that summered near the southern boundary of Yellowstone National Park, wintered near Jackson. Two marked pairs summered in the Henry's Fork drainage along the Idaho-Wyoming line; one pair wintered in Idaho, while the second pair wintered near Jackson each year.

Swans left the Jackson Hole winter areas as soon as warming temperatures began to melt nearby ponds. Since 1982, the departure of Wyoming breeders from the wintering areas has varied from about mid-March to mid-April. Experienced breeding pairs, family groups, and territorial pairs without breeding experience moved directly to ice-free habitats within a few km of their nesting territories. Territorial pairs utilized the same spring feeding sites each year, limiting their movements to occasional visits to the territory. Constant territorial occupancy did not occur until the territories were partially or completely ice-free. For one pair, territorial occupancy was not constant until 5-10 days prior to the initiation of laying each year.

Cygnets often left the wintering grounds with their parents. In two instances, cygnets remained with their parents until the pair occupied their territory. In another instance, the adults and cygnet group separated upon arrival at the spring use area. The siblings were subsequently observed on three occasions, about 10 km from the parents. In late May, about 50 days after the family had split up, the family was again observed together on the nesting territory. The siblings left the area by the time incubation began, about 10 June. In four other instances, sibling groups separated from their parents soon after leaving the wintering area. Two sibling groups were observed traveling together in April and May 1986, moving between four sites in the Jackson Hole area.

Following the harsh winter of 1984-85, spring melt was early, and breeding pairs had an unusually long period of access to prenesting food supplies. Two pairs did not occupy their territories until well after ice-off, however, indicating that spring foraging took precedence over occupation of the nesting territory. One territorial pair left the Jackson wintering area and was observed at the wintering area in Teton Basin, Idaho, in mid-April. This was the only sighting of marked Wyoming swans using the Teton Basin habitat.

Unpaired two and three-year-old trumpeters often remained near the Jackson Hole winter area until mid to late May. These swans fed at sites that had been frozen during the winter, and thus contained food that was unavailable to wintering swans. The ponds and marshes at the National Elk Refuge were used as a spring staging area by subadult trumpeters, and courtship activity was intense there each spring. Interactions between potential mates were first observed to be initiated by young swans on the winter areas, beginning in mid to late February. Bonds appeared to strengthen through the spring. Since 1982, four enduring pair associations were observed to occur following spring courtship activities and departure from the Refuge. At least one of these associations has continued through 1986.

Dispersal Movements

The number of territorial pairs in the Snake River drainage has increased since 1982, and few suitable unoccupied territories were available in the core production area by 1986. Territorial searches by young pairs may lead to dispersal to peripheral areas or movement outside the flock's current summer range. Dispersal of swans produced within the lower elevation Wyoming flock was documented in 1985. A subadult male, produced in 1983 on the National Elk Refuge, was observed in courtship activities on the Refuge during the spring of 1985. He was last seen on the Refuge until 20 May accompanied by an unmarked swan. He was subsequently found during the summer molt on Sheridan Reservoir, Idaho, approximately 128 km (80 mi) to the northwest of the National Elk Refuge and observed during the following winter at Harriman State Park. In both instances he was accompanied by an unmarked swan.

Each spring, local adults disassociated from their cygnets after the Canadian migrants had left. Unmarked yearlings in recently independent sibling groups could be identified with reasonable certainty because of the number of sibling groups, group sizes, the areas occupied by each group, occasional reunions with parents, and in one case the presence of a leucistic cygnet. Preliminary observations indicated that yearlings tended to move out of the breeding range of the Wyoming flock by mid-June. Their summer location could not be determined but at least one sibling group, marked by the presence of a leucistic yearling, returned to the Elk Refuge in November.

In the summer of 1986, a group of five swans summered and molted near Farson, Wyoming. This area was about 120 km south of the known southern extent of the current swan range in the Green River basin. A group of five swans suggested a yearling sibling association. In 1985, six cygnets were fledged by a Green River pair on a territory about 165 km to the north, and after wintering in the Jackson area, this pair left the Snake River with five cygnets. Through the summer a yearling group of five could not be found within the Wyoming swan range or in peripheral areas. These observations further suggest pre-molt yearling dispersal could be occurring to areas peripheral to the current Wyoming summer range.

Although sample sizes were small, the Wyoming studies suggest that two dispersal periods exist: 1) a pre-molt yearling dispersal, and 2) a dispersal of 2 and 3-year-old swans following a courtship period in late winter and spring. The occurrence of yearling and subadult dispersal from the Wyoming flock is further supported by the annual concentrations of yearling and subadult trumpeters in the molting flocks on Sheridan, Island Park, and Lima reservoirs. Similar pre-molt movement occurs in Canada geese among subadults, nonbreeding pairs, and failed breeders. Lockman *et al.* (1987) suggested that young males were the dispersers and females remained within the flock's range, guiding new mates to vacant territories within relatively close proximity to their natal area.

In at least one instance, a marked pair occupied what was considered an unsuitable territory for two successive summers. A nearby territory that had been occupied and on which young had been produced became vacant during the third year and was immediately filled by the pair from the unsuitable territory. This observation suggested that "floating pairs" may comprise a portion of the breeding pairs in Wyoming, and may occupy unsuitable production habitats while they await the opportunity to fill vacated suitable habitats.

Summer Movements

Territorial pairs and family groups were more sedentary than subadults on all seasonal ranges and usually remained on their territories throughout the period of available open water. Flight was infrequent and usually occurred in response to intrusion by other swans, or most often, by humans. Failed breeders often left their territory for extended periods and in two instances they completed a molt on areas up to three km from their territory. Non-territorial swans were quite mobile before and after the molt.

Since at least the 1940s, flocks of molting swans have been observed on several of the larger water bodies in the Tri-state region (Banko 1960:151). In addition to use of these large reservoirs, on seven occasions we observed marked Wyoming trumpeters molting on territories of nonbreeding pairs and failed breeders. The yearling sibling groups that summered in the Snake River drainage often molted within 4 km of their natal lake.

Territorial pairs tended to displace subadults during the spring and early summer. The resultant mobility of subadults appeared to contribute to population dispersal and pioneering, as the young swans sought unoccupied habitats for feeding and territory establishment. These movements also increased the opportunities for subadults to find a suitable mate. Young pairs remained very mobile until they settled onto a territory. In two cases, young pairs found and established a territory during the summer of their second year, following spring pair-bonding. If a young pair occupied an area from June through early September, they were likely to return at ice-off the following spring and occupy the area until late September. Young pairs tended to have a stronger attachment to a territory after the first year of occupancy.

Fall Movements

After fledging, family groups remained within the summer area, feeding at sites that were familiar to the adults. These sites were often the same sites used by the adults in early spring. When sufficient food was available, the family often remained on their territory until ice conditions forced departure. Family groups remained on the summer areas after failed breeders and subadults had departed.

By late September, yearling and two-year-old sibling groups moved to the National Elk Refuge. Other nonproductive adults and subadults arrived during the month of October. Family groups were the last to arrive, generally about 1 November. Between 15 October and 15 November each year, at least 90% of Wyoming's marked swans were observed on the Refuge staging area.

Adult pairs utilized approximately the same travel corridors on fall flights to winter range and on spring flights to summer range. Swans moved along major drainages, crossing hydrographic divides in passes with gradual gradients and mild relief. Swans moved as family groups, pairs, sibling groups, and singles, and rarely in mixed flocks.

Tundra Swans also gathered on the National Elk Refuge during their migration. Use of the Refuge by Wyoming trumpeters during the fall and winter appeared to be influenced by the number of Tundra Swans present, and the length of time that they remained. In years of high Tundra Swan use in the fall, aquatic plant beds were almost totally consumed, leaving little food remaining on the Refuge for over-wintering trumpeters. Prior to the unusually heavy use of the Refuge by tundras in the fall of 1984, at least 50% of the swan use in the Jackson Hole area occurred on the Refuge in the winters of 1981-82 and 1982-83. Thereafter, Refuge use by wintering trumpeters has comprised only about 15% of the total for the Jackson area. This decline in use of the Refuge has coincided with the use of new wintering sites since 1982-83.

Winter Movements

Adults with cygnets used the same sites each winter, segregated themselves from other swans, and seemed to defend their feeding area. Often they were observed feeding in association with their offspring of previous years.

Adults with young did not move as frequently as did trumpeters without young. Subadults and pairs without young often loafed and rested on creeks or river banks, up to 7 km (4.2 mi) from a feeding area. Midday loafing concentrations of 25-60 swans were observed on a Snake River gravel bar and on a sand bar at Fish Creek ponds. These loafing swans moved to feed in the mornings and evenings at eight different locations within a 2.5 km (1.5 mi) radius.

Swans sought new feeding areas as the available vegetation was consumed or when ice formed during extended cold periods and reduced the availability of aquatic plants. The swans' feeding and swimming activity was sufficient to keep some sites from freezing over. During periods of extensive ice cover swans tended to move to warm-spring fed creeks and other sites that usually were not used (i.e. narrow channels with tree cover along the banks, higher velocity channels, and sites with higher human use and disturbance levels). Swans moved in midwinter between sites at Jackson Lake, the upper Snake River, and the Jackson area. In the extended cold periods of 1984-85, swans used sites where they had never before been seen.

Red Rock Lakes NWR Studies

Neck-banding studies were conducted at RRLNWR in 1966 and 1967 to determine the range of nonbreeding trumpeters, the timing and areas of seasonal use, and winter movements to and from the Refuge (Papike 1971). Three progress reports on these studies can also be found in the RRLNWR neck-banding files and were the source of the information presented below.

Between 1966-68, 85 nonbreeding trumpeters were marked at RRLNWR. Red neck-bands were placed on males and green on females; individuals were not distinguishable. Neck-band loss and fading hampered data collection. No marked swans were resighted after April 1970.

All but two resightings occurred within an 80 km (50 mi) radius of the Refuge, in the area bounded by Ennis, Hebgen Lake, the Railroad Ranch (Harriman State Park), and Lima Reservoir (Figure 3). No marked swans were seen in Yellowstone or Jackson Hole. The furthest documented movement was that of a trumpeter that was found crippled on 13 December 1968, near Ryegate, Montana, about 200 km (125 mi) northeast of Red Rock Lakes. Although its red neck-band was missing, its tarsal-band revealed that this male had been marked in 1967 on

Upper Lake as an after-second-year bird. Progress report No. 3 (RRLNWR neck-band studies files) noted that *"trumpeters have been seen wintering in the Helena area just over the mountain to the west [of Ryegate]"* but gave no documentation for this statement. We could not determine from the resighting record whether this male had summered in Canada and was injured while returning, or whether it had summered in the Tri-state Area. The Refuge was also notified of a green neck-banded swan wintering in January 1967 at Alamo, Nevada, but since Utah had marked wintering Tundra Swans with similar neck-bands, the origin of this bird could not be determined (Papike 1971).

In 1966, most marked nonbreeders left the Centennial Valley in August or September, after completing the molt. Only a few were found in the Island Park area in December, but ten to twelve were observed there in January and February. The number of neck-banded swans began to increase on the Refuge in February, with the sharpest increase occurring after 1 March. The movement of swans to the Refuge appeared to be influenced by the amount and availability of natural feed in off-Refuge wintering sites, winter severity, and the magnitude of human activity at wintering sites. By April, marked nonbreeders began moving to Lima Reservoir (Papike 1971).

A second neck-banding study began at RRLNWR in 1978 to clarify subadult movements and determine whether Refuge swans were moving out of the Tri-state Area. Unlike the 1966-67 studies, the 1978-85 studies used coded neck-bands, which made observations of specific individuals possible. Three progress reports, a paper by Sjostrom (1982), and two reports by McEneaney and Sjostrom (1983, 1986) summarized the results of this study and formed the basis of the following discussion.

Efforts to follow the movements of marked cygnets proved unsuccessful because of apparent high mortality and neck-band loss. Of 10 cygnets neck-banded in September 1978, eight were located on the Refuge prior to freeze-up in early November. By December, one family had moved to Harriman State Park and one family was on the Madison River south of Ennis, MT. By January 1979, no marked cygnets could be located. Only two marked cygnets were subsequently resighted; one was seen in the molt on Upper Red Rock Lake in July 1979, and the second was resighted in November 1980 at Harriman State Park. Ten cygnets and twenty yearlings were neck-banded in 1979. Yearlings were resighted at Harriman State Park, Ashton, Hebgen Lake, Island Park Reservoir, Albino Lake (Gallatin National Forest, MT) and in the Centennial Valley. Fourteen yearlings and two cygnets were resighted at the Refuge feeding areas in early January 1980. One cygnet was resighted on 22 December 1981. Eight cygnets and six yearlings were never resighted after the initial marking. Due to concern that the neck-bands might be contributing to the apparent high mortality of cygnets, only yearlings were marked after 1979. Success improved in 1980, when seventeen of the eighteen marked yearlings were resighted during the 1980-81 winter. Most were resighted several times during 1981 and 1982. Only two yearlings were marked in 1981, making a total of 60 swans marked between 1978-81 (RRLNWR files: RRLNWR-9 Progress Repts.).

Between 1978-82, 240 relocations were made, of which 73% were at the Refuge (McEneaney and Sjostrom 1983). Sjostrom (1982) found no evidence that the marked Refuge trumpeters moved north with Canadian trumpeters or migrated south with Tundra Swans. Sjostrom observed that approximately 20-30% of the Centennial Valley flock moved out of the Valley in early November. As winter progressed, the swan flock at the Refuge increased to early autumn levels (300+), indicating the return of trumpeters from adjacent wintering areas. Resightings were similar to the 1966-67 findings, being concentrated on the Refuge and within a radius of about 80 km (50 miles). The furthest movement observed was to the Madison River in the western portion of YNP, Albino Lake on the Gallatin National Forest north of West Yellowstone, and Ashton Reservoir, ID. Centennial Valley swans were not detected to use the eastern portions of the Tri-state region, Yellowstone NP, or the Snake River drainage of Wyoming (RRLNWR files: RRLNWR-9 Progress Repts.).

Lack of coordination between swan marking programs has sometimes made interpretation of potentially important resightings impossible and seriously reduced the value of marking studies. McKelvey *et al.* (1983) reported sightings of a green neck-banded swan in the Yukon at Crooked Lake on 25 April 1980, and between North and South Toobally Lakes on 20 August 1980. The identifying alpha-numeric code could not be read, however. Because green neck-bands were simultaneously in use on both Tundra Swans wintering in California (D. Paullin, pers. comm.) and on trumpeters at RRLNWR, Malheur NWR, and Turnbull NWR, the origin of this bird, or birds, could not be determined. Green neck-bands were also used on Wyoming swans in 1982-86; however, in this project the alpha-numerics contrasted greatly from those of previous studies.

1982-1986 Marking Studies

Study RRLNWR-9 was amended in December 1981 to permit the neck-banding of 40 additional subadults because of the earlier problems with neck-band loss. The 60 trumpeters marked between 1978 and 1981 were fitted with rigid green, 1/16 inch thick neck-bands. As in the 1966 study, poor neck-band retention was a problem, and thicker neck-bands were used starting in 1982 (McEneaney and Sjostrom 1983). Thirteen molting yearlings were neck-banded in 1982 (Sjostrom 1982) and 24 in 1983, on Lima Reservoir and Upper Red Rock Lake (RRLNWR Ann. Narr. Rept. 1983). Observations from 1983-86 substantiated the findings of the previous studies, e.g., Refuge trumpeters usually remained within a radius of 80 km (50 mi) of RRLNWR, and did not move into the eastern Yellowstone or Jackson Hole portions of the Tri-state region. One swan was observed to visit Teton Basin (McEneaney and Sjostrom 1986). Studies of these marked individuals into the years of pair formation and territory establishment did not continue.

After completing the molt on Lima Reservoir, trumpeters generally moved eastward to feeding sites along the Red Rock River corridor. Some began using the Refuge wintering ponds by September. Others moved out of the Centennial Valley to the Henry's Fork, the Madison River in YNP, Sheridan Reservoir (Idaho), and the South Fork of the Madison River at Hebgen Lake. Virtually all of the neck-banded swans were observed at the Refuge wintering ponds sometime between October and December; Culver Pond was especially important during this period (McEneaney and Sjostrom 1986).

As noted by Papike (1971), McEneaney and Sjostrom (1983, 1986) and by many observations in the Refuge Narrative reports dating back to 1938 (see Chapter 8), varying numbers of swans leave the Centennial Valley in late autumn and return to the Refuge in late February and March. The number that leave the Valley has varied from the entire flock in 1941 and 1950 (RRLNWR Ann. Narr. Repts. 1941, 1950), to only a few in years when heavy feeding began early to deliberately keep swans from moving out. Some swans use the feeding ponds regularly throughout the winter, but other leave the Refuge intermittently. One swan was resighted at Culver Pond in January 1984 and found again on the following day at Teton Basin, 115 km (72 mi) to the southeast (McEneaney and Sjostrom 1986).

By late November to mid-December, some neck-banded subadults and family groups moved to the Henry's Fork and to the Ennis, MT, area. In January and February, neck-banded swans were observed at Harriman State Park, Teton Basin, Madison River, Ashton, Sheridan Reservoir, and Hebgen Lake. The Refuge ponds received heavy use by marked swans in February and March, and resightings were restricted to the Refuge area in March and April (McEneaney and Sjostrom 1986).

The subadults dispersed during April. Many moved westward toward Lima Reservoir, making use of the Red Rock River habitat en route. In May some marked individuals were observed on the Madison River, Harriman State Park, Albino Lake, Hebgen Lake, and the Henry's Fork near Ashton, ID. Most of these swans molted on Upper Lake, although some were resighted on Lima Reservoir. From June through August, swans were on their respective molting areas. In addition to the three main molting areas, (Lima Reservoir, Upper Lake, and Sheridan Reservoir), neck-banded swans were found in July at Albino Lake, Ashton Reservoir, and Harriman State Park (McEneaney and Sjostrom 1986).

Just as described by Banko (1960:148), 30 years later Lima Reservoir was still an important molting area, providing relatively undisturbed habitat for over 100 trumpeters in some years. Although tarsal and neck-banding data have shown that an individual trumpeter may use several different molting areas during its lifetime, all the neck-banded swans in the 1982-86 study molted at Lima Reservoir during at least one summer (McEneaney and Sjostrom 1986).

Telemetry Studies

In 1983, McEneaney began further study of swan movements using radio transmitters attached to alpha-numeric coded neck-bands. Telemetry was chosen because previous resightings of neck-banded swans had been primarily restricted to the more easily accessible areas (RRLNWR and Harriman State Park). Three adults from pairs on Lima Reservoir, Elk Lake, and Lower Red Rock Lake were fitted with neck-band radios in August 1983. The transmitters failed within a few months, presumably due to water leakage (RRLNWR Trumpeter Swan Rept. 1983).

Because other researchers were also using green neck-bands, RRLNWR used red bands in 1984 (RRLNWR Ann. Narr. Rept. 1983). McEneaney fitted seven adult trumpeters with radios in 1984, but all radios failed within five months. All but one or two radio-equipped swans moved out of the Centennial Valley in late September. One (R2) was visually relocated on the Madison River in YNP on 23-26 September 1984. McEneaney flew two flights in November and December 1984, but could not locate any of the radio-equipped swans outside of RRLNWR. Three radioed swans were seen at the Refuge feeding ponds in December 1984, but could not be located later in the winter (RRLNWR Trumpeter Swan Rept. 1984).

After moving to the Madison River in September, R2 was reported again in late November 1984 near Grande Junction, CO, where she was observed several times throughout the winter, at least once with a second swan. R2 was an after-second-year (ASY) female, marked on 18 July 1984 on Lower Red Rock Lake. She was seen in the Grande Junction area from late November 1984 to 6 April 1985. She was next seen near Vernal, UT, on 11 April 1985, and observed several times during the summer of 1985 at RRLNWR. A red neck-banded swan seen near Big Piney, WY on 19 April 1985 was probably R2, but the code was not observed. The sightings of R2 provided the first evidence that some Refuge trumpeters migrate to western Colorado and revealed the migration route that was followed (McEneaney 1986a).

McEneaney fitted six adult trumpeters with radios in 1985. One swan died soon after handling from a combination of elevated lead levels, poor physiological condition, and capture stress. The remaining five radio-equipped swans stayed on the Refuge until the lakes froze. Two wintered on the Refuge, one was found dead on the Buffalo River, and two could not be found during the winter. At least one swan lost its radio during the first winter. No sightings were recorded after 29 April 1986 (McEneaney 1986a).

Tarsal-banded Swans

Although 1,050 swans were tarsal-banded on or near RRLNWR between 1945-84 (Anderson *et al.* 1986), only two recoveries from outside the traditional Tri-state range have been reported. As previously mentioned, one was recovered in 1968 near Ryegate, MT. The second recovery was of a female banded on 29 July 1966, and found dead southeast of Calgary, Alberta on 10 September 1975. The recovery of this swan provided the only evidence that any Tri-state swans have moved north into Canada. Tarsal banding has provided evidence that Centennial Valley trumpeters occasionally disperse into Yellowstone and the lower elevations of Wyoming. In the late 1970s, Shea observed that at least two of the nesting pairs in YNP contained a banded swan. At that time, the only banded Tri-state trumpeters were from the Centennial Valley. Lockman *et al.* (1983) captured a bachelor male trumpeter in the Jackson area in 1983. This bird hatched and was banded at RRLNWR in 1971, and to date has been the only marked Centennial Valley trumpeter resighted in the lower elevations of Wyoming.

Grande Prairie Studies

In order to identify the wintering grounds of the Grande Prairie flock, R. Mackay of the Canadian Wildlife Service marked 9 adults and 59 cygnets with colored tarsal-bands during the summers of 1954-56. Results from this study provided the first evidence that most Grande Prairie trumpeters wintered with the Tri-state Subpopulation, although at least one family migrated much further eastward. Marked trumpeters were resighted in the Tri-state Area on the Buffalo River, at the Railroad Ranch (HSP), at RRLNWR (Mackay 1957), and at Yellowstone Lake and River (RRLNWR files 1957). The breeding pair and two broods, fledged in consecutive years at Lowe Lake, Alberta, were recovered after being shot in the fall of 1957 near Cody, WY and in Nebraska (Mackay 1957). These sightings provided the last evidence of use of these areas by Grande Prairie trumpeters.

The CWS again marked Grande Prairie trumpeters during the summers of 1973-78, using neck-bands with individually identifiable codes. A total of 232 swans (33 adults, 18 yearlings, 181 cygnets) was marked, including 56 broods. Cygnets were neck-banded in early September just prior to fledging. Most of the yearlings and adults were marked during the midsummer molt in 1976 and 1978.

Of the marked swans, a total of 29 adults (88%), 12 yearlings (67%) and 107 cygnets (59%) were resighted and individually identified at least once after leaving the marking area. Marked Grande Prairie trumpeters have been observed in the Tri-state Area each winter from 1973-86, although very few neck-bands have been observed since 1981. Records of all reported resightings are kept by the Canadian Wildlife Service, Edmonton, and provided most of the data referred to in the following discussion.

172

Fall and Winter Movements

By mid-September, family groups made short flights from their nesting territories to nearby lakes. Neck-banded adults were also observed to separate briefly from their cygnets in September and feed at nearby lakes. After the molt, most nonbreeders departed from their molting lake in early August and moved among several lakes until early September. Prior to migration they congregated on two or three lakes not occupied by breeding pairs (G. Holton, pers. comm.).

Migrants began their departure from Grande Prairie in the second week of October and most swans left by early November. Migration appeared to be direct and rapid; no marked swans were observed in southward migration during 1973-78 (Turner 1982, 1987), however migrant trumpeters were heard at Benton Lake NWR, MT, in 1983 (T. Turnow, pers. comm.).

Marked migrants began arriving at Yellowstone Lake between 14 and 20 October each year. Nonproductive trumpeters, in groups of three to eight, arrived first. In 1976-78, over 100 adults and less than a dozen cygnets were present prior to about 30 October, when most family groups began to appear (Shea 1979).

The peak of swan migration at Yellowstone Lake occurred during the second week of November in 1976-78. The highest single count of 312 trumpeters was made on 9 November 1977, during a census of the north end of Yellowstone Lake and River from the Lake outlet to Chittenden Bridge. Other trumpeters, including marked Grande Prairie swans, were known to use the south and southeast arms of Yellowstone Lake, but these inaccessible areas could not be surveyed from the ground. Based upon observed ratios of marked to unmarked swans, Shea (1979) estimated that the total number of migrant trumpeters using Yellowstone Lake was about 325 in 1976, and 470 in 1977. Of the 148 neck-banded Grande Prairie swans that were resighted at least once, 81 individuals (55%) were observed at Yellowstone Lake. The Yellowstone Lake area provides a very heavily used and important autumn habitat. This area is relatively free from human disturbance and totally free from hunting and thus the deposition of lead shot. Yellowstone Lake and River protect hundreds of trumpeters during the height of the Tri-state Area's waterfowl hunting season.

The fidelity of individuals to particular sites was apparent. Many individuals were relocated in consecutive years and often observed using the same sand bar or river bend each year. Of the 30 neck-banded adults observed at Yellowstone Lake in either 1976 or 1977, 25 returned to the same area and were resighted the following year. Six of the 16 adults found in 1976 returned to the same sites both in 1977 and 1978. Cygnets apparently suffered high mortality or wandered more widely as yearlings: of the 20 observed in 1976 or 1977 at Yellowstone Lake, 14 were never again seen alive. Marked families also followed rather predictable movement patterns, using the same sites year after year. These patterns appeared to be learned by the cygnets and followed by them in subsequent years. Marked yearlings and subadults were occasionally observed to arrive at their family's traditional autumn feeding site, where they were soon joined by their parents with their most recent brood (Shea 1979).

As the Yellowstone River and Lake gradually froze in December, swans moved south and west, primarily to Harriman State Park where protection from waterfowl hunting was also available. By early January, marked swans were observed on the Teton River in Idaho. A few swans moved west from Yellowstone to Hebgen Lake and the Madison River. Only one marked trumpeter was observed to use both Yellowstone Lake and RRLNWR. This bird was seen on Yellowstone Lake on 9 November 1981 and was resighted at Culver Pond four days later. Some trumpeters remained in Yellowstone throughout the winter. Others moved back into the Park in late February or March, if ice conditions moderated.

Although marked Grande Prairie swans were first observed at RRLNWR in March 1956 (Mackay 1957), the resighting data indicate that relatively few Grande Prairie swans use the Refuge feeding ponds. Ten neck-banded Grande Prairie trumpeters were resighted at the Refuge between 1977-83 (McEneaney and Sjostrom 1983). Banding records show that eight of these swans were the 1977 and 1978 broods from Flyingshot Lake. A ninth individual was a breeding adult from Flyingshot Lake that led unmarked broods of 6, 5, and 1 cygnets to the Refuge in 1979, 1980, and 1982 respectively. Apparently many of the Grande Prairie trumpeters that used the Refuge grain in the late 1970s were an extended family group from Flyingshot Lake.

Winter counts of swans on the Refuge also indicated that few Canadian trumpeters used the Refuge in the 1970s and early 1980s. Unusually high use of the Refuge by some 300-360 swans occurred in 1986-87 (C. Mitchell, pers. comm.), even though the Centennial Valley flock included only some 200 swans in September 1986. This winter influx of swans may represent a recent shift in the pattern of use by Canadian swans; however, this cannot be determined conclusively because few Canadian trumpeters in the wintering population are currently marked.

Shea (1979) identified 50 marked Grande Prairie trumpeters at HSP during the winter of 1977-78, and at least thirty-three of these came to Harriman via Yellowstone Lake. The remaining 17 apparently moved directly to the Henry's Fork from Grande Prairie. The HSP area was the single most important wintering site for marked Grande Prairie trumpeters in the 1970s. Of the 15 families marked in 1977, 10 were resighted at HSP. Of the 148 trumpeters which were ever resighted after marking, 90 (61%) were observed at least once at HSP.

Twelve marked Grande Prairie trumpeters were located at least once on the Teton River between Driggs and Victor, ID, and movement of marked swans between Harriman SP and the Teton River has been observed. Use of the Teton Basin area by marked swans was probably under-reported because fog, poor access, and lack of cover made viewing conditions difficult.

Few swan surveys have been made at Hebgen Lake. Sporadic observations indicated low swan use until January, when duck hunting ended and swan numbers increased. Neck-banded swans have been observed to move from Yellowstone to Hebgen Lake in December, and then on to HSP.

Marked Grande Prairie trumpeters occasionally have been observed in the Jackson Hole area. J. Weaver (pers. comm.) reported at least 10 marked trumpeters in two family groups that moved through Grand Teton National Park in November 1978, but remained for only a few days. One marked swan was seen on the National Elk Refuge in January 1980 and a second was seen at the Refuge in November 1983.

Both the Tri-state and Canadian marking programs have shown that swans from all the known breeding flocks intermingle in winter in the Tri-state Area. Many move from one site to another during the winter. If a disease outbreak should occur among the wintering waterfowl in the Tri-state Area, it could quickly spread to all the key wintering sites and could impact all of the Canadian and Tri-state trumpeter flocks. An outbreak of avian cholera occurred in March 1987 among Lesser Snow Geese (*Anser caerulescens*) at Market Lake Waterfowl Management Area, less than 75 miles from the Henry's Fork and Teton River wintering areas (D. Wright, IDFG, pers. comm.). Such an outbreak in late winter among the trumpeters could prove catastrophic to the restoration of the species in the Rocky Mountains.

Spring Migration

Most migrants leave the Tri-state Area by 15 March in most years. In 1986, February was unusually warm and most swans had left the Teton and Henry's Fork rivers by 1 March. In the latter half of February 1987, temperatures were again unusually mild at HSP. A group of 100-150 trumpeters congregated on less than 0.4 ha (1 acre) of open water at Golden Lake, approximately 3.2 km (2 mi) from the Henry's Fork. The swans were quite vocal and appeared to be staging prior to migration. Seventeen migrants were observed heading north over Island Park Reservoir on 1 March in a V-formation.

Few sightings of migrants have been made between early March and mid-April. After departing the wintering area, some trumpeters move northward into southwestern Alberta, where they begin arriving by the third week of March (Turner 1987). Marked swans have been observed at Mountain View, Lethbridge, Cardston, and Waterton Lakes NP between 1-14 April. These are the only areas where trumpeters have been regularly observed during spring migration. The migrants remain for about two weeks, then depart for Grande Prairie, where they arrive about 20-25 April. Breeding pairs arrive first, but by the end of April subadults, including yearlings, have also arrived (Turner 1982, 1987).

Holton (1985) reported trumpeters returning north through areas west of Calgary. One observation of a neck-banded swan in the Columbia River Valley of British Columbia during spring migration also indicated some movement along the Rocky Mountain trench (Mackay 1978). Observations of unmarked trumpeters at Hauser Lake and the Coeur d' Alene area also showed that some spring movement occurred along the west slope of the Continental Divide (Appendix I).

Summer movements

Breeding pairs molted on their nesting territories. Some pairs occupied and defended lakes in early spring, but made no apparent attempt to nest. Four of these pairs remained on their lake to molt, while one pair left its territory (G. Holton, pers. comm.). Nonbreeding flocks moved among three to four lakes which were not occupied by breeding pairs, prior to remaining at one lake to molt. In two cases, nonbreeders moved onto lakes occupied by breeding pairs.

Some subadults dispersed from Grande Prairie. One male which wintered at Red Rock Lakes in 1978-82, was observed as a 3-year-old to summer southeast of Fort St. John, B.C., and a 3-year-old female was observed summering near Fort Nelson, B.C. At least two marked Grande Prairie trumpeters spent one summer in the U.S. A yearling male spent the summer of 1977 at Ninepipe NWR near Moise, Montana, and was last resighted at Hamilton, MT, on 11 November 1977. Another Grande Prairie subadult summered at Red Rock Lakes in 1980 (RRLNWR Ann. Narr. Rept. 1980). This bird was most likely one of the three marked sibling subadults from Flyingshot Lake that wintered at Red Rock Lakes in 1979-80. In both cases these solitary swans showed no obvious signs of injury or ill health, however, neither was ever subsequently resighted.

Use of Wintering Sites Outside of the Tri-state Area

Virtually all of the marked Grande Prairie trumpeters which survived winter and were resighted in Canada were observed to winter in the Tri-state Area. Of the 159 trumpeters marked during 1976-78 while the winter area searches were most intensive, only one swan that had not been found in the Tri-state Area was resighted in Canada. There was no evidence to suggest that Grande Prairie trumpeters wintered successfully in other locations. Grande Prairie trumpeters have been observed migrating and wintering south of the Tri-state Area in Utah and Colorado (Appendix I), but no regularly used wintering areas have been located. These pioneers were primarily subadult swans: of six known-age individuals observed to winter outside the Tri-state Area, all were less than 3 years old. None of the subadults that were observed wintering in unusual locations were subsequently relocated alive. Apparently the young birds which migrated to new wintering areas experienced high mortality rates, or subsequently used summer and winter habitats which were not searched.

Unmarked migrating trumpeters have been regularly observed during spring and fall in the Coeur d'Alene area of northern Idaho (F. Bear, pers. comm.), but their winter destination is unknown. Several other sightings indicate infrequent movement of Grande Prairie swans through the Flathead and Bitterroot valleys of western Montana. These sightings suggest that some trumpeters are regularly wintering outside the Tri-state Area. It is possible that these trumpeters are moving with Tundra Swans to wintering grounds in California, but data to confirm their winter whereabouts are lacking.

Other Canadian Marking Studies

Saskatchewan

In conjunction with the Grande Prairie studies, one adult and four cygnets were banded in 1972, and 2 adults and six cygnets were neck-banded in 1973 in the Cypress Hills area of Saskatchewan (Nieman and Isbister 1974). Two of the neck-banded swans were resighted at Red Rock Lakes NWR on 12 February 1974, and were observed at the Refuge and on the Yellowstone River in 1975 and 1976 (Killaby 1987).

Yukon

In 1980, an adult female and her two cygnets were neck-banded on the Smith River, between North and South Toobally Lakes. On 19 December 1980, both cygnets were observed near Moise, Montana, apparently separated from the adult. The adult was resighted near the marking location during July 1981 (McKelvey 1983). One cygnet was subsequently resighted at Red Rock Lakes on 25 November 1981 (McEneaney and Sjostrom 1983). Based upon the relocations of the two cygnets, which apparently became separated from their parents, and the movement of other swans along the east slope of the Rockies, it has been assumed that the Toobally Lakes trumpeters are part of the Rocky Mountain Population, rather than the Pacific Coast Population (USFWS 1984; R. McKelvey, pers. comm.).

Northwest Territories

The first effort to mark trumpeters from the rapidly expanding NWT flock occurred during the summer of 1986 when the Canadian Wildlife Service neck-banded 20 adults/subadults. From this recent small sample it is evident that some NWT trumpeters are exploring new migration routes and wintering sites, while others migrate to the Tri-state wintering areas. During the winter of 1986-87, at least 7 marked individuals were located in the Tri-state Area, with resightings most frequent at the Teton Basin wintering site. NWT trumpeters were also observed at Yellowstone Lake, Harriman State Park, and at Red Rock Lakes. Other NWT swans were observed in migration at Coeur d'Alene, ID, at Goose Lake on the Oregon-California border, and near Alamo, Nevada (L. Shandruk, pers. comm.; Appendix I).

Evidence of Population and Subpopulation Interchange

The summer range of some 500 migrants that winter in the Tri-state Area is unknown. This raises the possibility that some Pacific Coast trumpeters winter with the Rocky Mountain trumpeters (Chapter 2). However, no resightings of marked PCP trumpeters have occurred on the Tri-state wintering area. Likewise, no marked Interior Canada or Tri-state trumpeters have been resighted within the Pacific Coast Population's range. To date, there is no evidence of population intermixing.

Only one case of a PCP swan recovered within RMP summer range has been documented. McKelvey (pers. comm.) reported that a trumpeter banded as a wintering cygnet on the Powell River, B.C., winter habitat on 10 January 1984 was presumably from Alaska. This swan apparently died during autumn 1984 along Highway 97, 160 km (100 mi) west of Dawson Creek, B.C., near Grande Prairie, where it was found in a snow bank in spring 1985.

As the numbers of wintering Canadian migrants increase in the Tri-state Area, the odds of pair formation between Tri-state and Canadian trumpeters will increase if such pairing is not prevented by subtle social or temporal barriers. At present, however, there is no evidence of pair formation and genetic exchange between the Interior Canada trumpeters and the Tri-state trumpeters; these groups so far appear to be reproductively isolated. Although about 1,050 trumpeters have been banded at Red Rock Lakes since 1945 (Anderson *et al.* 1986) and about 300 have been marked at Grande Prairie since 1954 (Mackay 1957, Turner 1987), there have been only four observed cases of marked swans summering outside the range of their natal subpopulation. As previously mentioned, one marked Grande Prairie subadult spent the summer of 1977 at Moiese, MT, and a second marked Grande Prairie subadult spent the summer of 1980 at RRLNWR. Both of these swans disappeared after their summers in the U.S. One trumpeter marked as a yearling female at Red Rock Lakes in July 1966, was found dead in September 1975, southeast of Calgary, Alberta. A male, marked as an after-second-year bird at Red Rock Lakes in 1967 was found crippled near Ryegate, Montana, in December 1968. The green neck-banded swan observed in the Yukon in 1980 (McKelvey *et al.* 1983) could have been from Red Rock Lakes, but this cannot be established because several other simultaneous studies were also using green neck-bands. The Yukon swan was more likely from Turnbull NWR, since other Turnbull swans have wandered into Canada. One Turnbull neck-banded trumpeter was observed at Revelstoke, B.C., on 13 May 1980, and again in the Calgary area in October 1980. On 6 April 1985, McKelvey observed a Turnbull-marked swan on Risky Creek near Williams Lake, B.C. (Turnbull NWR files).

Summary

Studies of marked trumpeters of known sex and age have recently produced much information regarding seasonal movements and associations, and clarified similarities between the Grande Prairie and Tri-state trumpeters. Like many other waterfowl species (Hochbaum 1955), trumpeters learn from the behaviors of their experienced companions. The objects, places, or actions thus learned become traditional. With pair bonds that may endure for life, and family bonds that endure for several years, family traditions are strong.

Wyoming trumpeters usually remained within the Green River and Snake River drainages of northwestern Wyoming. Movement beyond these drainages included a pre-molt dispersal by yearlings, dispersal after late winter/spring courtship and pair bonding by subadults, and spring foraging in Teton Basin, ID, by one territorial pair. Subadult and yearling dispersal occurred in late spring, and was aided by the territorial defense and aggression of breeding pairs.

Family groups were more sedentary than all other age classes, both in winter and summer. Experienced breeding pairs were the most consistent and traditional in their use of sites and flight paths. Breeding pairs with cygnets segregated from other wintering swans on the National Elk Refuge, and appeared to defend their traditional feeding sites. Breeding pairs were sometimes joined by their yearling and subadult offspring during the winter and permitted them to feed in close association.

Subadults remained on the National Elk Refuge in April and May. Courtship activity was intense during this period, and pair-bond formation was observed. Breeding pairs left the wintering grounds by about 20 March, and moved to traditional spring feeding sites within 10 km of their territory. Pairs made occasional flights to their territories, but did not begin constant territorial occupancy until their lake was at least partially ice-free. The parent-cygnet bond weakened in late winter and early spring. While some cygnets remained on the Refuge when their parents departed in March, others moved to the spring feeding sites with their parents. One brood of cygnets was observed on the nesting territory prior to incubation.

After remaining on their territory throughout the summer, successful breeding pairs occasionally led their newly fledged young to feed in the same sites which the pair had used in the spring. Almost all Wyoming trumpeters staged at the National Elk Refuge annually, between 15 October and 15 November. Nonproductive adults and subadults arrived first, followed by family groups. Swan movements to and from seasonal ranges followed major river courses, and movements over hydrographic divides occurred through natural passes, with gradual relief.

Swans sometimes depleted the available feed at wintering sites. Lack of feed, or its reduced availability due to ice formation during extended cold periods, caused swans to increase their movements to alternate, less preferred feeding sites.

Grande Prairie trumpeters also began to arrive at the wintering areas about 15 October, and like the Wyoming trumpeters, nonbreeders were the first to arrive, followed within two to three weeks by family groups. Grande Prairie trumpeters were also quite traditional in their annual movement patterns. Pairs regularly brought their broods to the same wintering sites each year, and occasionally associated with their yearling and subadult offspring.

Yellowstone National Park provided important fall habitat for hundreds of migrant trumpeters. Harriman State Park was the most important wintering area, and also provided fall habitat for Grande Prairie swans. Heavy use of these two areas, which are closed to waterfowl hunting, protected a large portion of the Canadian trumpeters during the waterfowl hunting season. Trumpeters moved within the Tri-state Area during the winter. Movements were probably influenced by varying weather and ice conditions. Because swans from all the known breeding flocks share the same wintering sites, and movement of swans between these sites occurs throughout the winter, the RMP trumpeters remain quite vulnerable to heavy losses should a disease outbreak occur among wintering waterfowl in the Tri-state Area.

Grande Prairie trumpeters usually left the Tri-state Area about 1-15 March. Earlier departure was observed during unusually mild winters. After a slow migration through southwest Alberta, the breeding pairs arrive in Grande Prairie about 20-25 April. Nonbreeders, including yearlings, arrived by the end of April.

Studies of marked Red Rock Lakes trumpeters have shown movement out of the Centennial Valley by nonbreeders, beginning in August and September. Some family groups also moved out of the Valley in October and November. Swans movements ranged within an 80 km (50 mi) radius of the Refuge. Marked swans moved back to the Refuge in January and February, with a noticeable influx occurring in March. In April the swans dispersed from the feeding ponds, and many nonbreeders moved west along the Red Rock River to Lima Reservoir. Other nonbreeders moved out of the Centennial Valley to Hebgen Lake, the Henry's Fork River, and other nearby sites.

Although most marked Grande Prairie and Tri-state trumpeters wintered in the Tri-state Area, a few swans migrated further south. A marked swan and its mate from Red Rock Lakes wintered near Grande Junction, Colorado, in 1984-85, and another marked Refuge trumpeter was shot at Great Salt Lake in November 1986. In recent years, marked Grande Prairie swans have been seen in Utah and Colorado. Swans marked in the Northwest Territories in 1986 were found during the winter of 1986-87 in northern Idaho, along the Oregon-California border, and in southern Nevada, as well as in the Tri-state Area. If the current increase of the Interior Canada Subpopulation continues, further exploration of new migration routes and wintering areas is likely. To date, however, this exploration has been primarily by subadults, and mortality has been high. No regular use of any wintering area outside the Tri-state Area has yet been documented.

Observations of marked swans so far indicate that there is little overlap between the ranges of the Pacific Coast Population and the Rocky Mountain Population, although the summer range of some 500 trumpeters that winter in the Tri-state Area is still unknown. Even though the Tri-state and Interior Canada trumpeters share a common wintering area, observations of some 1,300 marked swans have not yet produced any evidence of pair-bonding between members of the two groups. With the lack of evidence to the contrary, there is a real possibility that the subpopulations are reproductively isolated.

CHAPTER 10. BIOENERGETICS AND FOOD HABITS
J. Bradley Bortner and Ruth E. Shea

Bioenergetics

Few quantitative data have been gathered regarding the bioenergetics or the physiological condition of wild trumpeters due to the concern that the necessary handling would result in nest failure or mortality. Recent research, however, on Tundra Swans and arctic nesting geese has described some bioenergetic relationships that may be similar to certain aspects of the trumpeter's ecology.

Tundra Swans

Bortner (1985) studied migratory Tundra Swans wintering in North Carolina where climatic conditions were much milder than in the trumpeters' Tri-state wintering habitat. Minimum daily temperatures in North Carolina occasionally fell below 0 °C (32 °F) but daily maxima were often above 10 °C (50 °F). In addition to feeding on aquatic vegetation, the wintering Tundra Swans fed in agricultural fields on soybeans, wheat, and corn.

Periodic sacrifice of swans, subsequent carcass analyses, and time-activity budget analyses showed that Tundra Swans lost weight throughout the winter despite available food and that the swans could have consumed enough soybeans and wheat in several hours to meet their daily energy needs. Weight losses were independent of both food supply and ambient weather conditions and were due primarily to use of lipid reserves. Energy from stored reserves provided less than 10% of the swans' daily energy expenditures, while energy derived from food provided the remainder. Catabolism of stored lipids continued throughout the winter, despite warmer late-winter weather. Adult males lost 25% of their body weight overwinter. Adult females, juvenile females, and juvenile males lost 15%, 12%, and 17% respectively. Swans appeared to lose weight throughout the winter in an endogenous rhythm and left the wintering area at their lowest weight and with low reserve levels (Bortner 1985).

Swans arrived in North Carolina with fat reserves that were sufficient to supply approximately two weeks of energy if activity patterns were maintained and no food was acquired. Swans catabolized energy stores as the winter progressed and the probability of encountering harsh weather decreased. Tundra Swans conserved energy by feeding less, remaining in sleeping or resting postures later in the day, refraining from maintenance behaviors and seeking protected habitats during periods of low temperature accompanied by high winds and/or precipitation. These behaviors probably also minimized convective heat loss (Bortner 1985).

Tundra Swans have apparently evolved a pattern of nutrient dynamics which decreases tissue maintenance costs through weight loss, while providing sufficient reserves with which to survive a spring storm. Bortner concluded that this weight loss was adaptive, in that it minimized wing-loading during northward migration. As Tundra Swans stop to feed along the northward migration, food availability at spring staging locations may be critical to the accumulation of lipids for successful nesting and incubation. Bortner stated that *"swans apparently have reduced the energy requirements of migrational flight by acquiring fat reserves on the more northerly staging areas instead of the wintering grounds."* Tundra Swans probably carry as little weight as possible as they begin spring migration, and then add reserves for reproduction as close as possible to the breeding grounds. If trumpeters share a similar pattern of catabolism and accumulation of energy reserves, then the availability and quality of spring foods will directly influence their survival and productivity.

Arctic and Prairie Nesting Geese

Arctic and prairie nesting geese also share a similar pattern of seasonal weight dynamics and energetics. Greater and Lesser Snow Geese (*A. caerulescens atlanticus* and *A. caerulescens caerulescens*, respectively), and Canada Geese accumulate nutrient reserves in the fall, slowly lose weight throughout the winter, and then accumulate reserves for reproduction during northward migration. (Gauthier *et al.* 1984, Ankney 1982, Wypkema and Ankney 1979, Raveling 1979, McLandress and Raveling 1981). These studies have indicated that the energetics patterns are based on an endogenous rhythm of weight gain and loss. Giant Canada Geese (*B canadensis maxima*), which in some respects could be viewed as an evolutionary correlate to Trumpeter Swans, also have a typical energetics pattern that includes the slow loss of lipid reserves during the winter and the accumulation of energy reserves in the spring prior to nesting (McLandress and Raveling 1981). Ankney and MacInnes (1978) and Raveling (1979) have stated that the amount of energy or nutrients which geese bring to the breeding grounds proximately controls the clutch size and nesting success of individual geese. Yearly reproductive success, therefore, may be dependent on food availability, food quality, and the ability of geese to accumulate nutrient reserves on the wintering and spring staging areas.

Trumpeter Swans

Hampton (1981) studied the winter ecology of Trumpeter Swans on the Henry's Fork. While he did not directly quantify the energetics or physiological condition of trumpeters, he did make several predictions about these factors based upon his observations of trumpeter behavior. Hampton observed that the amount of time that swans spent feeding each day decreased from November to December. It then increased by late January, and continued to increase into March. He concluded that the trumpeters' increased feeding throughout the winter reflected their deteriorating body condition and suboptimal feeding habitat. Hampton's behavioral observations on the Henry's Fork included no observations in November and were limited to one day during December. Observations in Yellowstone included two days in November and 8 days in December. We conclude that the validity of the comparison of monthly changes based on such a small sample is suspect.

Based upon his assumption that the amount of time spent feeding was indicative of a swan's physiological condition, Hampton (1981) concluded that the local Tri-state swans arrived first on the winter range and were in poor physical condition, spending over 75% of the daylight hours feeding in mid- to late-November. Hampton attributed the lower feeding rates observed in December to the arrival of Canadian migrants which were in better condition.

Hampton's hypothesis that the lowered December feeding rates were due to the arrival of Grande Prairie swans was not supported by our observations. Shea (1979) documented that marked Canadian migrants regularly begin arriving in the Tri-state region by 15 October, and migrants were among the swans using the wintering areas when Hampton made his November observations. We suggest an alternate hypothesis, i.e., that Hampton observed intense feeding bouts in November that were indicative of birds building energy reserves for the winter (i.e. autumnal hyperphagia) but that did not necessarily signify poor physiological condition.

Hampton compiled a time and energy budget for wintering trumpeters. We find, however, that some important factors were overlooked. First, Hampton did not document weather conditions because he assumed that a swan's metabolic rate varied more with the weight of the bird than as a function of air temperature. However, the ambient air temperatures during winter along the Henry's Fork are frequently below -18 °C, which is probably below the swans' lower critical temperature. When ambient air temperatures fall below the lower critical temperature, swans must increase their metabolic rate to maintain body temperatures.

Second, underlying all his energy expenditure calculations were the assumptions that the mean body weights for cygnets and adults were 10.6 kg and 11.5 kg respectively, and that these weights remained constant throughout the winter. In apparent contradiction, however, Hampton noted that the swans' body condition appeared to decline throughout the winter, to the extreme that emaciation was the leading cause of death. If trumpeters wintering on the Henry's Fork experienced weight losses as Hampton asserted, then his calculations of energy expenditures were increasingly inaccurate as the winter progressed. We conclude that overwinter declines in body weight of trumpeters are probable, based up on data from Tundra Swans.

Third, Hampton did not investigate relationships between weather and swan behavior. Bortner (1985) observed no behavioral response to low temperature alone but did note energy conserving behavioral responses to high winds and precipitation. Daily temperature minima in North Carolina, however, often exceed the daily maxima in the Henry's Fork area, and we have observed that trumpeters also attempt to conserve energy when they encounter severe cold. During winter storms on the Henry's Fork, trumpeters noticeably altered their behavior. Swans gathered in groups of 5-40 in sites where they were sheltered from the wind and blowing snow. Almost all swans could be found on islands or ice shelves in energy-conserving postures with their head, necks, and feet tucked under their wing or body feathers. Typically the coldest, stormiest weather in the Island Park area occurs in December and early January, and conditions gradually moderate in February and March. Weather influences on swan activities could provide an alternate explanation for the changes in feeding rates observed by Hampton.

Food Habits

Banko (1960:122-130) presented a detailed review of Trumpeter Swan food habits, however no quantitative food habits studies have been made. The existing information is an accumulation of isolated observations, fecal analyses, and stomach contents of dead swans collected in the Tri-state Area. No data exist which quantify the importance of invertebrates, particularly to young cygnets, the existence of seasonal food preferences, or the relative importance of various aquatic plant species.

Several observers (Page 1976:44, Hampton 1981, Smith 1985) have reported that adults feed primarily on submerged aquatic vegetation and that cygnets consume both invertebrates and plant material. Trumpeters were also observed to capture and swallow small fish on four occasions (Hampton 1981), and to feed during the spring on cow manure in the Centennial Valley (RRLNWR Ann. Narr. Rept. 1963). Trumpeters at MacDonald Pond in March 1987 were observed grazing in the sagebrush/grassland adjacent to the ponds at a time when all available aquatic vegetation in the pond had been consumed (C. Miller, pers. comm.).

During the summer of 1966, 16 trumpeters were captured and held in a pen at Red Rock Lakes. At first the swans were fed only grain, but after about one week various aquatic plants were supplied. The swan were fed *Potamogeton pusillus*, *P. pectinatus*, *P. richardsonii*, *Ranunculus aquatilis*, *Chara vulgaris*, *Sagittaria cuneata*, *Myriophyllum exalbescens*, and *Elodea canadensis*. The swans showed a preference for *S. cuneata* and *E. canadensis*. Utilization of these two plant species by swans was also observed on Upper Lake (RRLNWR Ann. Narr. Rept. 1966). Observations of three caged adult swans at Red Rock Lakes in 1971 showed that they preferred *E. canadensis* over eight other species of aquatic plants. Each bird consumed nearly 9 kg (20 lbs) wet weight of plants daily (RRLNWR Ann. Narr. Rept. 1971).

Hampton (1981) examined the digestive tract contents of two adult female and one adult male trumpeters which died from unknown causes while wintering on the Henry's Fork. The three swans contained an average of 1,425.4 gm wet weight of food material. Individual contents ranged from 1,374-1,467 gm. The animal portion of the contents totaled less than 0.5%. Plant material consisted of *M. exalbescens*, *P. richardsonii*, *P. pectinatus* and *E. canadensis*, which were the dominant species growing at Harriman State Park.

In Grande Prairie, trumpeters fed at sites where the biomass of *Chara sp.*, *Lemna trisulca*, *P. richardsonii*, *P. strictifolia*, and *P. zosteriformis* was greater than the lake average (Holton 1982).

Cygnet food habits have not been quantitatively studied. Page (1976:39) observed one-week-old cygnets feeding on dead insects, and small bits of vegetation and seeds floating on the water surface. Two-week-old cygnets took both floating animal and plant material, and small bits of vegetation from under water. Holton (1982) studied cygnet feeding sites at Grande Prairie and determined that invertebrates greater than 2 mm in length such as *Gammarus limnaeus* , and *Notonectus sp.* were probable foods. Two and three-week-old cygnets were also observed to select seed capsules of *P. richardsonii* which grew above the water surface. *P. praelongus*, *L. trisulca*, and *L. minor* were recovered from the bills of Grande Prairie cygnets during banding (Holton 1985). Studies of nesting habitats in Wyoming indicate that the low survival of cygnets at some territories may be related to inadequate supplies of invertebrates. Further studies to investigate the importance of invertebrates will begin in 1987 (D. Lockman, pers. comm.).

Food Plant Analysis

In August 1972, eleven of the main aquatic plants eaten by Trumpeter Swans at Red Rock Lakes were analyzed along with wheat to determine their nutritional value. All foods were high in protein (7.2 - 19.0% air dried weight) and in a normal range for phosphorus (0.10 - 0.58% air dried weight); the aquatic plants, particularly *C. vulgaris*, were high in calcium (1.8 - 23.7% air dried weight) and contained less crude fat and more crude fiber than did wheat (Page 1976:45).

Discussion

Several biologists have suggested that quantitative and/or qualitative inadequacies in the Tri-state trumpeters' winter and spring diet contribute significantly to their small clutch sizes, reduced incubation constancy, failure of full-term embryos to hatch, and low viability of cygnets during their first weeks of life. No data regarding the adults' actual condition prior to nesting have been gathered due to concern that data collection would cause swans to abandon nesting.

Possible dietary deficiencies were first mentioned in the 1964 RRLNWR Annual Narrative Report. Of 20 eggs collected for incubation studies at the Refuge, only one hatched and the cygnet quickly died. While egg viability may have been reduced by the incubation techniques, eggs left in the nests fared little better. Less than 30% of the eggs which were left in the nests hatched. Fifty percent of those unhatched eggs showed no embryo development. Most of the remaining eggs contained embryos which died between the 27th-33rd day of development. The 1964 Annual Narrative Report suggests that the dietary intake of the adults prior to and during the 1964 laying period did not provide the proper protein nor vitamins necessary for embryo development and was the main cause of the poor reproduction. Unusual weather conditions prevailed that year; the spring was late and cold, and the Refuge lakes were not ice-free until 20 May. The report explained that normally the swans stop using the grain and return to a diet of aquatics as the lakes open up in April. Due to the late thaw, no aquatics were available. Feeding of wheat and barley continued into the laying period, and apparently provided an inadequate diet.

Page (1976:103) suggested that the reduced amounts of grain fed during the winters of 1969-70, 1970-71, and 1971-72 had caused Refuge swans to be in poor physiological condition during the breeding season. Page suggested that the consequences of poor condition were: inability to acquire territories, small clutch size, unviable eggs, weakened cygnets, reduced parental care, and high over-winter mortality rates. During the winter of 1972-73, feeding returned to a schedule of twice per week and the volume of grain fed was increased by about 50%. The number of nests, the percentage of eggs hatching, cygnet survival, and total cygnet production all increased in 1973. No changes in clutch size or territory size were detected, and parental attention and over-winter mortality were not measured. Page (1976:104) noted that the 1973 increase in number of nests might also have been influenced by the cessation of the removal program in the late 1960s and the increased recruitment of young breeders into the breeding flock.

Shea (1979) observed that most broods in the Yellowstone area died within a few weeks of hatching; cygnets were commonly found dead on or near the nest, sometimes dying within hours of hatching. Necropsies typically could not determine the cause of death and Shea suspected that nutritional inadequacies were the underlying cause. Twenty-two percent of the surviving cygnets showed retarded development, and three cases of leg and foot deformities were noted. Banko (1960:140) reported similar deformities at Red Rock Lakes in the 1940s and 1950s.

King (1973) summarized the energetic requirements of reproduction in birds, indicating that egg production is particularly sensitive to caloric deficiencies. We suggest that reduced clutch size in Tri-state swans is the result of inadequate accumulation of prebreeding energy reserves. Thirty-two percent of clutches in the Yellowstone area contained ≤3 eggs. Although small clutches are common among first-time breeders (Lumsden 1987b), Shea's observations indicated that most of these small clutches were laid by established pairs and that these pairs laid small clutches year after year. Cygnets from these small clutches suffered higher rates of pre-fledging mortality than did cygnets from large clutches (Shea 1979).

Another symptom of inadequate energy reserves may be reduced attentiveness during incubation. Nesting trumpeters in Yellowstone spent more time feeding and less time incubating than did the well-fed captive trumpeters monitored by Cooper (1979). Yellowstone swans which laid below average size clutches, had lower hatching success, and incubated with lower constancy than swans with average or larger size clutches (Shea 1979). Shea hypothesized that the small clutch size, increased time spent feeding, and low cygnet viability resulted from the inadequate energy reserves of the incubating female. Yellowstone trumpeters nest at the highest elevations occupied by swans in the RMP range. Winter and spring weather is usually severe; in many years, the higher elevation territories remain ice-bound until mid-May or early June. Food supplies during the prebreeding period are normally limited by ice cover on the ponds. Females may lack the opportunity to accumulate sufficient energy reserves to incubate with the high constancy typical of captive swans.

Little information is available regarding trumpeter food habits and habitat selection during the prenesting period in March and April. Yellowstone swans are often observed occupying ice-bound territories and using limited sites at lake inlets and outlets (Shea 1979). Centennial Valley swans make use of creeks, creek outlets, seeps, wet meadows, and ephemeral ponds in April as the thaw proceeds (McEneaney 1986a). Breeding swans in the lower elevations of Wyoming move to traditional feeding sites during the prebreeding period. These sites usually are within a few km of the nesting territory and provide food and cover while the territory remains icebound (Lockman *et al.* 1987). Lockman *et al.* (1987) attributed the high cygnet production in 1985 to the unusually early ice-off, which allowed the adults almost six weeks of access to spring habitats prior to egg-laying.

In contrast, the Canadian migrants leave the Tri-state winter ranges in early March, and move north to milder, lower elevation habitats. Migrants have been observed moving slowly through southwestern Alberta in early April, feeding in a variety of ponds and flooded agricultural sites (Turner and Mackay 1982). Northward migrating trumpeters often mixed with flocks of Tundra Swans and fed on submerged macrophytes including *P. pectinatus* and *P. richardsonii*. Both species of swans fed in stubble fields of barley (*Hordeum vulgare*) early in the spring, particularly in 1979 when ice breakup was late (Holton 1985).

Shea (1979) suggested that the Canadian migrants encountered more varied and higher quality spring habitats than those available to Tri-state resident trumpeters and that the migrants' higher nutritional plane during the prenesting season was reflected in their higher reproductive performance. Compared to Yellowstone swans, the Grande Prairie trumpeters laid larger clutches, incubated with a higher constancy, had higher nest success, higher percent hatch of eggs, and higher cygnet survival to fledging (Holton 1985).

Red Rock Lakes trumpeters encounter spring habitats and climatic conditions intermediate in severity between those of Yellowstone NP and Grande Prairie (see Chapter 7) and also receive an energy subsidy from winter grain feeding. The reproductive performance of Refuge trumpeters is intermediate to that of the Yellowstone and Grande Prairie swans (Appendix IX), particularly when nest losses at the Refuge due directly to flooding are removed from the comparison.

The best reproductive performance among Tri-state trumpeters is shown by the lower elevation Wyoming and Idaho swans, which usually have access to more varied and earlier thawing spring habitats than the Refuge or Yellowstone swans. D. Lockman (pers. comm.) has found that the chronology of nesting in Wyoming is related to the severity and duration of winter weather, and the length of time which breeders have to feed on good quality prebreeding habitat.

As discussed in Chapter 8, less winter habitat existed in the Tri-state Area prior to grain feeding at Red Rock Lakes NWR and the construction of Island Park and Hebgen dams. In addition to the limited areas available, the harsh climate of the Tri-state Area was marginal for wintering and prenesting trumpeters. Prior to the man-made modifications, the area provided only a few scattered warm spring sites where slow-moving waters remained ice-free during even the coldest weather.

Trumpeters winter in large numbers in the Tri-state Area today because they have in effect been "short stopped" by the elimination of other migratory traditions, by grain feeding, and by the creation of new ice-free areas below dams. The swans' normal endogenous pattern of weight loss, coupled with limited access to aquatic vegetation due to ice cover and the caloric demands of surviving air temperatures which may be as low as -40 °C (-40 °F), may exhaust a swan's energy reserves when winter conditions last from November to April. The limited availability of late winter-early spring feeding areas in the Tri-state region compounds the energetic restrictions facing nonmigrants by not allowing the swans sufficient opportunity to store the energy reserves needed to adequately provision a clutch of eggs and incubate with the proper constancy.

Summary

Studies of Tundra Swans and arctic nesting geese have shown that weight is lost throughout the winter in an endogenous rhythm, and that the energy needed for successful reproduction is gained during spring migration or at the nesting area. Tundra Swans wintering in North Carolina lost up to 25% of their body weight despite abundant food supplies. We conclude that trumpeters likely share a similar annual cycle of nutrient reserve and body weight changes.

We also conclude that use of the Tri-state Area by wintering trumpeters has increased due to the elimination of other migratory traditions, and the "short stopping" of swans by grain feeding and the creation of ice-free habitat downstream from dams. Trumpeters wintering in the Tri-state region endure a long, cold winter with subzero temperatures. The trumpeters' normal endogenous rhythm of winter weight loss, coupled with the marginal climatic conditions of the Tri-state region for wintering and breeding swans, have combined to depress the productivity of nonmigrants. In contrast, the Canadian migrants which leave the Tri-state Area have a greater opportunity to accumulate reserves necessary for breeding.

Few details are known about trumpeter food preferences or the nutritional value of the various food items. Captive trumpeters consumed up to 9 kg (20 lbs) per day of aquatic vegetation. Invertebrates may also be an important part of their diet, particularly during the first weeks of cygnet development.

In summary, numerous observations during the past 20 years support the theory that prenesting food resources are key to successful reproduction by trumpeters. Although the migrant Canadian trumpeters share the same wintering areas with the Tri-state residents, they have access to more varied and milder prenesting habitats as they migrate northward through lower elevations, while the Tri-state swans remain locked in winter. We conclude that these differences in access to transitional spring habitats are a major factor contributing to the superior reproductive performance of the Canadian migrants.

CHAPTER 11. GENETIC CONSIDERATIONS IN TRUMPETER SWAN MANAGEMENT
Peter F. Brussard and Ruth E. Shea

Random changes in a population's genetic makeup can threaten the viability of small populations. Such changes may have deleterious effects on the ability of individuals to survive and reproduce, as well as on the population's capacity to respond adaptively to changes in its environment. The two major genetic factors which cause deleterious effects are inbreeding depression and genetic drift.

Inbreeding depression results from the expression of deleterious alleles as a result of the mating of close relatives. Most populations carry a genetic load of deleterious alleles which are usually rare and not particularly harmful in the heterozygous state. However, when related individuals mate, the chance of inheriting such an allele in the homozygous state is increased. When deleterious alleles appear as homozygotes, they often result in the decline of fitness traits such as fertility, fecundity, and viability. If population size is large enough, natural selection can eliminate the individuals that express the deleterious recessive alleles and the population may recover. If the population is too small, its fitness remains impaired and the population will decline, possibly to extinction.

Genetic drift refers to random changes in gene frequencies in populations. These changes sooner or later result in the loss of genetic variation, the rate of loss being dependent on population size. In chronically small populations, this loss can be quite severe and rapid. For example, in a population with a genetically effective size (see below) of 10 individuals, 40% of the heterozygosity (one measure of genetic variability) in the population will be lost within 10 generations. There is ample evidence from the literature on animal breeding and from laboratory experiments with *Drosophila* and other organisms that populations with higher levels of genetic variation have greater fitnesses than those with less variation. Less evidence exists from natural populations, primarily because of the difficulties involved in estimating both fitness and genetic variability under natural circumstances (Hedrick *et al.* 1986). The evidence that does exist, however, usually confirms the results from domestic or laboratory populations.

As described in Chapter 1, both the Interior Canada Subpopulation and the Tri-state Subpopulation were reduced nearly to extinction by the early 1900s. Censuses of the Tri-state summer residents in the 1930s located only about 50 adults and their cygnets. By the mid-1950s, the descendants of those few families had increased to about 600 birds (Appendix III). The first question that we will address is whether or not the effects of inbreeding and drift during the bottleneck phase of the Tri-state Subpopulation's history, or subsequently, might be an underlying cause of the recent poor cygnet production.

Genetically Effective Population Size

In order to maintain its fitness, a population must be large enough to prevent intense inbreeding and serious reductions in genetic variation by random drift. However, the number of individuals actually involved in contributing progeny, and therefore genes, to subsequent generations is generally only a small fraction of the total population size. A population containing several hundred individuals (its census size, or N) may have a genetically effective size (N_e) of only a small fraction of that number.

A genetically ideal population is one in which all individuals in the population have an equal probability of being the parents of any individual in the next generation (Wright 1931). Thus, the genetically effective size of a population is reduced by a variety of factors that represent departures from the ideal situation. These include the presence of nonbreeding individuals, skewed sex ratios, non-random tendencies for inbreeding or outcrossing, non-Poisson variation in progeny survivorship, the loss of genetic variability which may have occurred during previous periods of low population size (the "bottleneck" effect), and the tendency of the individuals of some species to mate with individuals from an area which may be smaller than the area occupied by the population as a whole (the population structure effect). The cumulative effects of these factors are multiplicative, and are usually expressed as the ratio of N_e/N.

Although simple mathematical formulas for calculating N_e/N exist for most of these factors individually (e.g., Hedrick 1983), the necessary data do not exist to apply them to Trumpeter Swan populations. However, N_e/N ratios have been calculated for the Spotted Owl (*Strix occidentalis*) (Barrowclough and Coats 1985; Salwasser *et al.* 1984), the grizzly bear (*Ursus arctos*) (USFWS 1986a), and the black-footed ferret (*Mustela nigripes*) (Groves and Clark 1986; Brussard and Gilpin 1987). The N_e/N ratios resulting from these analyses range from 0.1 to 0.5; that is, the genetically effective sizes of these populations range from one-tenth to one-half their census sizes.

Effective Size of the Tri-state Subpopulation

The N_e/N ratio in Trumpeter Swans may lie toward the low end of this range, at 0.1, due both to the "bottleneck" through which the Subpopulation passed in the early 1900s, the subsequent failure of many nesting pairs to fledge cygnets (see Chapter 3), and the existence of several age classes of subadult nonbreeders. In 1985, the Tri-state Subpopulation consisted of about 500 individuals. If the N_e/N ratio is 0.1, this implies an N_e of 50. The expected rate of increase in inbreeding in each generation for an N_e of this size is about 1.0% (Frankel and Soule 1981). This is just about the level where natural selection for performance and fertility can balance inbreeding depression. With an estimated N_e of 50, 99% of the genetic variation in the Tri-state Subpopulation will be conserved each generation. Thus on the basis of genetic considerations, an N_e of 50 is probably adequate for the short-term survival of the Subpopulation, provided that the Subpopulation's total size is large enough to reduce the risks of extinction from demographic and environmental factors to acceptable levels.

Effective Size of the Centennial Valley Flock

It is not at all clear from banding data (see Chapter 9) that the various flocks within the Tri-state Subpopulation actually interbreed. A few individuals have been observed to disperse from the Centennial Valley flock to Yellowstone and Lower Wyoming, but it is not obvious that the other flocks have ever produced enough cygnets to maintain their own stability and provide emigrants to the Centennial Valley. The Centennial Valley flock could be a relatively isolated unit. If so, it could be in serious trouble genetically. The flock now consists of about 40 breeding pairs, or 80 breeding adults. If we conservatively estimate its N_e to be 20 (we use an N_e/N ratio of 0.25 instead of 0.1 because we are considering breeding birds only), inbreeding could be increasing and genetic variation could be decreasing at a rate of 2.5% per generation.

The number of genetically effective individuals which must be exchanged among flocks within the Tri-state Subpopulation to eliminate the effects of isolation and maintain genetic continuity is between one and two per generation (Lande and Barrowclough 1986). Thus, rather low levels of gene flow could alleviate most problems with inbreeding and loss of variability in the Centennial Valley flock. However, it is impossible to determine from the available data the amount of gene flow that is actually occurring between the flocks in the Tri-state Subpopulation as a whole. It could well be none, particularly during periods of Subpopulation decline.

Long-term Considerations

Even if sufficient gene flow exists among the Tri-state flocks to prevent high levels of inbreeding and losses of genetic variability in any one flock, the estimated N_e value of 50 or so for the Subpopulation as a whole provides a comfortable margin only for short-term survival. Survival over longer periods depends on adaptability, the ability of the Subpopulation to change its genetic composition in response to long-term changes in the environment. This ability, in turn, depends upon a broader array of genetic resources than are required for short-term survival, implying a need for considerably larger N_e values. For example, at an average N_e of 50, roughly 64% of a population's neutral (i.e., currently unselected) genetic variation will be lost through drift in one hundred generations. Since evolutionary change depends on the presence of heritable variation, this amount of loss represents a serious erosion of adaptive potential.

Theoretical considerations suggest that an N_e of 500 is a satisfactory first approximation of the minimum N_e necessary for continuing evolution (Franklin 1980; Frankel and Soule 1981). Since the actual number of individuals necessary to maintain an effective population size this large is likely to be considerably greater than 500 (over 5,000 if the N_e ratio is 0.1), it is clearly possible that both the Tri-state and Interior Canada Subpopulations are in jeopardy of losing their potential for long-term adaptability, unless the subpopulation sizes or rates of gene flow between the subpopulations increase.

Population Differentiation, Gene Flow, and the Risk of Outbreeding Depression

One solution to the problem of loss of long-term adaptive potential in Trumpeter Swans would be to manage them by maintaining a trickle of gene flow between the populations, thereby increasing the N_es of the individual populations to that of the species as a whole. This gene flow could be either natural or artificially induced. However, it has been suggested on the basis of morphological measurements (Hansen 1973) that the Pacific Coast Population may represent a differently coadapted stock. If hybridization between differently coadapted populations takes place, an outbreeding depression may occur (Templeton *et al.* 1986). This has many of the same symptoms as inbreeding depression such as increased mortality, decreased fertility, etc., and it may be manifested in the F1 or delayed until the F2 or backcross generations. If the Pacific Coast Population does represent a differently coadapted stock, gene flow between it and the Rocky Mountain Population could make matters worse, not better, by trading a probable inbreeding problem for a definite outbreeding problem.

Thus, an informed decision on genetic management of the Tri-state Subpopulation should depend upon an assessment of the extent of genetic differentiation between the Pacific Coast Population, the Interior Canada Subpopulation, and the Tri-state Subpopulation. If the level of differentiation is high, an outbreeding depression may result from artificially induced gene flow.

If, on the other hand, little or no genetic differentiation is observed among these three stocks, the risks of outbreeding depression would be substantially reduced.

How can the level of genetic differentiation be determined? Banding returns, field observations, reproductive patterns, morphometrics, chromosome analysis, gel electrophoresis, and DNA mapping can all play a part. None of these tools is adequate to do the job alone; each has varying ranges of uncertainty. Gel electrophoresis and morphometrics probably constitute the most cost-effective tools currently available for estimating genetic structure in bird populations.

Morphological Differences

Differences in morphological traits have been used to assess the degree of genetic differentiation of Trumpeter Swan populations. Hansen (1973) stated *"From the detailed studies conducted both at the Red Rock Lakes and in Alaska, it appears that these two widely separated populations of trumpeters may be distinct subspecies. There are significant physical differences, the Alaska trumpeters being larger, starting from the egg."* Measurements of eggs from the RMP have been made by Banko (1960:201), Shea (1979), Maj (1983), and Holton (1985). Hansen (1971) provided measurements of eggs from south-central Alaska (Table 35). Comparison of these measurements shows that the eggs from Red Rock Lakes were significantly smaller than those from Alaska, Grande Prairie, the Targhee National Forest, and Yellowstone. Eggs from Grande Prairie, Yellowstone, and the Targhee were intermediate between those from Alaska and Red Rock Lakes.

It is also possible to compare weights and bill lengths of swans from Alaska (Hansen *et al.* 1971; Barrett and Vyse 1982), Grande Prairie (Holton 1985) and Red Rock Lakes (Barrett and Vyse 1982) (Table 36). Our analysis of these combined data sets shows that female swans from Alaska, measured by Hansen, weighed significantly less than females from all other localities. There are no other significant weight differences among males or among females. Bill lengths of female trumpeters from Red Rock Lakes are significantly smaller than those from Alaska measured by Hansen. We find no significant differences in bill lengths between males from any localities, nor among the Alaskan females measured during the two different studies.

Thus, contrary to Hansen's assertion, Alaskan trumpeters do not appear to be consistently larger than those from Red Rock Lakes or from any other locality, at least in parameters other than egg size. Alaskan swans measured by Barrett and Vyse do not differ significantly from those at Red Rock Lakes in weight or bill length. Furthermore, since most morphological characters are affected to an unknown extent by environmental influences, the occasional differences observed between swans from Red Rock Lakes, Alaska, and Grande Prairie may not reflect underlying genetic differences at all.

Electrophoretic Differences

Barrett and Vyse (1982) compared trumpeters from Alaska ($N = 43$), Grande Prairie ($N = 58$), and Red Rock Lakes ($N = 128$) by electrophoretic examination of plasma and erythrocyte protein products of 19 presumptive gene loci. Five of these were reported to be polymorphic.

Table 35. Measurements of Trumpeter Swan eggs.

Site	Number of Eggs	Diameter			Length		
		Mean + SD	Min.	Max.	Mean + SD	Min.	Max.
Red Rock Banko (1960)	109	72.4 + 2.0	68.0	77.5	110.9 + 4.5	104.0	123.0
Yellowstone Shea (1979)	45	73.8 + 2.1	68.6	76.9	114.2 + 3.7	107.5	123.5
Targhee NF, ID Maj (1983)	54	75.3 + 1.8	71.0	79.0	115.7 + 4.4	107.0	126.0
Grande Prairie Holton (1985)	97	73.2 + 2.0	68.9	77.8	116.4 + 5.9	101.0	126.8
Alaska Hansen *et al.* (1971)	146	75.0 + 2.0	69.8	81.0	117.4 + 4.0	109.8	125.0

Table 36. Morphological measurements of adult Trumpeter Swans.

Site	Mean Weight (kg + SD)		Mean Bill Length (mm + SD)	
	Male	Female	Male	Female
Red Rock Lakes[a]	11.4 + 0.7 *n* = 27	10.3 +1.2 *n* = 47	54 + 2.0 *n* = 27	50 + 3.6 *n* = 47
Grande Prairie[b]	11.7 + 0.7 *n* = 3	10.0 + 0.7 *n* = 7	No Data	
Northwest Territories[c]	12.1+1.6 *n* = 9	10.2 + 0.7 *n* = 11	54.0 + 2.6 *n* = 9	52.1 + 2.1 *n* = 11
Alaska[d]	11.7 +1.2 *n* = 18	9.5 + 0.9 *n* = 20	55 + 2.6 *n* = 21	53 + 2.3 *n* = 19
Alaska[e]	11.8 + 0.6 *n* = 7	10.2 + 0.7 *n* = 15	54 + 2.6 *n* = 7	51 + 3.8 *n* = 8

[a] Barrett and Vyse 1982
[b] Holton 1985
[c] McCormick and Shandruk 1986.
[d] Hansen *et al.* 1971.
[e] Barrett and Vyse 1982.

Three loci (s-MDH, m-MDH, and Est-1) had extremely low levels of polymorphism in the Alaskan sample, and were monomorphic elsewhere. The CAR locus showed variation in all three samples, but the phenotypes observed may or may not represent genetic variation since two of the three samples (Red Rock Lakes and Grande Prairie) were not in Hardy-Weinberg equilibrium and showed large heterozygote deficiencies (our analysis). The locus TRF showed virtually identical frequencies of an alternate allele and a slight heterozygote excess in all three samples. Thus, overall heterozygosity was low (H = 0.01), but within the range of values found in other bird species (Corbin 1978; Avise *et al.* 1980; Avise and Aquadro 1982).

Barrett and Vyse found no significant differences in mean heterozygosity between the three populations, but suggested that the presence of alternate alleles at s-MDH, m-MDH, and Est-1 in the Alaskan sample indicated that some unique variability exists in the Pacific Coast Population. It seems more likely that the Alaskan trumpeters have retained these rare variants which have been lost elsewhere as a result of past population bottlenecks. Bottlenecks of short duration will have little effect on heterozygosity, but will severely reduce the number of rare alleles (Allendorf 1986, Fuerst and Maruyama 1986).

The overall genetic distances between these three stocks are very small. The average Nei (1972) distance between the three is 0.002; the average Rogers' (1972) distance is 0.0043. This is very little differentiation at the population level, even in birds (Avise *et al.* 1980). Thus, Barrett and Vyse's data suggest that there is negligible differentiation between the Alaskan, the Red Rock Lakes, and the Grande Prairie trumpeters. The results of the morphometric analyses tend to support this conclusion.

Summary and Management Implications

Judging from considerations of N_e/N ratios in other birds and mammals, there is a good chance that the Centennial Valley flock is suffering from the effects of inbreeding and loss of genetic variation in addition to its many other problems. The genetic effects can be counteracted by implementing one-way gene flow with other trumpeter stocks, preferably through transplants from the relatively genetically diverse Pacific Coast Population. However, there is some chance that this might result in an outbreeding depression in the F1 or F2 generations after the transplants, if the Red Rock Lakes and Alaskan trumpeters represent differently coadapted stocks. The evidence, albeit sketchy, suggests that this is unlikely.

Nevertheless, a higher level of certainty of the degree of differentiation between the extant trumpeter stocks should be obtained prior to making such transplants. Thus, we recommend that a second electrophoretic and morphological study be conducted so that a more adequate assessment of population boundaries can be made. Additional electrophoresis is recommended so that more polymorphic loci can be found. An element of chance exists in such a search; although researchers now routinely resolve at least twice as many loci as Barrett and Vyse did, the larger samples do not guarantee that any more of these loci will be polymorphic. Thus, we suggest that the first step in a new study should be to survey 40 to 50 loci in a sample of 50 to 100 trumpeters, collected from the Pacific Coast Population only. If variation exists in the species, it will be found there. Blood samples, fertile eggs, or freshly dead swans would provide adequate material. If the survey proves successful and additional useful loci are discovered, then the study can be expanded to include samples from the Interior Canada and Tri-state subpopulations. On the other hand, if no more useful loci are found, it may be necessary to consider looking at mitochondrial or ribosomal DNA, a much slower and more expensive procedure.

Additional morphological measurements involving many additional characters from adult birds (e.g., length of fifth primary, forearm, neck, tarsus, bill nail, width of bill and foot web, etc.) should also be taken and analyzed by multivariate techniques. Significant variation between populations and strong concordance between the electrophoretic and morphological data sets would suggest genetic differentiation. Lack of concordance would tend to confirm the results of the earlier studies.

As a final caution, however, we must emphasize that electrophoretic and other molecular techniques assay only a very small proportion of the loci in the swan genome. Furthermore, these loci probably have little, if anything, to do with population fitness. Likewise, one cannot tell how much morphological variation is genetic and how much is environmentally induced without heritability studies. Although reasonable correlations exist between genetic similarity as measured by electrophoresis or DNA analysis, patterns of morphological variation, and overall genetic similarity, absolute proof of potential outbreeding depression can only come from controlled matings between individuals from the populations of interest. Such experiments would be a very tedious and time consuming enterprise in a bird with such a long generation time. However, molecular techniques combined with morphological data can provide an adequate level of precision with which to make informed judgments, and such a data base would provide valuable guidance on how best to incorporate genetic structure into management plans.

CHAPTER 12. POPULATION-HABITAT RELATIONSHIPS

Previous Analyses of the Tri-state Subpopulation

Banko (1960:144-164) made the first analysis of the population dynamics and habitat relationships of Tri-state trumpeters, for the period from 1931 through 1958. Banko showed that during these years of protection and intensive management the descendants of the few surviving families increased at an annual rate of about 10%, from about 70 swans in 1932 to 630 in 1954. From 1955-58, the number of Tri-state trumpeters leveled off and fluctuated between 560-640. Unexpectedly low census results in 1957 led Banko to conclude that about 100 trumpeters moved outside the Tri-state survey area but apparently returned prior to the 1958 census.

Using census data from the late summer USFWS Trumpeter Swan Surveys, Banko (1960:146-151) analyzed the relationships between the numbers of cygnets, mated pairs, total adults, total nonbreeders, and total swans. Banko found that cygnet production was inversely related both to the total number of adults and to the number of mated pairs. Banko also observed that by 1950-54, the proportion of flocked nonbreeders was much greater than in the early years (1939-41).

To explain the inverse relationship between cygnet production and population size, Banko (1960:164) suggested that cygnet production was regulated by density-dependent mechanisms that affected cygnet hatching and survival rates. Although Banko was not able to determine how these mechanisms worked, he suggested that they possibly involved social competition, because he found no evidence of habitat deterioration. To explain the rapid increase in nonbreeders in the 1950s, Banko (1960:148) suggested that as the number of swans increased, a shortage of suitable breeding habitat prevented the rising number of breeding-age swans from nesting. The shortage of available territories prolonged the swans' flocked nonbreeding status and in effect, created a reservoir of surplus eligible nonbreeders. Banko (1960:163) further concluded that the Tri-state Trumpeter Swans should no longer be considered an endangered species because after 20 years of near constant increase, they showed signs of leveling off at the maximum number that their environment could support.

Red Rock Lakes NWR Flock

Banko (1960:151) showed that the RRLNWR flock played a major role in determining the status of the total Tri-state Subpopulation. The inverse relationships between cygnet production and mated pairs, and between cygnet production and total numbers were even more pronounced for the Refuge flock than for the entire Subpopulation. As in the case of the total Tri-state Subpopulation, the production decline in the Red Rock Lakes flock also was attributed to the failure of eggs to hatch and young to survive. Banko concluded that the Refuge habitat appeared to be saturated when it contained about 40 breeding pairs and greater breeding numbers depressed cygnet productivity.

Changes in the size of the Refuge flock during the years from 1932-73 were subsequently examined by Page (1976), using two graphic methods: semi-log plots and serial correlations. Page (1976:31-39) reached the following conclusions:

1) Prior to 1954, the adult portion of the flock grew and declined in a density-independent manner, but the growth of the whole flock (all age classes) showed density-dependent controls.

2) After 1954, the flock reached an upper maximum and fluctuated widely. The increases in the adult segment were density-independent, but the decreases were strongly density-dependent.

3) The equilibrium point for the adult segment of the Refuge flock was near 155 in the late 1950s and 1960s, but declined to about 135 by 1973 (Page 1976: Fig. 9).

We question the validity of Page's analysis due to several problems with his methods. First, the serial correlation technique that Page used to reach conclusions one and two was criticized by Brockelman and Fagen (1972), who showed that the technique is of little usefulness because the results depend on the amount of a population's initial displacement from equilibrium and on the number of serial samples. We also question Page's (1976:34) justification for calculating separate axes for years of increase and years of decrease after 1954, in order to calculate an equilibrium point. We disagree with Pages' application of the concept of equilibrium point. Rather than being a central point around which the flock numbers fluctuated, Page used the equilibrium point merely to represent the lower levels to which the flock declined during a period of years when census results varied widely. In effect, Page merely documented that the number of swans on the Refuge was indeed declining.

Page (1976:73-92) also constructed a deterministic computer model to analyze the relative importance of survival rates, nesting habitat limitations, and age at first reproduction in the maintenance and stability of the RRLNWR flock. His main conclusions were:

1) When no limits were set on the number of available territories, all increases in cygnet and subadult survival rates caused the flock to expand. All increases in cygnet and subadult survival rates, when compared by percent increase, caused similar increases in the flock growth rate. In marked contrast, small increases in the adult survival rate caused much higher rates of flock growth, most likely due to the compounding effect of the many adult age classes (Page 1976:Fig. 13).

2) When the computer simulations restricted the number of available territories to 50, increases in survival rates caused the flock to expand and then stabilize at an upper maximum, which was determined by the amount of the survivorship increase. As survival rates were increased, the number of surplus eligible breeders increased (Page 1976:85).

3) As adult survival rates were increased, the size of the flock that could be maintained with a particular number of nesting territories increased. Correspondingly, with higher survival rates, fewer territories were required to maintain a given size flock (Page 1976:85).

4) Small increases in adult mortality caused severe flock declines, due to the compounding effect of the many adult age classes. Increasing the adult mortality rate from 15 to 20% caused the extinction of the flock, with the flock size halving every 20 years.

5) In an unlimited habitat, lowering the age at first reproduction from five years to four years caused the flock to grow at a slightly more rapid rate, however the difference was not great.

Like Banko, Page assumed that the molting flocks contained a large number of eligible breeding-age swans that were prevented from nesting due to habitat limitations. He did not, however, actually demonstrate that excess eligible breeders existed in the Tri-state flocks. Rather, Page used his model to demonstrate that if breeding habitat limitations were set on a simulated population, then excess eligible breeders resulted.

Page (1976:89) also used his model to examine the possible effects of swan removals from

RRLNWR. Simulations showed that removals approximating the actual levels of removals in the 1950s and 1960s could have caused the flock to decline by the late 1960s, if the removals were additive to natural mortality. Page (1976:125) found that from 1954-68, the number of swans removed from the Centennial Valley equaled almost 61% of the known recruitment to the Refuge flock. In 5 of 15 years, the number of swans removed actually exceeded recruitment. While concluding that the flock's downward trend in the late 1960s could have been caused by the removal program, Page did not address the question of whether the removals were in fact equivalent to additive mortality, or whether the removals were compensated by some type of density-dependent increase in cygnet production or adult survival.

Neither Page nor Banko analyzed the records of the Refuge winter feeding program in detail. Both authors mentioned the effects of this program only briefly, and implied that they did not consider winter grain supplies to have been a limiting factor. Page (1976:102) erroneously stated that prior to the winter of 1969-70, the Refuge winter feeding program provided the trumpeters with 1,200-1,400 bu of wheat each winter. As we showed in Chapter 8, the feeding program actually varied widely in the amount fed, the timing, and methods of feeding. Page (1976:111) hypothesized that winter feeding resulted in decreased mortality rates and allowed the Refuge flock to increase by the 1950s to the level where a shortage of nesting habitat limited the number of breeders, and survival rates dictated the number of excess eligible breeders. The flock then declined below this level due to increased removals and/or increased mortality.

Page (1976:100) also stated that the Refuge flock had been maintained at an artificially high level since 1954 by winter feeding and these high numbers of swans had overgrazed and damaged *Elodea*, one of their preferred spring foods, causing the carrying capacity of the Refuge to decline. He suggested that early spring was the most critical time of year for Refuge swans, because many aquatic plants deteriorated during winter and spring food supplies were limited. Based upon his assumption that the artificially large trumpeter flock had damaged *Elodea*, Page (1976:138) concluded that the removals of swans had probably been beneficial because they reduced the flock which had apparently overused its food resources. Thus, Page's studies supported Banko's conclusion that the Refuge flock exceeded the carrying capacity of its habitat. Page, however, suggested that spring habitat, rather than the number of suitable breeding territories, was the limiting factor in the 1970s.

Yellowstone National Park

Banko (1960:155-162) suggested that when the number of swans in the Centennial Valley was high, CV swans moved into Yellowstone NP and thus depressed the apparent ratio of cygnets per mated pair in the Park. When the number of swans in the CV decreased, movement to Yellowstone did not occur and the number of swans in the Park declined. Prior to 1949, a relatively constant flock of paired swans in the Park produced cygnets at a fairly constant rate. In most years from 1949 to 1955, greater numbers of paired swans in Yellowstone coincided with reductions in cygnets per mated pair, although cygnets per brood and total cygnets did not decrease significantly. Breeding habitats in Yellowstone appeared to be saturated when occupied by about 15 pairs; greater breeding populations seemed to depress productivity. Banko suggested that in the discontinuous Park habitat, environmental factors other than population density were more variable and influential in regulating annual production than in the continuous Refuge breeding habitat. As we will explain in the following section, we disagree with the method that Banko used to quantify relationships between cygnet production and reproductive effort.

Although Page (1976:136) made no studies outside of Red Rock Lakes, he concluded that much of the breeding habitat in the Tri-state Area, particularly in Yellowstone and Grand Teton National Parks, was unproductive due to high levels of human disturbance. As we discussed in Chapter 3-5, although human disturbance has definitely contributed to reduced cygnet production at particular sites, it by no means is the dominant factor involved in nest failures and early cygnet mortality. Many of Yellowstone's unproductive nesting territories are virtually undisturbed by humans.

Discussion of Population-Habitat Relationships

Management of the Tri-state trumpeters, as recently as 1983 (USFWS Trumpeter Swan policy memo, January 1983), has been dominated by the assumptions that the Tri-state Subpopulation exceeded the carrying capacity of its summer habitat, a surplus of breeding age swans existed that were nonproductive due to a shortage of breeding habitat, and cygnet production was depressed by density-dependent social interactions. Further efforts to increase the number of swans were thought to be futile, since higher numbers of swans were expected to further depress cygnet production. These ideas were accepted and expanded by several subsequent authors (Banko and Mackay 1964, Erickson 1969, Denson 1970, Hansen 1973). Borrowing also from Page's ideas, managers concluded that cygnet production could be stimulated, and the Tri-state trumpeters would benefit, by the harvest of surplus swans through increased removals (Erickson 1969; USFWS Trumpeter Swan policy memo, January 1983). In conjunction with the confirmation of a robust population of trumpeters in Alaska, the assumption that the Tri-state swans were at, or above, the carrying capacity of their habitat provided the impetus to remove the species from the rare list in 1968 (Erickson 1969).

These assumptions have persisted, even though the Tri-state Subpopulation decreased both in total number and number of nesting pairs in the 1970s and 1980s, without exhibiting any evidence of increased cygnet productivity and recruitment.

With the advantage of hindsight formed by analysis of the massive 50-year data base, we found that Banko's conclusions were not as straightforward as subsequent writers and managers have assumed, due to the interactions of the following factors:

1) Banko overestimated cygnet production prior to 1954 compared to production after 1954 because the USFWS summer surveys were flown two to six weeks earlier in the years prior to 1954 (Appendix III). In 1954 and 1955, cygnet production was also underestimated because a total of eleven cygnets were removed prior to the survey (Appendix IV).

2) As Page (1976:95) pointed out, Banko's use of the term "cygnets per mated pair" was not an accurate measure of cygnet production per unit of reproductive effort. For want of a better term, Banko (pers. comm.) used "mated pairs" in all of his analyses of production and populations. However, the number of mated pairs recorded in a given year equaled the number of groups of two white-plumaged swans observed on the late summer USFWS survey. Recent neck-banding studies have revealed that nonbreeding trumpeters such as subadult siblings, unrelated unpaired subadults, paired three and four-year-olds, as well as actual nesting pairs, may associate in groups of two during late summer. Groups of two swans from all of these nonproductive categories were lumped in Banko's category "mated pairs". Therefore, the number of "mated pairs" increased in the 1950s as the number of adults and subadults increased,

but certainly did not represent an accurate measure of increasing reproductive effort. The inclusion of subadults in the category "mated pairs" caused Banko to further underestimate cygnet production per breeding pair in the 1950s when the rapidly increasing Subpopulation included a large proportion of subadults.

3) Banko's (1960:149) analyses of the swans that he referred to as "Total Population", did not deal solely with Tri-state trumpeters, but also included swans from Malheur and Ruby Lakes National Wildlife Refuges, and several zoos. The data were from the USFWS Annual Trumpeter Swan Surveys and prior to 1971 these surveys included all trumpeters in the lower 48 states, including those in zoos and transplant flocks. Most of the swans transplanted in the early years failed to breed in their new environments (Banko 1960:181): therefore, their inclusion in the analyses increased the proportion of nonproductive adults, and further depressed calculated cygnet/adult ratios in the later years.

In retrospect, it is apparent that the data used in Banko's many calculations were quite imprecise, and we therefore have little confidence in the calculations of the annual relationships between cygnet production and population size. Despite the problems with the details of the analyses, however, we agree with Banko's conclusions regarding the general trends in the proportion of nonbreeders and in cygnet production. The increase in both the number and proportion of flocked nonbreeders in the 1950s was shown quite clearly by the Tri-state Area survey data (Banko 1960:147). No data were gathered for the number of nest attempts in the total Tri-state Subpopulation in any year prior to 1979. Therefore, cygnet production per nest attempt could not be calculated. However, our analysis of cygnet production per nest attempt at RRLNWR (Chapter 3) showed that a significant negative correlation ($P = 0.001$) existed between cygnet production per nest and the number of nesting pairs, thus supporting Banko's observations.

As both Banko and Page recognized, a shortage of nesting territories could theoretically restrict the number of pairs that nest, and result in a surplus of eligible nonbreeders. Although this may be occurring locally in the lower elevations of Wyoming (see Chapter 9), several recent observations indicate that suitable breeding habitat was vacant in the 1950s and remains vacant today. Since the 1950s, when Banko considered the Tri-state breeding habitat to be filled by trumpeters, swans have pioneered a total of at least 20 new nesting territories in the Centennial Valley, and in the lower elevations of Idaho and Wyoming. We could find no evidence to suggest that these territories were not suitable and available to swans in the 1950s. Also since the late 1950s, the number of pairs nesting at RRLNWR has declined by about 50%, and the number of nesting pairs in Yellowstone NP has shown a comparable decline since 1978. As a result of these declines, at least 30 historically occupied territories are currently vacant. We conclude that several factors other than saturated breeding habitat were responsible for the leveling off of total swan numbers, the increase in the number and proportion of nonbreeders, and the inverse relationship between cygnet production and population levels.

Between 1950-54, the number of adults censused in the Tri-state Subpopulation rose from 279 to 548 (Appendix III). Between 250-300 of these birds occurred in the large molting flocks at Lima Reservoir and Upper Red Rock Lakes. Banko suggested that many were eligible breeders that were prevented from obtaining territories because the breeding habitat was saturated (Banko 1960:147-8). We conclude that the sharp increase in nonbreeders in 1951-54 was the likely result of:

1) a rapid increase in the number and proportion of subadults in the Tri-state Subpopulation

due to several years of high cygnet production between 1948-53, and increased overwinter survival rates in the early 1950s. Between 1942-48, 311 cygnets or an average of 44.4 cygnets/year (after removals) were censused in the Tri-state Subpopulation. Between 1948-53, 562 cygnets were censused after removals, or an average of 93.7 cygnets/year. Overwinter survival of Centennial Valley trumpeters also was probably unusually high in the early 1950s. Our analysis (Chapter 5) showed a highly significant positive correlation between the overwinter survival rate of Centennial Valley trumpeters and the amount of water released at Island Park Dam. Winter water releases at Island Park Dam between November 1950 and March 1953 (Table 33) were much higher than in preceding or subsequent years. This unusual availability of winter habitat on the Henry's Fork, coupled with moderate levels of grain feeding at Red Rock Lakes, would have promoted relatively higher winter survival. Thus, a period of combined high recruitment and high winter survival resulted in a pulse of young white birds in the population. We suspect, however, that it was the young age of these birds, rather than lack of available nesting habitat, that prevented them from nesting.

2) the increased visibility of nonbreeders in the early 1950s due to several years of drought (RRLNWR Ann. Narr. Repts. 1953, 1954, 1956). As ponds dried up by midsummer, both subadults and failed breeders had little choice but to congregate on the larger water bodies where they were easily censused.

Inverse relationships between population size and cygnet production, and between the number of nesting pairs and cygnet production were found both by Banko and by our analyses (Chapter 3). These inverse relationships probably arose from several factors that are inherent in the population dynamics of a long-lived species having several subadult age classes and a delayed age of first breeding. During periods of rapid population increase, cygnet production ratios would be expected to decline due to shifts in the Subpopulation's age structure and differences in age-specific productivity. As the number of swans increased following several years of high cygnet recruitment, the average age of the breeding segment would decrease as these relatively large, young cohorts first nested. Young, inexperienced breeders are less productive due to smaller average clutch size (Lumsden 1986b) and inefficient care of small cygnets (Leach 1977). Increases in the proportion of young breeders in the breeding segment at high population levels would therefore contribute to the observed inverse relationship between cygnet production and number of nests. The presence of a high proportion of nonproductive subadults (age classes two to four) during periods of rapid population increase and high population levels would also contribute to the observed inverse relationship between cygnet production and the total number of adults/subadults. Thus, we maintain that although reduced per-capita recruitment accompanied high numbers of paired swans and of total white birds, the relationship was not one of cause and effect in a density dependent sense.

Reduced cygnet production per nest at RRLNWR after about 1950 coincided with many factors other than the increasing numbers of swans. Review of the Refuge files clearly indicated that human disturbance increased on the Refuge during this period. Throughout the 1930s and 1940s, managers maintained a virtual ban on human activity in the Refuge marshes during the nesting and brood rearing period. These restrictions were relaxed in the late 1940s as swan numbers increased and concern over the possible extinction of the species lessened. Numerous studies of the swans and other marsh resources were conducted beginning in the late 1940s, and visits to swan nests during incubation and immediately after the hatch became routine.

Disturbance increased substantially in 1952 with the acquisition of an airboat, which was used in subsequent years for research and management activities including the annual capture of flightless swans. Page (1976:58) noted that the noise of the airboat engine was extremely disruptive; swans that were several miles distant hurriedly left their nests and hid when the engine was started. In the continuous nesting habitat of the Refuge, use of the airboat had the potential to interfere with incubation and the care of small cygnets on many territories simultaneously. Other human disturbances, including boating and fishing, also increased on the Refuge as managers grew more confident that the trumpeter populations were secure.

In addition to increased levels of human disturbance, cygnet production was also depressed by an increase in the frequency of adverse weather conditions and higher water levels in the years since 1954 (Chapter 3). We found a highly significant negative correlation between cygnet production and July rainfall ($P= 0.02$) and a negative correlation between July water levels and cygnets fledged per nest ($P= 0.08$). Both water levels and July rainfall were higher after 1954. In addition, we showed that although none of the years prior to 1954 combined both high July rainfall and high water levels, 7 of the 31 years since 1954 have been of this worst type. We concluded that both weather patterns and water management on the Refuge contributed to the poor cygnet production after 1954.

The leveling off of swan numbers in the mid-1950s was likely caused by a combination of decreased cygnet production and increased overwinter mortality due to the shutoff of winter water releases at Island Park Dam, upstream from the Harriman State Park wintering sites. In our analysis of mortality factors (Chapter 5) we found no evidence of density-dependent mortality in the Centennial Valley flock. Instead, we found that losses of swans were significantly correlated with the availability of winter food sources, specifically the amount of grain fed per swan at the Refuge ($P= 0.016$) and the duration of low water releases at Island Park Dam ($P= 0.009$). As we discussed in Chapter 8, both of these winter food resources have varied widely and abruptly over the years. After three years of higher water releases in 1951-53, during which time the Tri-state Subpopulation grew rapidly, winter water releases were curtailed completely beginning in the winter of 1954, while the number of swans was nearing record high levels. We suspect that this loss of habitat at Harriman State Park in effect reduced the carrying capacity of winter habitat that was essential to the Centennial Valley trumpeters.

Winter water releases were cut off completely for at least a month in most winters until 1968. At the same time, the amount of grain fed per swan at RRLNWR also varied annually from about 3.5-7 bu. During these years, the number of swans in the Centennial Valley flock and the Tri-state Subpopulation fluctuated widely. We suspect that the continued decline of the Centennial flock since 1968 was due to its worsening cygnet production, coupled with a reduction in its winter food supplies. Although the water flows in the Henry's Fork have been generally higher and more constant since 1968, they were again curtailed during the winters of 1977-78 and 1980-81, resulting in extensive ice formation. The number of Canadian trumpeters using the river also has increased dramatically and the potential for competition at key feeding sites has increased. At the same time, the amount of grain fed per swan at RRLNWR decreased sharply during the 1970s. By the early 1980s, the amount of grain fed per swan was the lowest in the history of the feeding program (Figure 15) and the feeding methods used tended to restrict the availability of grain to subordinate swans.

We conclude that throughout most of the last 50 years, the availability of winter foods in the Harriman State Park area and at the RRLNWR feeding ponds have been the primary factors limiting the growth of the Centennial Valley flock. Page (1976:100) concluded that the RRLNWR trumpeters had been maintained at an artificially high level since 1954 due to the influence of the winter feeding program. We strongly disagree. The Red Rock Lakes/Centennial Valley trumpeters (and thus the Tri-state Subpopulation) have been maintained at an artificially high level since almost the first years of management, when grain feeding began in 1935. Without the energy subsidy provided by supplemental grain, trumpeters could not winter in the Centennial Valley. The "natural carrying capacity" of the Centennial Valley for nonmigratory trumpeters is virtually zero because it cannot meet the trumpeters' winter or spring habitat needs.

As we described in Chapter 8, prior to supplemental feeding and the impoundment of reservoirs that created downstream areas of ice-free habitat, the Tri-state region provided only very limited natural areas near warm springs, where both the lack of ice and sufficiently slow moving water provided reliable winter habitat. Except in unusually mild winters, we doubt that these sites were able to sustain more than a few hundred wintering swans. The management of recent decades, which had a goal of maintaining a nonmigratory Tri-state Subpopulation in balance with its natural habitat, faced a perplexing task. Without realizing it, management itself almost annually varied the winter carrying capacity of the Tri-state Area by manipulating water flows and the amounts of grain fed.

While the survival of swans through the winter is related to the availability of winter foods, the quality and availability of spring foods appears to be critical to the swans' reproductive performance (Chapter 11). Page (1976:100) observed that while the Refuge swans did not appear to be limited by breeding habitat in the 1970s, spring food sources were in short supply, particularly due to the observed decline of *Elodea* on the Refuge. Page suggested that the artificially high numbers of trumpeters might have overgrazed this preferred plant and contributed to its decline. *Elodea* has declined on the Refuge since the 1950s. There is no evidence, however, that its decline was caused by overgrazing; several other factors were probably more important. Since the mid-1950s, increased turbidity, increased water depths, decreased light, and decreased water temperatures have probably contributed to the decrease in biomass of *Elodea*, as well as overall plant biomass on the Refuge (Chapter 7). Any decrease in spring food availability, due either to long-term habitat changes, or annual climatic conditions, would be expected to decrease the nonmigrants' reproductive success by reducing their opportunity to accumulate adequate prenesting energy reserves.

We conclude that the winter and spring carrying capacity of the Tri-state region for nonmigratory trumpeters is considerably lower than the current size of the Tri-state Subpopulation, and that the nonmigrants will further decline if they are not subsidized with increased amounts of supplemental grain. At the same time, we conclude that the Tri-state Area likely supported many more migratory summering trumpeters, which were virtually eliminated as the species was brought near extinction. The Tri-state Area could provide numerous suitable nesting territories for trumpeters that wintered further to the south and stored the energy reserves necessary for successful reproduction as they journeyed northward in the spring. Based upon nest location records for the last 50 years, we estimate that at least 100-150 suitable nesting territories still exist within the current breeding range of the Tri-state Subpopulation.

Removals

In addition to losses due to mortality and occasional dispersal, swans have also been deliberately removed from the Tri-state Subpopulation. The initial success in protecting and enlarging the Tri-state Subpopulation led to a decision by the USFWS in 1938 to extend the trumpeters' remnant breeding range in order to insure against any catastrophic loss in the Tri-state Area (Banko 1960:178; Marshall 1968). Over the next 45 years, approximately 160 eggs, 327 cygnets, and 200 adults were removed from the RRLNWR flock. Appendix XVII documents the history of transfers from RRLNWR, based upon information reported in the Refuge Annual Narrative Reports, 1938-86.

In the first transfer, four cygnets were moved from RRLNWR to the National Elk Refuge near Jackson, WY, in October 1938. The transfer program expanded to include Malheur NWR in southeastern Oregon in 1939, and Ruby Lake NWR in northeastern Nevada in 1949 (Banko 1960:178). From 1938 to 1949, cygnets were captured and transferred in September just prior to fledging. Successful nesting occurred at the National Elk Refuge in 1944 (Banko 1960:179) and a single breeding attempt was recorded near Ruby Lake NWR in 1953 (Banko 1960:181). These initial efforts failed to create breeding flocks at Malheur and Ruby Lakes. Banko (1960:181) attributed these failures to the practice of pinioning and confining the swans to a single pool, and to the loss of swans to accidents and disease. In 1954, an attempt was made to capture and transfer successful breeding pairs and their cygnets. Of 25 pairs periodically chased with the airboat to check for flight capabilities in mid-July, only 6 pairs were simultaneously flightless and could be captured. Even poorer capture success resulted in 1955 and 1956, and the mated pair transfer program was suspended (Banko 1960:73). Cygnets and nonbreeding adults were transferred in 1955-57 and successful nesting by three-year-old swans occurred at both Malheur and Ruby Lake in 1958. The most successful transplanting technique was to wing-clip the transferred cygnets and hold them flightless and semi-captive on a pond while they slowly became acquainted with their new environment. When they regained flight in the next summer's molt, they were free to explore further (Monnie 1966, Marshall 1968, Hansen 1973).

Beginning in 1959, trumpeters were transferred to zoos as well as to Refuges. Between 1959-69, 18 cygnets and 123 nonbreeding adults were sent to zoos. Most often each zoo received one male and one female of unknown age and breeding history.

The effort to create new flocks continued and 57 cygnets were sent to Lacreek NWR, SD in 1960-62; the first successful breeding occurred there in 1963 (Marshall 1968). Thirty-seven cygnets were transferred to Turnbull NWR between 1963-66. The Hennepin County Park Reserve district near Minneapolis, MN, received 20 cygnets and 12 adults between 1966-70 to form the nucleus of a second new flock in the Midwest.

In 1969 the swan loan program to zoos was discontinued and transfers declined greatly. Only 21 cygnets and 7 adults were transferred in 1970-86.

Beginning in 1977, trumpeter eggs were made available to private propagators. Up to six permittees annually were allowed to take either a clutch of eggs or a pair of cygnets for either propagation or display purposes (USFWS Trumpeter Swan Policy Memo 1983). A total of 113 eggs were transferred between 1977-82. Twelve eggs were transferred to the Minnesota restoration project in 1983.

Because of overall concern regarding the long-term effects of the removal program, heightened by the decline in the number of adults and acute production problems at RRLNWR in 1980 and 1982, the future of the removal program came under considerable debate. The Refuge staff made a preliminary assessment of the transfer program in 1984 and recommended that all removals be curtailed immediately (RRLNWR report to Pac. Flyway RMP Trumpeter Swan Subcommittee, March 1984). On March 16, 1984 the Pacific Flyway Study Committee adopted a recommendation for a three year moratorium on the removal of swans and eggs from the Tri-state Subpopulation. No removal would be allowed until the RMP Subcommittee developed guidelines and criteria for future removals.

We developed a population model that incorporated the environmental variables that were significantly correlated with production and survival rates (Chapters 3 and 5). When we ran simulations with actual removals and without removals, our model showed that:

1) prior to 1954, removals only slightly slowed the growth of the Centennial Valley flock;

2) between 1955 and 1975, removals caused the Centennial Valley flock to decline. Without removals, the simulated flock size remained relatively stable;

3) after about 1975, the low rate of removals has depressed the size of the flock by about 7%.

We conclude that although removals contributed to the leveling off of the Tri-state Subpopulation in the mid-1950s and to its fluctuations in the 1960s, the net effect at present is slight after over a decade of only minor removals.

Age/sex Distribution of Molting Nonbreeders

Discussions of the Tri-state Subpopulation by Banko (1960) and Page (1976) assumed that the flocks that molt on the larger water bodies contained significant numbers of surplus eligible nonbreeders that were prevented from nesting due to an inadequate supply of breeding habitat. Very little is known, however, about the swans that comprise the molting flocks, and we suspect that the use of the term "nonbreeders" to describe these swans has been misleading. In addition to nonbreeders, the flocks also likely contained breeding pairs that failed during the early stages of incubation or brood rearing and joined the flocks prior to the molt (Page 1976:107). Swans in the Centennial Valley molting flocks have been banded since 1945. We reviewed the banding and recapture data to determine the sex and age structure of the banded birds (Table 37).

Between 1978-83, 375 (156 male and 219 female) adults were captured in the molt at Lima Reservoir and Upper Lake. Due to recognizable yearling plumage characteristics or previous banding at a known age, the age of 94 males and 137 females was determined. Approximately 32% of the known-age molting swans were yearlings. The records also showed that 22% of the known-age males, and 33% of the known-age females were breeding age birds (age classes 6+), despite the vacancy of some 30 historically used territories in RRLNWR and Yellowstone. Because suitable vacant breeding habitat was available, we conclude that the breeding age swans in the molting flocks were most likely either failed breeders, or individuals that were physically unable to reproduce due to a variety of miscellaneous factors. One factor that could disrupt normal reproductive behavior is chronic sublethal lead accumulation. Recent studies of lead levels in the blood of 22 molting swans from the Centennial Valley found that 61% contained detectable lead (L. Blus, letter to RRLNWR, 6 August 1986). These findings indicate the need for further investigation to determine whether chronic sublethal lead poisoning may lead to the reproductive failure of some breeding age swans.

An imbalanced sex ratio could also prevent some eligible swans from breeding, however, the Refuge records do not indicate whether the observed excess of females among the captured swans represented a real imbalance, or was due to sampling bias. Slight sex-specific differences in molt timing or catchability could bias the sex ratio of captured swans. Based upon a larger sample of 1,050 banded Centennial Valley swans, Anderson *et al.* (1986) found a sex ratio of 48 males:52 females.

Table 37. Age and sex distribution of known-age trumpeters captured at Lima Reservoir and Upper Lake, 1978-83[a].

Age Class	Males		Females		Both Sexes	
	Number	Percent	Number	Percent	Number	Percent
2	34	36	41	30	75	32
3	17	18	16	12	33	14
4	13	14	14	10	27	12
5	9	10	21	15	32	14
6	8	9	15	11	23	10
7	6	6	8	6	14	6
8	2	2	7	5	9	4
9	2	2	6	4	8	3
10-19	2	2	9	7	11	5
>20	1	1	0	0	1	0
Total	94	100	137	100	231	100

[a] These data were extracted and summarized from the Red Rock Lakes NWR banding files.

Future Flock Stability

Predictions regarding future flock growth or decline can be made by calculating whether breeding pairs are currently producing cygnets above or below the rate necessary for self-replacement. To make these calculations we used the age specific survival rates that have been derived from neck-band data (Chapter 5) and the number of cygnets fledged per active nest in each of the Tri-state flocks (Appendix IX). We calculated the number of years in which a pair of swans must nest in order to replace themselves, and then compared this value to the average life expectancy of an adult trumpeter.

When survival rates intermediate between those derived from RRLNWR and lower elevation Wyoming data are used, and swans are assumed to first breed as five-year-olds (age class six), the calculation is as follows:

Number cygnets fledged per nest attempt X first winter survival (.60) X yearling survival (.66) X age class three through five survival (.90)(.90)(.90) = Number of cygnets fledged per nest attempt which will survive to age class 6.

For swans in the RRLNWR flock, which in recent years have fledged about 0.7 cygnets/nest attempt, our calculations indicated that only about 0.2 cygnets/nest will survive to age class 6. Therefore, a pair of Refuge swans must nest for 10 years (2/0.2 = 10) to replace themselves with two 5-year-old offspring.

Our estimates of the number of nest attempts required for pair replacement in the other flocks are given in Table 38. After establishing the number of years needed to replace the pair unit, we then proceeded to estimate the life expectancy of a five-year-old trumpeter as it begins reproduction. The expectation of further life for adult birds can be calculated as

$$e = (2-D)/2D$$

where D is the annual mortality rate (Botkin and Miller 1974). Therefore, if the annual adult mortality rate is as low as the 0.07 estimated for lower elevation Wyoming swans (Lockman *et al.* 1985), then $e = 13.9$.

As annual mortality rates increase, the expectation of further life declines rapidly. For example if the annual mortality rate for RRLNWR swans is closer to 0.12 (Anderson *et al.* 1986), then $e = 7.8$. If we use an intermediate annual mortality rate of 0.10, then $e = 9.5$ years. These calculations confirm that trumpeters nesting at the Refuge are not replacing themselves, even if adult survival is 90%.

We conclude that the decline of the Centennial Valley flock over the past 20 years has not been due to dispersal, but reflects the failure of nesting swans to produce sufficient cygnets for self-replacement. If we assume that the annual adult and subadult mortality rate for RRLNWR swans is about 12%, a nesting pair must fledge an average of about one cygnet per nest attempt in order to replace themselves within their 7.8 year life expectancy. With the potential for healthy trumpeters to lay from 6-8 eggs per clutch, trumpeters need only fledge roughly 2 broods of 3.5 cygnets during a 7 year breeding life to replace themselves. With their relatively long life span and potential large clutch size, trumpeters seem well adapted to nest in harsh environments where weather conditions may permit only sporadic nest success.

Table 38. Number of years required for replacement of a breeding pair that begins breeding at 5 years of age. Expectancy of further life for a 5-year-old adult would be 13.9 years if annual survival rate (s) = .93; 9.5 years if s = .90; and 7.8 years if s = .88

Flock	Cygnets fledged per nest[a]	Number of years required for pair replacement
Yellowstone NP	0.5	14.0
Red Rock Lakes NWR	0.7	10.0
Idaho	1.4	5.0
Lower elevation Wyo.	1.8	3.9

a. Sources of data for calculation of cygnets fledged per nest are shown in Appendix IX.

Production at RRLNWR has been less than one cygnet per nest attempt in 23 of the last 33 years (Figure 9). Production has been effectively reduced below one cygnet per nest attempt in four other years due to the removal of cygnets in September. We conclude that the decline of the Centennial Valley flock will continue unless current rates of cygnet production or adult survival increase.

Trumpeters that nest in Yellowstone are also not replacing themselves at their recent production rate of about 0.5 cygnet/nest attempt. The Tri-state Surveys indicate that the Yellowstone trumpeters have only rarely produced more than 0.5 cygnet/nest attempt since the mid-1950s. In view of this low productivity, we conclude that the maintenance of a nesting flock in the Park in recent decades has depended upon immigration, most probably of birds dispersing from the Centennial Valley as Banko (1960:162) suggested. Recent observations support this theory. During 1977-79, at least two of the nesting pairs in Yellowstone studied by Shea (1979) contained banded swans that most likely had been marked in the Centennial Valley banding program. In addition, coinciding with the recent poor cygnet production and decline in the Centennial flock, fewer subadults were available to disperse from the Centennial Valley, and the Yellowstone nesting population abruptly declined from about 20 pairs in the late 1970s to less than 7 pairs in the 1980s.

In contrast to the RRLNWR and YNP flocks, the Idaho and lower elevation Wyoming flocks are producing cygnets at rates that should result in either flock growth or dispersal, unless mortality rates are underestimated. The rates, in fact, appear to be reasonable, as both flocks have grown and shown a tendency to pioneer peripheral habitats in the 1980s.

Interior Canada Subpopulation

Detailed analyses of the population dynamics of most of the Canadian flocks have not been possible because long-term annual census data are lacking. Over 25 years of data from the Grande Prairie flock (Appendix VII) permit more detailed discussion, although emigration and variations in census effort complicate any analysis.

Grande Prairie Flock

As previously discussed in Chapter 2, the Grande Prairie flock remained remarkably stable from 1959-77 (Turner and Mackay 1981). Since 1978 the flock, as well as the area surveyed, has increased substantially (G. Holton, pers. comm.).

Turner and Mackay (1981) used the model described by Page (1976) to project the growth of the Grande Prairie flock, starting with a population containing 19 reproductively mature females and using survival rates determined from neck-banding studies. Clutch size and prefledging survival data were from Burgess (1972). The model predicted that under average fecundity and survival, the flock would grow at an exponential rate of 0.036 if habitat did not become limiting.

Using the survival rates shown in Table 15, Turner and Mackay (1981) also calculated the expectation of further life and potential natural longevity for Grande Prairie trumpeters. When the annual mortality rate for adults is 0.176 (1-survival rate of 0.824) an adult's expectation of further life is 5.2 years. Potential natural longevity, i.e. the calculated maximum lifespan of one swan in 100, was 24 years. Based upon their observed reproductive potential and estimated survival rates, a pair of Grande Prairie trumpeters must breed for 3.8 years in order to fledge enough young such that at least two will survive to age five (Turner and Mackay 1981). Clearly, the current fecundity of the Grande Prairie flock will result in continued flock growth, unless adult mortality rates increase. Actual mortality rates were probably lower than estimated from the neck-band studies because of the effect of neck-band loss, and thus the expectation of further life is probably somewhat greater than 5.2 years.

Birth rates and death rates were found to be interrelated; fluctuations of the birth rate were usually followed by similar variations of the death rate. Both rates decreased in the early sixties and subsequently increased in the late sixties. Population growth after 1978 resulted from a tendency toward a stabilized death rate, coincident with the maintenance of a high birth rate in the 1970s (Turner and Mackay 1981). Our analysis of Centennial Valley data also found a similar interrelationship between birth and death rates. We concluded that this was due to the higher mortality rates of first winter cygnets and yearlings compared to older swans. When the birthrate was high and thus the proportion of young birds in the wintering flock was high, the average mortality rate for all age classes increased.

The lack of growth of the Grande Prairie flock in the 1950s and 1960s led to the conclusion that the flock's growth was restricted by a shortage of breeding habitat (Banko and Mackay 1964). This theory was reassessed by Turner (1978) and Turner and Mackay (1981) who concluded that winter habitat, rather than breeding habitat, limited the flock's growth.

Turner and Mackay (1981) also used Page's model to test the effects of removals from the Grande Prairie flock. Various levels of simulated removals resulted in a flock that decreased slightly and then quickly stabilized. The authors emphasized that if the flock was limited by density-dependent winter competition, then the effects of the removals might be offset by increased survival rates. They concluded that eggs and cygnets sufficient for a transplant project could be removed with marginal detriment to the flock.

Holton (1982) analyzed lake occupancy in the Grande Prairie region and also concluded that the numbers of nesting trumpeters in the Grande Prairie region was not limited by the number of suitable lakes in at least five years between 1973-80. Although Holton's conclusion might seem to support Turner's suggestion that density-dependent regulation occurred on the winter range, Holton pointed out that Turner's analysis assumed that the number of swans in the Grande Prairie autumn flock was correlated with the total number of trumpeters that shared the Tri-state winter range. Holton (1982) showed that there was no such correlation.

Holton (1982) suggested an alternate hypothesis, that pairs from the Grande Prairie region were moving into suitable habitat outside the flock's known breeding range and represented an undetected expansion, rather than losses due to winter mortality. This dispersal may have contributed to the recently detected swan activity near Cardston, Edson, and the Peace River region, and even in northern British Columbia, the Yukon, and the Northwest Territories.

Holton (1982) also suggested that if the number of suitable lakes does not limit the size of the Grande Prairie flock, then the removal of swans for transplant programs could potentially reduce the flock. He recommended that one or a few breeding pairs be transplanted and the speed of reoccupation of the vacant territories be monitored to determine whether "replacement" pairs of breeding age existed.

Perhaps the most puzzling aspect of the population dynamics of the Grande Prairie flock was the failure of the flock to expand between 1945-75, despite high cygnet production that exceeded 35 cygnets per 100 adults during the years of record in the 1960s. Prior to the mid-1970s, no other significant flocks of trumpeters were detected in Canada, and thus it seems unlikely that dispersal alone could account for the high loss of swans from the Grande Prairie flock. We found that the annual loss of swans from the Grande Prairie flock showed a significant correlation with winter severity in the Tri-state Area ($P = 0.0003$). The Grande Prairie swans, at least until 1985, had made little use of the supplemental grain at RRLNWR. In contrast to the grain-fed Centennial Valley swans, whose survival rate showed only a slight relationship to winter severity, the Grande Prairie swans were quite vulnerable to severe cold weather which reduced the availability of ice-free feeding sites and increased energy demands.

The recent growth of the Grande Prairie flock and the increase of the other Canadian flocks followed the increase in winter swan use of the Harriman State Park area that was made possible by the restoration of higher winter water flows in 1968. Prior to 1968, water releases were completely curtailed in January and February, after the swans had settled in for the winter and during the coldest months when little suitable habitat was available elsewhere. We suspect that the Grande Prairie trumpeters, which used the Harriman State Park area extensively, suffered high winter mortality when the water flows were curtailed. The increased water flows of the 1970s and 1980s increased the availability and stability of the Harriman Park winter food resource and probably reduced the mortality rate of Grande Prairie swans, allowing this flock to grow and disperse by the mid-1970s. From this primary wintering site, the increasing number of swans began to expand their area of use to Hebgen Lake, Teton River, and other portions of the Henry's Fork (Chapter 8).

Summary

Banko concluded that cygnet production in the Tri-state Area was inversely related to population numbers and was depressed at high population levels possibly by mechanisms involving social competition. He also concluded that breeding habitat limitations prevented further growth of the Subpopulation, and caused the proportion of nonbreeders to increase because eligible breeding-age swans were prevented from obtaining suitable territories.

Page suggested that artificially high numbers of swans were maintained on the Refuge after 1954 by supplemental winter feeding. He suspected that overgrazing by swans contributed to the decline of Elodea and reduced the carrying capacity of the Refuge. Page suggested that removals were beneficial because they helped to reduce a flock which was damaging its food resources.

Page used a deterministic model of the RRLNWR flock to show that small increases in adult mortality caused a severe decline of the flock. Also, he showed that if a limit were placed on the number of nesting pairs in a simulated population, then the swans would increase and stabilize at an upper maximum that was determined by adult survival rates, and surplus eligible nonbreeders would result.

We found that no shortage of breeding habitat currently exists in the Tri-state Area, except possibly in the lower elevations of Wyoming. We also conclude that breeding habitat was not limiting in the 1950s, because since that time trumpeters have pioneered at least 20 new territories in the Tri-state Area.

Our analysis showed that mortality rates of Centennial Valley trumpeters were not density-dependent, but were significantly correlated with the amount of grain fed per swan at RRLNWR and the reduction in water flows at Harriman State Park due to cutoffs of water released at Island Park Dam. Abrupt variations in the availability of the swans' two key winter food sources have effectively varied the winter carrying capacity of the Tri-state Area and exerted a strong influence on annual mortality rates.

The molting flocks increased in the early 1950s due to several years of high cygnet production and increased overwinter survival made possible by the unusual availability of winter habitat at Harriman State Park in 1950-53. Flocked subadults were also joined in the molt by older failed breeders. The existence of a large surplus of eligible breeding age swans that were assumed to be nonproductive due to a shortage of breeding habitat was never documented.

The leveling off of the growth of the Tri-state Subpopulation after 1954 was caused both by depressed cygnet production, removals, and increased overwinter mortality, due in part to periodic total curtailment of winter water releases from Island Park Dam. Mortality rates also increased as a result of the reduction in the amount of grain fed per bird at RRLNWR after the late 1960s. By the early 1980s, the amount of grain fed per swan was the lowest in the history of the Refuge and likely was a major factor in the flock's continued decline.

Refuge cygnet production was reduced after about 1950 by the combined influence of several factors including: increased human disturbance, the lower productivity of young inexperienced breeders, increased frequency of adverse weather, and increased water levels. The inverse relationships observed between cygnet production and population levels may be due in large part to factors inherent in the shifting age structure and age-specific fecundity of a long-lived species with several nonproductive subadult age classes.

The number of nonmigratory trumpeters in the Tri-state Area has been maintained at an artificially high level by supplemental feeding. However, the productivity of these birds is limited by their ability to obtain adequate spring foods in order to store sufficient energy reserves for successful reproduction. On the other hand, the Tri-state Area could probably support many more breeding trumpeters if they were migrants that imported the energy stores needed for reproduction from more southerly winter and spring habitats.

Nesting swans at RRLNWR and YNP are not fledging enough cygnets/nest attempt to replace themselves and we predict the further decline of these flocks unless productivity or adult survival increases. Yellowstone is now, and probably for many years has been, dependent upon the immigration of Centennial Valley swans to maintain its breeding flock. With the decline of the Centennial Valley flock, the Yellowstone flock has virtually collapsed. At the survival rates estimated from banding data, RRLNWR swans must fledge about one cygnet/nest attempt in order to maintain a stable nesting flock; currently they are only fledging about 0.7 cygnets per nest attempt.

Idaho, lower elevation Wyoming, and the Grande Prairie flocks are producing cygnets at rates that will allow flock expansion if post-fledging survival rates do not decrease.

The Grande Prairie flock was thought to be limited by breeding habitat in the 1950s and 1960s because the flock failed to increase despite high cygnet production. Subsequent analysis by Turner and Mackay (1981) indicated that winter habitat rather than breeding habitat was limiting. Holton (1982) also concluded that breeding habitat did not limit the Grande Prairie flock and suggested that dispersal of swans into expanding flocks in Alberta, B.C., Yukon and NWT may be confounding the estimates of winter mortality.

We found that the mortality rates of Centennial Valley trumpeters, which made regular use of the Henry's Fork and the supplemental grain, were closely linked to the availability of those two food sources but were only slightly influenced by winter severity. In contrast, the mortality rates of the Grande Prairie trumpeters were very highly correlated with winter severity in the Tri-state Area, probably reflecting their dependence upon the availability of natural open water habitats, rather than the Refuge grain. We conclude that the restoration of higher, regular water flows in the Henry's Fork after 1968 increased the availability of high quality winter habitat and resulted in higher winter survival of Grande Prairie trumpeters. We suspect that this increase in winter habitat was a key factor in the expansion and dispersal of the Grande Prairie flock in the 1970s.

212

CHAPTER 13. SUMMARY

Over the past sixty years, observations of Trumpeter Swans in the wild and in captivity have resulted in the accumulation of a wealth of information about the species. The population and habitat data gathered on the Rocky Mountain Population form a remarkable documentary of the problems and responses of a population as it recovered from near extinction. The biological data only make sense, however, when viewed within an historical perspective. We cannot understand the Rocky Mountain Trumpeter Swans today, without also understanding how their habitats and patterns of habitat use have changed, and why events that happened decades ago still affect the Population.

The task of reconstructing the patterns of habitat use prior to the species' decline is hindered by the lack of historic data. Trumpeters were eliminated from many areas before naturalists were able to identify the species and systematically record their occurrence. Where trumpeters remained, territorial pairs were widely dispersed and difficult to observe from the ground. In the absence of historical data, biologists can either leave important questions unanswered, or they can take what little historic information exists from other locations, and then examine the current behavior and biology of Trumpeter Swans in order to deduce how trumpeters probably functioned prior to their disruption by man. We chose the latter approach. The ideas expressed in the following summary represent a synthesis of information from many sources. Throughout this summary, we will refer to the chapters in which the specific ideas were developed. The reader should refer to the detailed discussions and citations in the preceding chapters in order to trace the origins of each idea.

Prior to the near elimination of the species in the 1700s and 1800s, trumpeters nested widely across most of Canada and the United States. They migrated south to winter on ice-free freshwater habitats scattered across the United States, and to many coastal bays and estuaries (Chapter 1). Then, as now, these long-lived birds likely were highly dependent upon prolonged family group associations and the accumulated experience of family members. Breeding pairs developed traditional annual movement patterns and passed the knowledge of their specific habitats and migration routes to their offspring. Thus, family traditions were passed from one generation to the next. Also as occurs today, dispersal and exploration of new habitat probably occurred mainly among subadults (Chapter 9).

Trumpeter Swans most likely maintained a continuous presence in the Tri-state Area throughout the period of decline, but no data exist to document whether most of the Tri-state trumpeters historically were migrants or year-round residents (Chapter 1). Today, however, both migrant and nonmigrant trumpeters occupy the region. Since the 1930s, observations have revealed that although the majority of trumpeters that summer in the Tri-state Area are nonmigratory, some individuals occasionally migrate to more southerly wintering areas. No evidence of regular migration by Tri-state summer residents has yet been detected however (Appendix I, Chapter 9). Canadian trumpeters, which summer in Alberta, British Columbia, the Yukon, the Northwest Territories, and Saskatchewan have been observed in recent decades to winter primarily in the Tri-state Area. Within the last fifteen years, however, their numbers have grown rapidly and the Canadian trumpeters have shown an increasing tendency to migrate to a variety of other more southerly wintering sites. Spring and fall migrants annually move through the Coeur d'Alene area of northern Idaho, but the winter range and numbers of these migrants are unknown.

The number of trumpeters from both Canada and the Tri-state Area that migrate south of the Tri-state Area probably is higher than records indicate. Most observations of trumpeters from more southerly wintering areas have been reported only because the individuals were marked. Few trumpeters presently are marked and no efforts have been made to search for trumpeters wintering alone or among Tundra Swans outside of the Tri-state Area. As a result, migrating trumpeters are likely to winter unnoticed due to the inability of most observers to distinguish between Tundra Swans and Trumpeter Swans (Appendix I, Chapter 9).

The variation shown by the movement patterns of the Rocky Mountain Trumpeter Swans today suggests that the trumpeter families that used the Tri-state Area in pristine times would also have followed a similar variety of movement patterns. As they do today, some families probably wintered in the Tri-state Area and migrated north to summer habitat in Montana and Canada. Other northern families likely passed through the Tri-state Area during migration, possibly spending several weeks at Yellowstone Lake as they do today, before continuing their migration to more southerly wintering sites. Some families probably nested in the Tri-state Area and wintered at a variety of more southerly locations. Still other families remained in the Tri-state Area year-round, finding scattered sites where geothermal activity provided small areas of ice-free winter habitat (Chapter 8).

By the early 1900s, Trumpeter Swans had been eliminated from most of their historic range outside of Alaska. The last survivors in the Rocky Mountains were also reduced to near extinction. Swans were slaughtered when their traditional migration patterns exposed them to the guns of the settlers and Indians, and they were subjected to some 125 years of commercial exploitation by the Hudson's Bay Company on their Canadian breeding grounds. Outside of Alaska and British Columbia, the only trumpeters known to have survived the decline were those that wintered in the Tri-state region. Truly unique on the continent, this remote geothermal area remained virtually unexplored until the 1870s. Despite the region's high elevation and severe winter weather, its isolation from human settlement and the availability of ice-free habitat, created by the runoff from warm springs, provided the last winter refuge for the trumpeters of both the United States and Canada (Chapter 1).

In the 1930s, the first summer surveys of the Tri-state Area found only about 50 adults and their cygnets (Chapter 2). Apparently these few survivors were mostly nonmigrants, whose traditional movements had kept them isolated in the most remote portions of Yellowstone and the Centennial Valley where human-caused mortality was very low. We suspect that trumpeters that nested in the Tri-state Area and migrated south to winter in less remote areas were more vulnerable to human-caused mortality than the nonmigrants and were virtually eliminated (Chapter 1).

Almost all of the trumpeters that nested across the interior of Canada were also eliminated. Only one small flock was known to survive in the vicinity of Grande Prairie, Alberta. When the Grande Prairie flock was first censused in 1946, only 77 adults and their broods were found (Chapter 2). Marking studies in the 1950s revealed that these Canadian survivors also wintered almost exclusively in the Tri-state Area. Like the trumpeters of the United States, any interior Canadian trumpeters that wintered in less remote locations were gradually eliminated (Chapter 9).

The descendants of those two remnant breeding groups now comprise the Tri-state Subpopulation and the Interior Canada Subpopulation; together they are known as the Rocky Mountain Population. Though they still continue to share the same Tri-state wintering range, marking studies have not yet provided evidence that these two groups interbreed. The data gathered so far indicate that these two groups are reproductively isolated and should be regarded as separate populations until evidence to the contrary is found.

The near extinction of the Trumpeter Swans probably caused at least two lasting impacts on the Rocky Mountain Population. First, genetic variability most likely was lost as the number of swans declined to somewhat less than 150 adults. Loss of genetic variation was even more likely to have occurred because trumpeters normally spend much of the year in parent/cygnet or subadult sibling associations, and thus many of the survivors were likely to have been closely related. Although the effects might not be obvious, a loss of genetic variability could render their descendants somewhat less fit to cope with environmental stresses and increase the likelihood of defects due to inbreeding (Chapter 11).

Second, as the vast majority of trumpeters were eliminated most of the Population's cumulative knowledge of traditional migration routes to other winter and spring habitats was also destroyed. The few surviving families retained only two basic annual movement patterns to pass on to their offspring: they either remained in the Tri-state Area year-round, or wintered there and migrated to Canada to nest (Chapter 1). **This loss of knowledge of other migration routes and alternate winter and spring habitats is the underlying cause of several of the problems now facing the Rocky Mountain trumpeters.**

Although management actions were able to increase the numbers of trumpeters, the migratory traditions that were destroyed during the species' decline were not restored. Therefore, as the remnant was protected and their offspring increased, growing numbers of nonmigrants and wintering swans became dependent upon the Tri-state habitat. Managers were able to increase the numbers of nonmigrants and wintering Canadian trumpeters to artificially high levels due to the influence of two factors:

1) the winter carrying capacity of the Tri-state habitat was substantially increased by supplemental feeding which began at Red Rock Lakes NWR in 1935. The feeding program was also intentionally used to hold trumpeters on the Refuge and discourage their traditional fall movements out of the Centennial Valley (Chapter 8).

2) since the 1930s the amount of available winter habitat has also been increased in various locations within the Tri-state region due to man-made warm spring impoundments and the construction of dams. Water stored in reservoirs retains significant amounts of heat and its release during the winter months creates ice-free river habitat downstream. Ice cover in the Harriman State Park area on the Henry's Fork is substantially less under the present water flow regime than it would have been prior to the construction of the Island Park Dam in 1938 (Chapter 8).

One consequence of the loss of traditional movement to more southerly winter and spring habitats has been that the increased numbers of nonmigratory Tri-state trumpeters have access to only very marginal, high elevation spring feeding areas. Recent studies have shown that the reproductive success of other species of swans and geese is directly influenced by the amount of stored energy reserves that the breeding female brings to egg laying and incubation. Energy reserves are catabolized during the winter months, and a female must therefore accumulate sufficient reserves in the weeks prior to egg laying to adequately provision a clutch and to sustain herself during the incubation period. Thus, the quality and quantity of available spring foods are critical to reproductive success (Chapter 10).

The Canadian trumpeters usually depart from the high plateaus of the Tri-state wintering area early in March, and migrate slowly through the lower elevations of Montana and southern Alberta, where they have access to a wide variety of ice-free habitats (Chapter 9). Upon their arrival in Grande Prairie in late April, migrant trumpeters have apparently stored energy reserves sufficient to provision clutches containing an average of 5.6 eggs and to incubate with constancies exceeding 94%. Their average clutch size and constancy are only slightly lower than those of well-fed captive trumpeters (Chapter 4). Their high rate of cygnet productivity has resulted in the rapid expansion of the migrants' numbers and summer distribution (Chapter 12). In contrast, the Tri-state trumpeters usually face several weeks of winter weather after the migrants depart for lower elevations. Often spring thaw in the higher elevations of the Tri-state Area does not begin until early April. In some years, the lakes at Red Rock Lakes NWR have not thawed until late May (Chapter 6). Access to spring food sources is highly dependent upon spring weather conditions and ice cover. Cold, late springs or unusually high spring runoff delay the development of aquatic plants and invertebrates and reduce their availability, both to the prebreeding adults and to the young cygnets in the critical first weeks of life (Chapter 5, 7). The relatively low availability of spring foods in most years in the Tri-state Area likely limits the amount of energy reserves that the swans can accumulate prior to egg laying in late April or early May and thus reduces their productivity (Chapter 10). Likely as a result of low energy reserves, the mean clutch sizes of the several nonmigratory flocks ranged only from 3.6-4.9 eggs (Appendix IX). The Tri-state females incubated with lower constancy and spent more time feeding compared both to the Grande Prairie migrants and to captive trumpeters (Chapter 4).

Several other examples of the relationships between productivity and spring habitat conditions have been shown. Clutch size at Red Rock Lakes NWR was positively correlated with increasing May temperatures ($P = 0.08$) and negatively correlated with water levels during the entire spring and early summer period ($P = 0.08$). Cygnet survival was positively correlated with mean minimum temperatures in June ($P = 0.07$), and negatively correlated with July water levels ($P = 0.06$). The number of cygnets alive in late summer was significantly lower in years when the peak of hatch was delayed ($P = 0.01$); a late peak of hatch was normally associated with adverse spring weather conditions (Chapter 3). Also coinciding with the relative timing of spring thaw and the availability of spring foods, clutch size, hatchability, and cygnet survival are lower in the high elevation Red Rock Lakes and Yellowstone flocks than in the lower elevation Idaho, Wyoming, or Grande Prairie flocks (Appendix IX, Chapter 6).

Another consequence of the population's loss of traditional knowledge of other wintering areas has been its continued high fidelity to the Tri-state wintering area, and thus the dependence of increasing numbers of trumpeters on this single wintering area (Chapter 8). Although the tendency of subadults to explore new wintering habitat is becoming more apparent as the Rocky Mountain Population increases in numbers, over 95% of the trumpeters marked in the Canadian and Tri-state flocks have wintered in the Tri-state Area (Chapter 9). The 1,600 trumpeters that wintered in the Tri-state region in 1986 represented Tri-state residents and all the known Canadian breeding flocks, as well as approximately 500 swans from unknown summer ranges (Chapter 2, Chapter 9). Despite the fact that the main concentrations of swans are scattered between about 8-10 key sites, throughout the winter some trumpeters regularly move from one site to another. If a disease outbreak should occur in the Tri-state region, it could spread rapidly between the wintering sites and be disastrous to the recovery of the Trumpeter Swan both in Canada and the U.S. (Chapter 9).

Very little is known about the relationship between the number of wintering swans and the carrying capacity of the Tri-state winter habitat. Studies of the Wyoming wintering sites suggest that they are presently at their carrying capacity, and this may be the reason that the number of swans wintering in the Jackson area has stabilized. Preliminary data suggest that the carrying capacity of the Henry's Fork has not yet been exceeded, however the vegetation/swan relationships have not been adequately investigated and the potential for the wintering swans to alter aquatic plant communities in the Henry's Fork is unknown.

The future of key wintering habitats is not secure. Private lands along the Teton River in Idaho will likely be developed in the near future and human disturbance is likely to increase in most habitats. The water flows that are essential to maintaining ice-free habitat at Harriman State Park are not guaranteed. Although the Bureau of Reclamation attempts to provide adequate flows when water is plentiful, releases during years of low water storage will likely be insufficient to prevent freezing unless changes in water management are made (Chapter 8).

The artificially high numbers of swans that winter in the Tri-state region must endure some of the most severe winter weather in the lower 48 states, with temperatures regularly declining to -40 °C. Winter conditions can last from November until April and cause trumpeters to deplete their energy reserves. In addition to increasing the amount of energy that swans must expend in order to survive, the subzero winter temperatures also create ice cover that severely reduces the amount of available food (Chapter 8). The Canadian migrants appear to be particularly vulnerable to severe winter weather. Mortality rates in the Grande Prairie flock showed a highly significant positive correlation with winter severity in the Tri-state region ($P = 0.0003$) (Chapter 5).

The availability of winter foods also appears to have a large influence on the mortality rates of the Centennial Valley trumpeter flock and was an important factor in the flock's lack of increase after the mid-1950s and decrease since the late 1960s (Chapter 5, 12). Although a few marked Centennial Valley swans have been observed at other Tri-state wintering sites, most of these nonmigratory trumpeters appear to depend on the supplemental grain at Red Rock Lakes NWR and aquatic vegetation at Harriman State Park (Chapter 8, 9). Both of these food supplies have varied widely and somewhat erratically over the last 50 years as a direct result of management actions (Chapter 8). The mortality rate of the Centennial Valley swans increased in years when either one of these food resources was in short supply. The amount of grain fed annually per swan at Red Rock Lakes has varied from about 1-7.5 bu, with a general decline in the 1970s and 1980s to the lowest volumes fed, per swan, in the Refuge's history. A significant correlation between hatching success and the amount of grain fed during the previous winter suggests that the decline in the amount of grain fed in recent years also contributed to decreased productivity (Chapter 3). This decrease in winter feeding was a key factor in the recent decline of the Centennial Valley flock and thus the Tri-state Subpopulation (Chapter 5, 12).

The amount of vegetation available to swans on the Henry's Fork has also varied abruptly, due to the drastic reduction of winter water flows at Island Park Dam during most years between 1938-50, and 1954-67. When winter water releases were curtailed, the resulting low flows at Harriman State Park allowed the river to freeze during periods of normal or subnormal winter temperatures (Chapter 8). Annual mortality rates of the Centennial Valley swans were negatively correlated with the amount of grain fed per swan at Red Rock Lakes NWR ($P = 0.016$), and positively correlated with the duration of reduced water flows on the Henry's Fork ($P = 0.009$).

In effect, the wide, unpredictable fluctuations in these two key winter food sources caused the winter carrying capacity of the Tri-state Area to vary considerably over the last 50 years (Chapter 5, 12). The curtailment of water flows on the Henry's Fork for several weeks or months each winter in most years prior to 1968 apparently also caused high mortality among the wintering Grande Prairie trumpeters, and was an important factor in that flock's failure to increase until the mid-1970s (Chapter 8, 12).

In addition to the underlying influences of spring and winter habitat availability on the Population's productivity and mortality rates, several other factors have contributed to the decline of the Tri-state Subpopulation. Among the most important factors influencing productivity have been habitat changes at Red Rock Lakes National Wildlife Refuge. Increased water levels, due to the influence of the lower water control structure, have directly caused the flooding of numerous nests in the 1980s. Higher, more stable water levels have also probably decreased the productivity of the marsh by reducing the periodic aeration of marsh soils, and reducing average water temperatures, particularly during the critical early spring period. Erosion and the deposition of sediment accelerated in the Refuge lakes in the late 1950s and 1960s due to mining activities, heavy grazing, and increased water levels. The 1959 earthquake probably also altered patterns of erosion and deposition (Chapter 6). Although vegetation changes were poorly quantified, *Elodea* declined dramatically on the Refuge since 1956, and its loss as an important spring food may have further reduced the resources available to the nonmigrants. The decline of *Elodea* most likely was due to the increased turbidity of the refuge waters and other habitat changes, rather than overgrazing by swans (Chapter 7).

Human disturbance also increased substantially at Red Rock Lakes as research and management activities expanded. Use of an airboat in the nesting areas after 1952 caused a dramatic increase in the level of disturbance in the marsh. Several researchers reported that the noise from its engine was very disruptive to the swans and use of the airboat has now been drastically reduced. Although long-term habitat changes in other portions of the trumpeters range have not been studied, increased human disturbance has been found to reduce the rate of occupancy of breeding territories in Yellowstone National Park and Grande Prairie, and to interfere with successful reproduction in the lower elevations of Wyoming (Chapter 4).

Several other factors have contributed to increased adult mortality. Losses of swans to collisions with power lines and fences are the leading cause of death in Wyoming, and may increase as further human development occurs in Trumpeter Swan habitat. The loss of trumpeters to lead poisoning has become more apparent in recent years as managers have made a more determined effort to retrieve carcasses and test for tissue lead levels. Of the 34 swans whose cause of death was determined between 1980-86, 33% died from lead poisoning. Tests of blood lead levels in 32 nonbreeders found that 23 carried detectable levels of lead. Due to the compounding effect of mortality in the many adult age classes, a slight increase in adult mortality rates due to lead poisoning could easily hasten the decline of the Tri-state Subpopulation. Chronic sublethal lead levels could have long-term detrimental effects both on an individual's reproductive success and its ability to withstand other environmental stresses.

The recovery of Trumpeter Swans is at a crossroads. The Interior Canadian trumpeters are increasing rapidly, and large areas of potential breeding habitat still remain vacant and available in Canada. Their recovery is precarious, however, because of their continued dependence upon the Tri-state wintering area and the vulnerability of all the flocks to high mortality if disease should occur on the wintering ground. An outbreak of avian cholera occurred among Snow Geese at Market Lake in March 1987, less than 75 miles from the Henry's Fork, and was a reminder of the potential for disaster. Even if the trumpeters avoid such an unfortunate event, the restoration of the Canadian flocks will be slowed by occasional high mortality in the Tri-state Area during severe winters, as occurred in 1984-85.

Recent observations indicate that Canadian trumpeters are increasing their exploration of other migration routes and wintering areas as their numbers increase. If this natural tendency can be protected and enhanced by management efforts, Canadian trumpeters have the potential to end their dependency on the Tri-state Area and develop a varied network of winter and spring habitats. A key question will be whether alternate wintering traditions can be restored before high mortality occurs in the Tri-state Area.

The continued survival of the Tri-state breeding population is in doubt. There is currently no evidence that these swans interbreed with the Interior Canada trumpeters. Until evidence of matings between the two groups is found, the Tri-state trumpeters should be viewed as a significant breeding population whose continued existence is threatened, and managed as a threatened population. Unless the current downward trends in total numbers, numbers of nesting pairs, and cygnet production are halted, this breeding population will continue to decline.

Without supplemental feeding, a few pairs of nonmigrants will likely persist in the lower elevations of the Tri-state Area, but their numbers would be inadequate to maintain adequate genetic variation for long-term survival (Chapter 11). Their productivity would be very low and highly vulnerable to harsh spring weather. They would rapidly decline if factors such as lead poisoning or human development increased adult mortality rates (Chapter 12). The past decades of management have increased the number of nonmigratory trumpeters kept alive by supplemental feeding, but managers have not begun to restore the essential patterns of habitat use that could allow the Tri-state trumpeters to truly recover and become self-sustaining.

The immediate problem of stabilizing the Tri-state Subpopulation before further losses occur can be solved. The number of nonmigrant trumpeters can probably be increased fairly quickly if the water management problems and winter feeding program at Red Rock Lakes NWR are corrected, and if other factors, such as habitat loss, human disturbance, and lead poisoning are reduced. But these efforts alone are not sufficient. Increasing the number of grain-fed nonmigrants only buys time, and reduces the rate of loss of genetic variation. The end result however, will be an artificially sustained group of trumpeters, perpetually dependent upon winter feeding. These birds will be continue to have low productivity except in unusually mild years, and population growth will continue to be highly vulnerable to factors that increase adult mortality.

As long as the existence of wild breeding Trumpeter Swan populations in the United States, south of Canada, depends upon supplemental feeding and the birds are unable to survive on their own, the species should not be considered to be recovered. The true recovery of the Trumpeter Swan as a self-sustaining member of our waterfowl community is possible, but it will require the active intervention of managers to actively rebuild broken traditions and enhance the birds' own efforts to migrate to other wintering areas. Migration patterns that were disrupted decades ago must be restored and the Tri-state trumpeters must gain access to less severe and more productive winter and spring habitats. Trumpeters have a high reproductive potential and could increase rapidly if they have access to adequate food supplies. If they can regain the use of a secure foundation of varied winter and spring habitats, the trumpeters' vulnerability to catastrophic loss will be also substantially reduced. With increased spring feeding opportunities and increased productivity, the potential is high that migrant trumpeter could once again return to large portions of their historic range in the United States and Canada.

CHAPTER 14. MANAGEMENT RECOMMENDATIONS

As we have previously discussed, the Rocky Mountain Population faces some chronic long-term problems that underlie the decline of the Tri-state Subpopulation and the vulnerability of the total wintering Population. In addition, however, the Tri-state Subpopulation faces an immediate crisis of declining numbers, declining nesting effort, and very low productivity that threatens its continued survival. Therefore, management requires the recognition of both long-term and short-term goals.

Perhaps we can best summarize the recommended approach as:
1) protect the existing subpopulations and prevent further losses while managers work to
2) rebuild the habitat use patterns that will allow the RMP to recover and no longer require intensive annual management to insure its continued survival.

A "quick fix" is needed to prevent the continued decline of the Tri-state Subpopulation and the further loss of its genetic variability. A long-term, coordinated management strategy will be necessary, however, to restore lost migratory traditions and provide the trumpeters with access to the habitats necessary for the birds to become self-sustaining without continued dependence upon supplemental feeding.

Management direction for RMP trumpeters is currently provided by the North American Management Plan for Trumpeter Swans (1984). We present the following management suggestions:

Short-term Management Goals:

I. Halt the decline of the Tri-state Subpopulation and increase it to contain a minimum of 105 nesting pairs, distributed as follows:

Red Rock Lakes NWR	40
Other Centennial Valley	15
Other Montana	5
Idaho	20
Yellowstone NP	15
Lower elevations, WY	10

A. Increase cygnet production
1. Reduce the loss of nests due to flooding at RRLNWR by remodeling the water control structure and actively managing water levels.

2. Enhance aquatic vegetation and invertebrate productivity at RRLNWR by manipulating water levels. Investigate methods of fluctuating water levels at other breeding sites where feasible to allow periodic soil aeration and prevent flooding.

3. Minimize human disturbance at nesting areas, particularly during incubation and early

brood rearing periods. Minimize airboat use at RRLNWR.

4. Continue supplemental feeding at RRLNWR with large quantities of high quality feed until alternate winter habitat has been identified and efforts are actively underway to restore migratory behavior.
    a. empty and clean grain bins each year
    b. disperse feed as widely as possible to minimize competition
    c. investigate feeds best suited to bring swans into good breeding condition.

5. Identify currently used spring habitats and investigate possibilities to enhance existing or create new spring food sources in the Tri-state Area.

6. Evaluate historic nesting territories and actions needed to maintain the long-term suitability of each site. Identify opportunities for site enhancement and protection needs. Nesting sites in the lower portions of the Centennial Valley downstream from RRLNWR appear to have relatively high potential to produce cygnets and management attention directed at these sites might be particularly effective.

7. Determine the extent of loss of heterozygosity in the Tri-state Subpopulation (see Chapter 11). If the loss is significant, develop methods to introduce genetic variation from the Pacific Coast Population.

8. Use floating platforms to save individual nests from flooding, particularly at ponds where suitable nest sites are limited, or where human disturbance prevents successful nesting at historic shoreline nest sites.

9. Salvage and hatch eggs that would otherwise be lost to flooding etc. Place hatched cygnets with natural parents if habitat is suitable or use in cross-fostering programs with lower elevation pairs.
    a. Until the objective of 105 nesting pairs has been met, removals from the Tri-state Subpopulation should be limited to the aggressive salvage of eggs and cygnets that otherwise would be lost.

B. Reduce mortality of adults and fledged cygnets

1. Feed large amounts of high quality food to those swans that remain after freeze-up at RRLNWR.

2. Minimize ice formation on the Henry's Fork at Harriman State Park by maintaining adequate winter water releases at Island Park Dam.

3. Protect other Tri-state winter habitats and increase their productivity where possible.

4. Eliminate lead shot and lead sinkers as soon as possible.

5. Increase hunter education and enforcement efforts, and work with landowners to remove or modify objects that pose direct collision hazards in important habitat.

II. Maintain the current distribution and abundance of the ICSP flocks

A. Protect key nesting habitats

    1. Develop site specific measures to insure the long-term suitability of currently occupied sites.

    2. Identify the summer habitat of the approximately 500 migrants that winter in the Tri-state Area but who were not accounted for in the 1985 rangewide survey.

B. Identify and protect key spring habitats

    1. Efforts are needed both in Montana and Canada.

C. Protect and enhance winter habitats

    1. Insure adequate water flows on the Henry's Fork and the long-term protection of other Tri-state wintering sites

    2. Explore possibilities of enhancement of habitat at Hebgen Lake and Teton Valley.

    3. Identify other sites used by Canadian trumpeters outside of the Tri-state Area. Develop strategies to insure adequate protection of these sites and to promote use by trumpeters.

D. Reduce vulnerability to catastrophic winter loss by actively developing a tradition of swan use of at least two other wintering areas by 1997 and 2 additional wintering areas by 2007.

    1. Identify suitable wintering sites and develop techniques to move trumpeters to these sites and establish new traditions of winter use.

Long-term Management Goal

Rebuild a predominantly migratory RMP that will use a diverse network of summer and winter habitats, and that will be able to maintain its stability without supplemental grain feeding.

A. Create at least four wintering areas outside of the Tri-state Area that will be used by both ICSP and TSP trumpeters.

    1. Identify suitable sites and develop and implement a strategy to establish regular winter use by trumpeters for each area. Attempt to build upon the current natural movement of

trumpeters to areas further south.

2. Techniques to move family and subadult groups to other wintering sites will be needed. Marking studies may be helpful in identifying individuals that would be most likely to utilize new areas.

B. Concurrent with the development of use of other wintering sites, gradually eliminate the Tri-state Subpopulation's dependence upon supplemental grain and increase the proportion of Tri-state trumpeters that migrate further south to winter.

1. As alternate wintering areas are developed, simultaneously begin to wean Centennial Valley Trumpeters from supplemental feeding at RRLNWR
   a. Begin the reconditioning process by withholding grain until one week following freeze-up at RRLNWR to encourage the movement of swans out of the Centennial Valley. Until the Tri-state Subpopulation's numbers have recovered, the birds that fail to leave must be fed in order to prevent unacceptable mortality.
   b. Each subsequent year, further delay the start of feeding by two additional days (based upon date of freeze-up). Terminate feeding entirely when start-up date has been gradually moved back to January 15, or when regular movements to new wintering areas have been restored and the small number of swans that refuse to leave the Refuge can be used as transplant stock.
   c. Concurrent with efforts to move trumpeters to other wintering sites and when grain is no longer used to "short stop" trumpeters on the Refuge at freeze-up, water management on the Henry's Fork could also be used to stimulate swan movement. Occasional cold weather in mid to late November provides the opportunity to temporarily freeze the Harriman State Park feeding areas by reducing flows as the swans first arrive. If this habitat was deliberately made unavailable early at the beginning of winter, trumpeters should be in good enough physical condition to move south to other sites.

We want to emphasize the apparent management paradox that exists. Supplemental feeding is currently needed to prevent the further decline of the number of nonmigrants. At the same time, the Tri-state nonmigrants will have little incentive to move to more southerly winter habitats and obtain the energy reserves necessary for self-sufficiency if the provision of supplemental grain continues to hold the swans in the Centennial Valley after freeze-up. Correction of this situation will require a coordinated management balance between teaching swans to move to new areas and the gradual phase-out of supplemental grain.

We also suggest that the following questions require further research:

1) To what extent has a loss of genetic heterozygosity reduced the viability of Tri-state or Centennial Valley trumpeters? Would it be desirable to introduce genetic variation from the Pacific Coast Population?

2) Although there is little past evidence of dispersal of Tri-state trumpeters, is this situation

changing as increasing numbers of Canadian migrants head north each spring? To what extent are subadults from the Tri-state Subpopulation pairing with Canadian migrants? Is such pairing underestimated due to the scarcity of marked individuals in recent years or do temporal or behavioral barriers keep the subpopulations reproductively isolated?

3) Are other migration routes and wintering areas currently being used by trumpeters whose presence is undetected? Could the restoration of the RMP be accelerated by identifying and protecting these migrants?

4) What is the summer range of the 500+ migrants that winter in the Tri-state Area but that could not be located on the 1985 rangewide summer survey? Does their summer whereabouts have implications regarding intermixing of populations or subpopulations? Do subadult trumpeters join in a molt migration to currently unknown habitats?

5) Do trumpeters wintering on the Henry's Fork have the potential to alter their habitat in ways that might be detrimental to the blue-ribbon trout fishery in the area? We need to determine the relationships between vegetation, invertebrate, sedimentation, water flow, fish, and swans on the Henry's Fork and understand the factors that are key to the long-term productivity of this resource.

6) Do trumpeter adults or cygnets have specific food preferences or needs which would have management implications?

APPENDICES

Appendix I. Observations of RMP Trumpeter Swans in locations outside their usual Tri-state range 1930-87.

Idaho

1932	Nest on island in Pend Oreille R. just below Idaho-Washington line, reported by R. James (Banko 1960).
4/17/37	Five seen at Marsh Cr., Portneuf R., 30 miles south of Pocatello
4/18/37	Pair on oxbow lake near Roberts, Jefferson Co. (Banko).
Fall/43	Trumpeter killed on Snake R. near Burley (Salter 1954).
3/30/51	Five seen on Elk Horn Res., Oneida Co. by E. Keppner (Salter 1954).
1/9/52	Eleven seen on Spring Cr., American Falls Res., Bannock Co., by W. Banko and R. Salter (Salter 1954).
1968-75	Nesting and occasional spring and fall migrants at Grays Lake NWR (Burgess 1986
1982-85	Annual spring and fall migration of groups up to 20, stay for a week or more at Lake Chatcolet. Usually intermingle with Tundra Swans. Vocalizations commonly heard (F. Bear, Heyburn State Park, pers. comm.).
4/85	Swan wearing yellow neck-band (from Elk Island National Park, Alberta) seen at Weiser Flat, Ada Co. (F. Bear, pers. comm.).
11/85	Three adults found dead in Couer d'Alene area by J. Nigh, Idaho Dept. of Fish and Game. Diagnosed lead poisoning, most likely from mining residues (R. Krieger, pers. comm.).
11/85	One adult found dead at McArthur Reservoir due to gunshot wound, reported by J. Nigh, Idaho Dept. of Fish and Game (R. Krieger, pers. comm.).
11/27/86	Neck-banded adult from Nahanni, NWT, observed at Round Lake near south end of Lake Couer D'Alene by W. Latshaw (G. Will, pers. comm).

Montana

10/10/50	Trumpeter found dead at Freezeout L., Fairfield, Teton Co. (Banko 1960).
10/31/53	Thirty trumpeters seen flying over Missoula by R. Hand (Banko 1960).
Spring/55	Pair attempted to nest on Mystic L. near the Rosebud R., one killed by hitting a powerline (Banko 1960).
4/19/56	Several trumpeters heard among tundras at Helena L. by Mrs. W. McKinney (RRL files 1956).
Spring/61	Pair present in Helena Valley, March to June, seen by Mrs. W. McKinney (RRL files 1961).
11/10/61	Five migrants joined with domestic geese a few miles south of Lolo along Rt. 93, observed by U.S. Forest Service employees (RRL files 1961).
3/14/62	Nine seen at Helena L. by Mrs. W. McKinney (RRL files 1962)

4/4/65	Four seen at Helena L. by Mrs. W. McKinney (RRL files 1965).
7/26/67	Pair on ranch 42 km west of Choteau (RRL files 1967).
12/13/68	Crippled adult found near Ryegate, banded at RRL as adult in 1967 (RRL files 1968).
2/72	20-30 on Dayton and Elmo Bays, Flathead Lake, Lake Co., identified by vocalization (M. Bishop)
1971-73	Nesting pair on Burdick Stone property near Silver Star (RRL files 1973).
4/73	Four on pothole near Ronan, Lake Co (M. Bishop)
12/30/74	Grande Prairie neck-banded yearling male (42TY) seen at Ravalli NWR, Stephensville (CWS Edmonton banding files).
4/26/75	Three on west lagoon by inlet canal at Pablo, Lake Co (M. Bishop)
1/28/76	Grande Prairie neck-banded male cygnet (29TA) found dead on highway in Broadwater Co. (CWS Edmonton banding files)
6/4-23/77	Grande Prairie neck-banded yearling male (3N) summered at Ninepipes NWR, Moise (CWS Edmonton banding files).
11/11/77	Grande Prairie neck-banded yearling male (3N) seen at Hamilton (CWS Edmonton banding files).
11/8/78	Grande Prairie neck-banded yearling male (8B) seen at St. Mary's, later found dead at Ravalli NWR (CWS Edmonton banding files).
Fall/80	Trumpeters on pond on Kelly Rd. near Columbia Heights reported by T. Stidham (Burgess 1986)
12/19/80	Two neck-banded swans (cygnets 01CA and 02CA) from Smith River, Yukon Territory seen at Moise (McKelveyt *et al.* 1983)
Spring/83	Trumpeters heard and seen around Glacier Bible Camp, Hungry Horse, by T. Stidham (Burgess 1986).
Fall/84	Trumpeters heard at Benton Lake NWR (R. Pearson and T. Turnow, pers. comm.)

Wyoming

Fall/39	Trumpeter killed at Dubois, reported by K. Roahen, Game Mgmt. Agent (RRL files 1939).
9/53	Pair with 5 cygnets on small lake near Pathfinder Migratory Bird Refuge by D. Condon (Banko 1960).
10/27/56	Two cygnets marked on 8/22/56 at Lowe Lake, near Grande Prairie Alberta were shot near Cody (Mackay 1957)
12/58	Subadult from Grande Prairie found dead near Sunlight basin (RRL files 1958)
Summer 1977	Pair nested and raised 3 cygnets on Trail L, Dunoir R near Dubois. Pair occupied the area for at least three years prior to nesting. Reported by J. Mumma, USFS (Lockman 1983).
12/15/79	Two trumpeters, Seedskadee NWR circle (American Birds 34(4)).
1980-81	Pair nested on Marsh Lake, Clarks Fork R. near Cody. Reported by M. Black, WGFD (Lockman 1983).
Spring/82	Yearling seen on Tongue R. near Sheridan by M. Maj (Burgess 1986).

Winter/83	Six trumpeters wintered on Salt R. near Afton (Lockman 1983).
3/84	Three trumpeters seen on Pool 6 near Graybull by M. Terry (H. Burgess, pers. comm.).
11/10/84	Four adults/subadults and three cygnets seen with Tundra Swans on Bear R. south of Cokeville (Lockman *et al.* 1985).
11/23/84	Two adults and one cygnet on Salt R. near Thayne (Lockman *et al.* 1985).
12/1/84-2/15/85	Four adults/subadults and three cygnets wintered on Salt R. near Grover (Lockman *et al.* 1985).
11/84-2/85	One wintered on Bull Lake Cr., Crowheart near Dubois (Lockman *et al.* 1985).
1/85	One seen on Popo Agie R. near Lander (Lockman 1985).
4/19/85	Swan with red neck-band (from RRL) seen on N. Piney Cr. near Big Piney by B. Johnson, WGFD (Lockman et al 1985)
1/86	One adult with one cygnet wintered near Grover on Salt R. (Lockman *et al.* 1987)
Summer 1986	Five swans summered near Farson, Wyoming (Lockman et al 1987)

Utah

6/14/32	Trumpeter seen at Bear River NWR. by A. Hull (Banko 1960).
7/40	Immature trumpeter, seen on Strawberry Reservoir by reported by D. Rasmussen and L. Couch (Banko 1960).
1949	"Formerly common, probably resident, now a rare straggler into northern Utah" (Woodbury *et al.* 1949)
11/23/59	One collected at Brigham Bird Refuge, Specimen at Utah Museum of Nat. Hist.
12/27/65	Trumpeter heard vocalizing in flock of Tundra Swans along east side of Great Salt Lake (Condor 68(5):521).
Summer 1968	One at Fish Springs NWR
7/15/69	Two molting nonbreeding males captured on Neponset Reservoir, 11 miles SW of Woodruff, Rich County, by J. Nagel. Wt. 26.5 and 24.5 lbs. (J. Huener, Utah Dept. of Natural Resources, pers. comm.).
1970s	SE Utah; trumpeters observed at confluence of Delores R and Colorado R (1); Colorado R. near Keg Springs Canyon south of Green R (1), and Desert Lake WMA south of Price (several) (L. Dalton)
Fall 74	Grande Prairie neck-banded female cygnet (16TA) shot 3 km west of Corinne, Box Elder Co. (CWS Edmonton banding files).
1975	Occasional in northern Utah, more common formerly (Behle and Perry 1975)
1977-78	One adult and five cygnets wintered at Fish Springs NWR, observed by R. Shea.
12/27/81	Four seen on Christmas Bird Count, Fish Springs NWR (Am. Birds 36(4))
3/83	Dead cygnet found by D. Lockman, and J. Deutscher, USFWS, on Big Creek about 3 km south of Laketown (Lockman 1983).
3/30/85	Two adults seen and heard along Hwy. 15, .5 km south of Riverside by T.

	Pogson, Malheur NWR (Lockman et al. 1985).
4/11/85	Neck-banded adult female trumpeter (R2) seen in Vernal by D. Condon. Marked at RRL in summer 1984 (T. McEneaney, pers. comm.).
11/26/85	Subadult female shot near Ogden Bay, Weber Co. Banded at RRL as a yearling on 7/16/84 (B. Reiswig, pers. comm.).
12/18/86-2/87	Marked yearling male from Wyoming wintered at Lake Powell, Glen Canyon National Recreation Area (D. Lockman, pers. comm.)

Nevada

12/2/52	One trumpeter east of Carson City (Banko 1960)
1/23/87	Two Nahanni, NWT trumpeters, 23AA (adult male) and 30AA (adult female) located near Alamo. 23AA was dead, 30AA flew off with a flock of Tundra Swans. (T. Redderer)

California

1935	One seen NE of Eagle Lake, between Termo and Grasshopper Valley, Lassen Co.; Others seen and heard years earlier at Honey lake, Lassen Co. (Ryser 1985)
1944	Grinnel and Miller (1944) "Believed to have been of regular winter occurrence, formerly, though in smaller numbers than Whistling Swan, south through interior of State. Reported but once since 1900.
1950s	Family of three in Marin Co (E. Krider)
1/57	"A banded Trumpeter Swan was found near Wards lake, Lassen County, in January 1957. It had been banded at Likely Lake, Terrace Skeena District, British Columbia, by an employee of the Canadian Wildlife Service. When banded on March 21, 1954 it was an immature bird. California Fish and Game employees report that they have seen and heard others at Honey Lake, but none of these reports have been verified as to exact place and date" (Ed R. Pickett)
2/8/69	Five at Santa Rosa (B.A. McLean in D. Fix notes)
11/27/77	Four adults at Mono Lake (D. Fix notes)
11/11/78	One at Lake Merced (D. Fix notes)
Late Jan-2/25/79	One at Tule Lake and Klamath NWRs (D. Fix notes)
12/23/84	Two at Tulelake NWR, Siskiyou Co. (Tule Lake CBC)
12/2/85-3/14/86	Four adults using bottomlands around Ft. Dick and Smith River (RAE, ADB, GSL)
11/15/86	One red neck-banded swan flying over Goose Lake just north of OR/CAL line, Lassen Co (J. Welch, Sheldon NWR)
12/30/86	Three adults near W. Butte Rd, Sutter (R. Ryno)
1/18/87	Two at Tule Lake NWR, Siskiyou Co. (P. Rostron)
2/2/87	Two seen at Lower Klamath NWR, including 1 neck-banded as cygnet at Turnbull NWR, WA in 1981 (NE, RE, MR)

Colorado

12/2/75	Grande Prairie neck-banded male cygnet (18TA) seen at Bowles L., Bow Mar, Jefferson County (CWS Edmonton banding files).
12/1/77-2/12/78	Seven trumpeters seen at Lake De Weese near Westcliffe by D. Griffith (D. Griffith, pers. comm.).
1/16/78	Two at Shadow Mtn., Grand Lake Village, Grand Co (D. Jasper)
Winter/80	Grande Prairie neck-banded swan at Denver City lagoon, observed by W. Grath (Burgess 1986)
3/27-4/11/80	Six seen at Wayne Seaton Pond, Buena Vista, Chaffee Co. (J. Porratta)
11/25/83	One adult at Dye Reservoir, Otero County reported by R. Bunn (CFO Journal 19(4), rept no #8-84-12)
12/13/83	Five adults and 1 imm. seen at Brush Hollow Reservoir, Freemont Co by R and J. Watts and W. Maynard (CFO Journal 19(4) rept no #8-84-12)
11/26/84-4/6/85	Two adults from Red Rock Lakes NWR (neckband R2) wintered near Grand Junction (McEneaney 1986)
12/23/84	One adult and one immature observed on Gunnison Christmas Bird Count, on a warm slough south of Gunnison by R. Meyer (Am. Birds 39(4))c

Oregon

11/15/86	One red neck-banded swan (from NWT) observed flying south in flock of unidentified swans at Goose Lake, south of Lakeview, OR, on the Oregon/California border by J. Welch (L. Shandruk, pers. comm.).

Nebraska

10/27/56	Three yearlings, marked as cygnets on 8/27/55, at Lowe Lake, near Grande Prairie, Alberta were found shot: one at Schoolhouse L.; one at Shoup L. near Valentine; one found 11/2 on Loup R. 12 miles west of Fullerton, Nance Co. (Mackay 1957).
10/30/56	One adult shot and crippled near Fullerton, a second adult captured wounded near Shelton, Buffalo Co. on 11/2 (Banko 1960).

Missouri

Apr./83	Two Grande Prairie neck-banded trumpeters observed with two other unmarked swans at Talley Bend on the Truman Res., St. Clair Co., by Talley, Sutlick, Wilson, and Smith (Burgess 1986)

New Mexico

Nov. 1931	Adult shot by hunter on Rio Grande 8 km south of Mesilla Park. Four other swans seen in group (Auk 1932:460) Questionable sighting according to Banko (1960:20)
Winter/75	Zahm photographed cygnet at Bosque del Apache NWR (Burgess 1986).
2/23/77	Adult shot in Bear Canyon Res., east of Silver City (Western Birds 9:90).
Dec./82	Zahm photographed cygnet at Bosque del Apache NWR (Burgess 1986).
Winter/84-85	Family of trumpeters at Elephant Butte Res., reported by H. Miller, J. Ward, J. Herring, B. Morrison (H. Burgess, pers. comm.).
12/14-21/85	One adult on Grays L. near Radium Springs, Dona Ana Co, reported by C. and J. Anderson, E. Wooten and T. Schulte (Trumpeter Swan Soc. News. 15(3):9.
Feb 1987	One adult observed on Rio Grande between Las Cruces and Truth or Consequences. Suspected to be a trumpeter but identity not confirmed (J. Hubbard, NMGF, pers. comm.).

British Columbia

1968	Successful breeding by a released pair on Swan L. near Vernon (Palmer 1976)
4/18/75	Grande Prairie neck-banded 2.5 yr old female (16TY) seen at Golden, also seen in Grande Prairie on 9/25/75 (CWS, Edmonton, banding files)
Summer 1981	Grande Prairie neck-banded 3 yr old male (7X) seen WE of Ft St John (CWS Edmonton banding files)
August 1981	Grande Prairie neck-banded 3 yr old male (70) seen 50 mi SW of Ft Nelson at Klowee L (CWS Edmonton banding files)

Yukon

4/25/80	A green neck-banded swan (from either Malheur NWR or RRL) seen at Crooked L (McKelvey *et al*. 1983)
8/20/80	A green neck-banded swan (from either Malheur NWR or RRL) seen on Smith R. between North and South Toobally Lakes (McKelvey *et al*. 1983).

Mexico

1/21/09	Trumpeter shot at Matamoros, Tamaulipas (Auk Vol. X, p. 72 in Coale 1915)

Appendix II. USFWS Midwinter Tri-state Trumpeter Swan Surveys, 1972-86[a].

	Montana			Idaho			Wyoming		Tri-state
	Centen. Valley	Hebgen Lake	Other[b]	Henry's Fk[c]	Teton River	Other[d]	YNP	Other[e]	Total
1972	209-14	ns[f]	ns	153- 14	149[g]	1-0	56	ns	419- 28
1973	212-28	ns	ns	198- 56	26- 2	2-0	61	ns	499- 86
1974	229-37	ns	4- 3	245- 93	37- 16	2-0	31- 7	7- 0[h]	553-156
1975	151-28	26- 1	15- 3	310- 88	23- 6	ns	30	40- 2	595-128
1976	203-18	48-14	2- 2	282- 55	26- 12	ns	32	62- 1	623-102
1977	290-34	15- 5	10- 4	361-111	34- 15	ns	43- 9	86	839-178
1978	141-48	47-18	6- 2	331- 69	61- 27	ns	46-11	47[i]	695-179
1979	253-20	37- 1	14- 5	290- 52	62- 26	1-3	71-13	15- 3	743-123
1980	304-53	60-25	10- 2	182- 29	68- 41	ns	80-16	63- 6	767-172
1981	292- 9	38-17	22-10	357-103	13- 7	ns	241-91	37-10	1,000-247
1982	249-25	97-48	44-17	358- 88	71- 49	ns	57-20	76-19	952-266
1983	166-12	157-39	40- 8	385- 67	108- 55	ns	88-14	81-12	1,025-207
1984	176-21	169-60	44-28	347- 79	147- 79	9-4	149-50	87-11	1,128-332
1985	212-17	164-13	17- 1	443- 91	249- 50	9-3	154- 7	76- 8	1,326-190
1986	215-45	158-26	7- 2	363- 72	367-103	14-8	89-18	91-25	1,304-299
1987	239-36	69-25	6- 8	524-170	164- 81	2-4	107-50	85-18	1,196-386

[a] Data are presented as adults - cygnets. Only one number is given when observers were unable to classify swans by age and reported only total numbers.

[b] Includes Madison River Valley and Ennis Lake area.

[c] Henry's Fork drainage above St. Anthony, including Buffalo River and Sheridan Reservoir.

[d] Henry's Fork River below St. Anthony, and South Fork of the Snake River.

[e] Primarily the National Elk Refuge, Snake River near Jackson, and Salt River; occasional swans on the Green River.

[f] Area not surveyed.

[g] Suspected to be migrant Tundra Swans, not included in survey total.

[h] Although not included in results of USFWS Midwinter surveys, between 1969-74 January waterfowl surveys conducted by the Wyo. Dept. of Game and Fish located between 43-47 swans annually in the Snake River drainage.

Appendix III. USFWS September Tri-state Trumpeter Swan Surveys (adults - cygnets)[a].

	Montana			Wyoming			Idaho		Tri-state
Date	RRLNWR	Other CV[b]	State Total	YNP	GTNP	State Total	HSP	State Total	Total
1931	ns[c]	ns	ns	14-12	6-3	0-15	ns	ns	20- 15 (35)
1932	19- 7	1- 2	20- 9	29- 2	2-1	31- 3	ns	7- 0	58- 12 (70)
1933	15- 9	2- 0	17- 9	27- 8	1-0	28- 8	ns	4- 0	49- 17 (66)
1934	16-26	ns	16-26	16-17	1-0	19-18	ns	13- 5	48- 49 (97)
1935	30-16	ns	30-16	16-11	ns	16-11	ns	ns	46- 27 (73)
7/13/36	29-26	2- 0	31-26	38-13	ns	43-15	ns	ns	74- 41 (115)
8/5/37	34-51	ns	36-51	38-26	ns	42-26	ns	3- 0	81- 77 (158)
8/38	28-42	18- 9	46-51	40- 4	ns	47- 4	ns	ns	93- 55 (148)
8/15/39	50-59	8- 0	58-59	47-17	ns	56-17	ns	12- 0	126- 76 (202)
8/15/40	58-48	5- 0	67-49	39-14	ns	46-14	ns	7- 5	120- 68 (188)
8/15/41	52-44	14- 4	74-54	44-15	ns	47-15	ns	19- 0	140- 69 (209)
8/20/42	45-43	20- 3	71-53	ns	ns	7- 5	ns	24- 0	102- 58 (160)
8/26/43	88-25	24- 1	126-34	ns	ns	3- 0	ns	8- 0	137- 34 (171)
8/12/44	106-58	21- 3	137-61	35- 8	ns	48-11	ns	22- 0	207- 72 (279)
8/31/45	113-50	25- 0	146-52	ns	ns	4- 3	ns	8- 0	158- 55 (213)
8/11/46	124-46	43- 1	181-62	43- 8	ns	51-10	ns	23- 0	255- 72 (327)
8/12/47	131-49	38- 3	179-52	45- 8	ns	60- 8	ns	24- 0	263- 60 (323)
8/16/48	121-73	67- 5	199-85	49-13	ns	63-21	ns	26- 0	288-106 (394)
8/3/49	132-61	87- 7	233-75	54-21	ns	72-23	5- 5	16- 5	321-103 (424)
7/31/50	106-40	73- 0	187-47	57-16	ns	68-21	4- 3	24- 5	279- 73 (352)
7/31/51	170-76	104- 5	285-89	63-11	ns	76-14	6-11	39-15	400-118 (518)
7/16/52	184-55	142- 1	340-67	58-10	ns	74-17	6- 5	54- 9	468- 93 (561)
8/3/53	211-38	132- 9	355-57	51-10	4-1	87-22	6-12	28-20	470- 99 (569)
8/31/54	352-28	49- 8	412-40	64-23	8-4	98-32	21- 5	38- 7	548- 79 (627)
8/29/55	242-41	109- 4	366-48	58-10	10-5	99-31	10-11	26-16	491- 95 (587)
8/27/56	293-39	68- 9	374-48	48- 9	10-6	81-19	6- 4	26-14	481- 81 (562)
8/20/57	159-45	78-10	247-57	44-16	13-6	85-28	5- 1	27- 4	359- 89 (448)
9/9/58	270-40	76-21	358-62	64-18	17-9	105-45	8- 0	48-23	511-130 (641)
9/8/59	271-40	92-12	379-59	62- 8	26-8	109-30	12- 0	44-10	532- 99 (631)
9/6/60	163-33	116-14	296-49	56- 7	36-6	125-25	20- 0	66-14	487- 88 (575)
9/26/61	155-14	91-10	257-29	69- 3	9-6	130-12	20- 5	43-18	430- 59 (489)
9/10/62	179-50	39-18	225-76	44- 7	13-0	96- 9	13- 5	45-18	366-103 (469)

[a] These data were assembled primarily from the original USFWS Tri-state Survey flight maps and data sheets which are on file at RRLNWR. Supplemental data from ground counts of areas not aerially surveyed were also included. In many instances these data differ from previously published data, primarily due to addition and transcription errors.

[b] Other CV includes the Centennial Valley downstream from RRLNWR. Elk, Conklin, Cliff, or Wade lakes are included in the category State Total for Montana.

[c] Not surveyed.

Appendix III, cont.

Date	Montana			Wyoming			Idaho		Tri-state
	RRLNWR	Other CV	State Total	YNP	GTNP	State Total	HSP	State Total	Total
9/3/63	145-122	78-12	229-138	49- 7	23- 4	111-16	4- 5	41-28	381-182 (563)
9/14/64	180- 22	217- 4	402- 31	61- 8	26- 1	106-10	9- 4	46- 7	554- 48 (602)
9/14/65	190- 16	157-20	354- 36	62- 5	24- 0	123-13	10- 4	62-12	539- 61 (600)
9/12/66	240- 54	104-11	351- 66	57-12	26- 4	101-28	18- 7	62-21	514-115 (629)
8/28/67	184- 20	143- 5	334- 25	55- 2	21- 5	100-18	8- 3	85- 8	519- 51 (570)
8/26/68	155- 90	79-24	242-123	57- 4	29-14	101-25	10- 3	88- 6	431-154 (585)
1969	ns- 7	ns-11	ns- 20	ns	14- 8	ns	ns	ns	ns
1970	ns- 50	ns	ns	ns	8- 6	ns	ns	ns	ns
8/23/71	146- 12	142-26	297- 49	30- 3	27-10	74-13	6- 1	60- 6	431- 68 (499)
8/20/72	ns- 20	ns	ns	ns- 3	19- 4	ns	ns	ns	ns
9/19/73	ns- 39	ns-26	ns- 65	ns	18- 6	ns	ns	ns	ns
8/26/74	139- 33	151-11	296- 49	50- 4	30-10	90-14	25- 8	71-17	457- 80 (537)
9/4/75	120- 22	ns-14	ns- 36	ns	ns	ns	ns	ns	ns
1976	ns- 23	ns-22	ns	52- 1	20- 0	ns	ns	ns	ns
8/29/77	137- 39	122-13	267- 64	51- 7	14- 6	76-15	25- 2	60- 7	403- 86 (489)
9/14/78	ns- 38	ns- 9	ns- 50	45- 2	ns	ns	ns	ns	ns
9/25/79	297- 52	18- 4	324- 63	38- 3	32- 4	80-15	34- 0	58- 9	462- 87 (549)
9/2/80	175- 5	128- 0	315- 6	39- 1	29- 1	74- 6	10- 1	73-11	462- 23 (485)
10/1/81	247- 37	40-13	ns	ns	ns	ns	ns	ns	ns
1982	ns- 4	ns-14	ns- 21	ns	ns	ns	ns	ns	ns
9/12/83	106- 19	112-12	228- 32	29- 6	23- 3	78-16	8- 0	92- 6	398- 54 (452)
9/11/84[d]	115- 4	131-13	259- 17	37- 4	20- 1	93-15	26- 7	72-21	424- 53 (477)
8/25/85[d]	108- 42	96-34	208- 83	27- 5	21- 5	73-25	20- 6	83-27	364-135 (499)
9/7/86[d]	87- 15	80-13	174- 28	24-12	26- 0	74-19	22- 0	83-14	331- 61 (392)

[d] For comparability with survey results of previous years, East Front Montana data have not been included. Since 1984 East Front data are: 1984 7-5; 1985 4-4; 1986 2-2.

Appendix IIIa. Explanations of differences between the survey data presented in Appendix III and data presented in previously published sources[a]. Data are presented in the format adults-cygnets.

Year	Previously Published Source	Survey Area	Previous Data		Changed To		Explanation for difference
1985	TTSS[b]	MT	212-	87	208-	83	Removed East Front, MT, data of 4-4 to conform to the format of previous surveys.
1985	TTSS	Tri-state	368-	139	364-	135	Same as above
1984	TTSS	CV	137-	13	131-	13	Changed Elk, Hidden, and Conklin Lakes from CV to Other Montana to conform to format of previous surveys
1984	TTSS	MT	268-	22	259-	17	Removed East Front, MT data to conform with previous surveys. Also grouped all YNP data in WY to conform with previous surveys.
1984	TTSS	YNP	27-	4	37-	4	All YNP grouped together in WY.
1984	TTSS	WY	83-	15	93-	15	Same as above
1984	TTSS	ID	80-	21	72-	21	Same as above
1984	TTSS	Tri-state	431-	58	424-	53	East Front, MT, data removed to conform with earlier surveys
1977	TTSS	HSP	23-	2	25-	2	HSP should include 2-0 on Fish Pond
1977	TTSS	Tri-state	406-	86	403-	86	Addition error in original records.
1974	TTSS	RRLNWR	125-	33	139-	33	Addition error of 14-0 in original records.
1974	TTSS	MT	282-	49	296-	49	Same as above
1974	TTSS	Tri-state	443-	80	457-	80	Same as above
1967	TTSS	CV	113-	5	143-	5	Addition error of 30-0 in original records.
1967	TTSS	MT	302-	25	334-	25	Addition errors of 30-0, 2-0.
1967	TTSS	GTNP	24-	5	21-	5	Transcription error from flight sheet.
1967	TTSS	WY	99-	12	100-	18	Add 2-1 in Pacific Cr., 2-5 at Pinto Ranch, less 3-0 in GTNP.
1967	TTSS	Tri-state	488-	45	519-	51	Addition errors in MT, transcription errors it in WY
1965	TTSS	YNP	58-	5	62-	5	Error noted by YNP staff.
1965	TTSS	WY	119-	13	123-	13	Same as above
1965	TTSS	Tri-state	535-	61	539-	61	Same as above
1963	TTSS	RRLNWR	148-	127	145-	122	Addition error in original records.
1963	TTSS	MT	232-	143	229-	138	Same as above
1963	TTSS	WY	89-	12	111-	16	22-4 in Squirrel Meadow should be added to WY
1963	TTSS	ID	63-	32	41-	28	Same as above, removed from ID
1963	TTSS	Tri-state	384-	187	381-	182	Addition error at RRLNWR.
1962	TTSS	WY	83-	9	96-	9	Addition error in original records.
1962	TTSS	Tri-state	353-	103	366-	103	Same as above
1961	TTSS	MT	255-	29	257-	29	Added 2-0 at Aldrich Lake.
1961	TTSS	YNP	71-	3	69-	3	Removed 2-0 at Aldrich Lake.
1961	TTSS	GTNP	21-	8	19-	6	2-0 changed from GTNP to other WY
1961	TTSS	ID	47-	19	43-	18	Deleted 2-1 duplicate record; moved 2-0 to WY
1961	TTSS	Tri-state	432-	60	430-	59	Deleted 2-1 duplicate record.

Appendix IIIa, cont.

Year	Previously Published Source	Survey Area	Previous Data	Changed To	Explanation for difference
1960	TTSS	RRLNWR	163- 34	163- 33	Addition error in original report
1960	TTSS	MT	292- 52	296- 49	Add 2-0 on Aldrich L., addition errors on and off Refuge
1960	TTSS	ID	95- 23	66- 14	29-9 in Squirrel Md. area moved to Wyoming
1960	TTSS	WY	98- 16	125- 25	29-9 added from ID, 2-0 on Aldrich L. shifted to MT
1960	TTSS	YNP	56- 7	54- 7	2-0 on Aldrich L. shifted to MT
1960	TTSS	Tri-state	485- 91	487- 88	Addition error in MT
1953	TTSS	WY	89- 23	87- 22	2-1 shifted from ID to WY
1953	TTSS	ID	26- 19	28- 20	Same as above
1952	TTSS	ID	60- 10	54- 9	6-1 shifted to WY from ID in Squirrel Mds
1952	TTSS	WY	68- 16	74- 17	Same as above
1951	TTSS	ID	46- 18	39- 15	Moved 7-3 to WY in Squirrel Md. area
1951	TTSS	WY	69- 11	68- 21	Same as above
1950	TTSS	ID	31- 7	24- 5	Moved 7-2 to WY in Squirrel Md. area
1950	TTSS	WY	61- 19	68- 21	Same as above
1946	TTSS	YNP	47- 10	43- 8	Removed swans on lakes adjacent to YNP
1944	TTSS	YNP	44- 11	35- 8	Removed swans on lakes adjacent to YNP
1942	TTSS	Tri-state	123- 76	126- 76	3-0 added from National Elk Refuge, WY

[a] The 1940-57 survey data in Appendix III differ from that published by Banko (1960:146). Banko's data included swans counted at Malheur NWR, Ruby Lakes NWR, and zoos, which were included in the USFWS Trumpeter Swan surveys. Those swans were not included in the data presented in Appendix III.

[b] TTSS = the USFWS September Tri-state Trumpeter Swan Survey

Appendix IV. Cygnet production at Red Rock Lakes NWR, adjusted for removals (not adjusted for variation in census dates)[a].

Year	Cygnets Censused	Cygnets Actually Produced	Cygnets Recruited	Comments
1935	16	16	16	No date given or comments on survival after census
1936	26	26	26	Early count, cygnets small.
1937	51	51	51	
1938	42	42	38	4 to National Elk Refuge (NER) on 10/24.
1939	59	59	53	3 to NER, 3 to Malheur NWR in October.
1940	48	48	48	
1941	44	44	35	6 to Malheur NWR on 9/19, 3 to NER on 9/23.
1942	43	43	43	
1943	25	[b]	[b]	New observer, only saw 4-0 in River Marsh, possibly an incomplete survey
1944	58	60	38	2 hatched in incubator and hand-reared; 22 to Malheur 9/4
1945	50	50	30	20 to Malheur NWR on 9/13
1946	46	46	46	
1947	49	49	49	
1948	73	73	53	20 to Malheur NWR on 9/12
1949	61	61	51	10 to Ruby L. NWR on 9/12.
1950	40	40	40	
1951	76	76	76	
1952	55	55	55	
1953	38	38	38	
1954	28	36	28	7 to Malheur NWR on 7/14: 1 to Ruby L. NWR on 7/28.
1955	41	44	8	3 to Delta Marsh on 8/7; 16 to Ruby L. NWR on 9/26; 17 to Malheur NWR on 9/27.
1956	39	39	27	12 to Ruby L. NWR on 10/10.
1957	45	45	25	20 to Malheur NWR on 9/25.
1958	40	55	35	55 seen on ground count after census; 20 to Ruby L. NWR on 9/22.

[a] Data were compiled from Tri-state Trumpeter Swan surveys, RRLNWR Annual Narrative Reports, and RRLNWR swan transfer files. Cygnets actually produced equals the number of cygnets censused plus the number of cygnets removed prior to the census. Cygnets recruited equals the number of cygnets censused minus the number of cygnets removed after the census.

[b] Narrative report indicates the possibility that the new observer overlooked a number of swans and data may be incomplete

Appendix IV, cont.

Year	Cygnets Censused	Cygnets Actually Produced	Cygnets Recruited	Comments
1959	40	40	40	
1960	33	33	13	20 eggs to Bowdoin NWR, 7 cygnets reared and released at RRLNWR on 9/1; 20 to Lacreek NWR on 9/12.
1961	14	32	14	1 hand-raised cygnet sent to zoo 7/14; 17 to Lacreek NWR on 9/21.
1962	50	50	30	20 to Lacreek NWR on 9/20.
1963	122	138	122	16 cygnets hand-reared -from 30 eggs: 10 to zoos on 8/26, 6 to Turnbull NWR after census
1964	22	24	22	2 to zoo on 9/11, not included in census.
1965	16	16	3	2 to Kellogg Sanctuary after 9/24; 11 to Turnbull NWR on 9/21.
1966 (Refuge)	54	65	53	11 to Turnbull NWR on 9/7 ;1 to zoo on 9/19 after census.
1966 (Cent.V.)	11	20	11	9 to Turnbull NWR taken from Blake Slough, CV off-Refuge.
1967	20	22	10	2 to Belgium zoo on 8/67; 10 to Hennepin County Pk. Res. on 8/31
1968	90	90	78	10 to Hennepin County Pk. Res. on 9/10; 2 to Portugal
1969	7	9	7	2 to zoos on 8/69
1970	50	50	40	10 to Hennepin County Pk. Res. on 9/9
1971	12	13	13	Actual count from Page (1976)
1972	20	20	20	
1973	39	39	39	
1974	33	33	33	
1975	22	22	22	
1976	23	23	23	
1977	39	39	39	
1978	38	40	38	2 To Logan Zoo on 9/9
1979	52	56	52	4 to private propagators in August.
1980	5	5	5	
1981	37	39	37	2 to private propagators on 8/31.
1982	4	4	4	
1983	19	19	19	
1984	4	4	4	
1985	42	42	42	
1986	15	15	15	

Appendix V. Summer counts of the Red Rock Lakes National Wildlife Refuge and Centennial Valley (CV) flocks. Data were compiled from USFWS Tri-state Trumpeter Swan Surveys and other ground observations recorded in the RRLNWR files. Observations are presented as adults - cygnets.

	Red Rock Lakes National Wildlife Refuge								
Year	Upper Lake	Lower Lake	Swan Lake	River Marsh	Other Areas	Total Refuge	Cygnets Removed[a]	CV (Off Refuge)	Total CV[b]
1935	2- 5	6-11	ns[c]	ns	ns	30- 16	0	ns	ns
1936	12-10	10-11	ns	8- 5	1-0	29- 26	0	2- 0	31- 26
1937	12- 9	10-21	2- 8	10-13	0	34- 51	0	ns	ns
1936	ns	ns	ns	ns	ns	28- 42	0	18- 9	46- 51
1939	19-12	14-13	6- 8	11-26	0	50- 59	0	8- 0	58- 59
1940	25-10	18-20	3- 0	12-18	0	58- 48	0	5- 0	63- 48
1941	26- 8	12-22	2- 0	10-14	2-0	52- 44	0	14- 4	66- 48
1942	20-13	13-19	3- 4	9- 7	0	45- 43	0	20- 3	65- 46
1943	52- 3	22-16	10- 6	4- 0	0	88- 25	o	24- 1	112- 26
1944	62-18	20-22	7- 5	17-11	0-2	106- 58	2	21- 3	127- 61
1945	76-10	14-26	8- 4	15-10	0	113- 50	0	25- 0	138- 50
1946	72-10	27-21	7- 6	18- 9	0	124- 46	0	43- 1	167- 47
1947	61-15	35-19	4- 2	31-13	0	131- 49	0	38- 3	169- 52
1948	59- 9	20-27	16- 6	26-31	0	121- 73	0	67- 5	188- 78
1949	60-15	30-24	11- 5	30-17	1-0	132- 61	0	87- 7	219- 68
1950	73-13	11-10	5- 4	17-13	0	106- 40	0	73- 0	179- 40
1951	109-12	16-21	8- 6	34-37	3-0	170- 76	0	104- 5	274- 81
1952	130-16	19-11	7- 5	27-23	1-0	184- 55	0	142- 1	326- 56
1953	145- 9	16- 5	9- 3	39-19		211- 38	0	132- 9	343- 47
1954	260- 3	16- 3	18- 6	57-16	1-0	352- 28	8	49- 8	401- 36
1955	147-14	14-13	16- 0	58-14	7-0	242- 41	3	109- 4	351- 45
1956	174-17	23- 5	17- 3	62-10	17-4	293- 39	0	68- 9	361- 48
1957	64- 6	15- 6	3- 6	63-27	14-0	159- 45	0	78-10	237- 55
1958	137- 7	25- 4	18- 5	88-14	2-0	270- 40	15[d]	76-21	346- 61
1959	166-12	32-11	20- 6	52-11	1-0	271- 40	0	92-12	363- 52
1960	30- 0	45- 9	30- 3	38-14	20-7	163- 33	0	116-14	279- 47
1961	65- 3	24- 4	20- 0	42- 7	4-0	155- 14	18	91-10	246- 24
1962[e]	ns-16	ns- 8	ns- 2	ns-22	ns-0	179- 50	0	39-18	218- 68
1963	39-17	27-29	22-26	53-49	4-1	145-122	16	78-12	223- 134
1964	86- 5	17-10	21- 7	48- 5	8-5	180- 22	2	217- 4	397- 26
1965	90- 2	26- 1	19- 1	46- 9	9-3	190- 16	0	157-20	347- 36
1966	152-15	22-23	14- 7	42- 5	10-4	240- 54	20	104-11	344- 65

[a] Cygnets removed from Centennial Valley prior to census.
[b] Centennial Valley does not include Elk, Conklin, Cliff, nor Wade Lakes.
[c] ns = no survey or data not recorded in Refuge files.
[d] Fifty-five cygnets were seen from the ground after the survey; at least 15 were overlooked on the survey.
[e] Census discrepancy, two cygnets were not reported in area breakdown.

238

Appendix V, cont.

	Red Rock Lakes National Wildlife Refuge								
Year	Upper Lake	Lower Lake	Swan Lake	River Marsh	Other Areas	Total Refuge	Cygnets Removed[a]	CV (Off Refuge)	Total CV[b]
1967	100- 5	15- 3	9- 3	49- 4	11- 5	184-20	2	143- 5	327- 25
1968	53- 6	29-24	22-16	37-37	14- 7	155-90	0	79-24	234-114
1969	ns- 0	ns	ns- 0	ns	ns	ns- 7	2	ns-11	ns- 18
1970	ns- 2	ns-15	ns-10	ns-23	ns- 0	ns-50	0	ns	ns
1971	62- 8	9- 0	16- 1	25- 0	34- 3	146-12	0	142-26	288- 38
1972	ns- 2	ns- 1	ns- 4	ns-10	ns- 3	ns-20	0	ns	ns
1973	ns	ns	ns	ns	ns	ns-39	0	ns-26	ns- 65
1974	52- 8	13- 2	9- 3	19- 7	6-13	139-33	0	151-11	290- 44
1975	ns	ns	ns	ns	ns	ns-22	0	ns-14	ns- 36
1976	ns-10	ns- 0	ns- 4	ns- 9	ns- 0	ns-23	0	ns-22	ns- 45
1977	54- 6	10- 7	10- 1	30-12	33-13	137-39	0	122-13	259- 52
1978	ns- 5	ns- 9	ns-15	ns- 9	ns- 0	ns-38	2	ns- 9	337- 47
1979	ns- 1	ns- 7	ns-13	ns-22	ns- 9	297-52	4	18- 4	315- 56
1980	43- 0	9- 1	16- 0	42- 1	65- 3	175- 5	0	128- 0	303- 5
1981[f]	ns-11	ns- 4	ns- 7	ns-12	ns- 3	247-37	2	40-13	287- 50
1982	42- 0	5- 0	12- 0	23- 0	36- 4	118- 4	0	120-17	238- 21
1983	8- 5	10- 0	12- 4	29- 1	47- 9	106-19	0	112-12	218- 31
1984	24- 3	10- 0	10- 0	29- 0	42- 1	115- 4	0	131-13	246- 17
1985	40- 9	4- 5	16- 9	35-14	13- 5	108-42	0	96-34	204- 76
1986	19- 0	4- 1	17- 6	27- 8	20- 0	87-15	0	80-13	167- 28

[f] Survey made on 1 October

Appendix VI. Canadian Wildlife Service late summer surveys of the Grande Prairie flock, 1959-1986[a].

	Total No. Lakes Surveyed	Pairs With Cygnets	Total Pairs	Single & Flocked Adults	Total Adults	Total Cygnets	Total Flock
1959	37	10	18	51	87	40	127
1960	36	9	14	42	70	38	108
1961	38	12	16	57	89	41	130
1962	39	8	19	35	73	36	109
1963	41	9	14	62	89	27	116
1964	38	7	16	58	90	14	104
1965	42	2	23	18	64	5	69
1966	42	7	21	19	61	24	85
1967[b]	42	7	20	4	44	24	68
1968	47	11	22	32	75	31	106
1969	43	6	23	47	73	13	86
1970	54	9	14	48	76	24	100
1971	55	11	24	31	78	36	114
1972	57	10	23	21	67	37	104
1973	60	19	29	11	68	55	123
1974	71	13	28	43	98	49	147
1975	79	12	31	22	84	37	121
1976	103	14	36	8	80	41	121
1977	113	25	31	26	88	80	168
1978	141(14)	20(0)	36 (3)	59(0)	133 (6)	72 (0)	203 (6)
1979	123(13)	17(1)	41 (4)	15(0)	97 (8)	58 (3)	155(11)
1980	107(13)	21(2)	36 (3)	55(5)	127(11)	64 8)	191(19)
1981	110(14)	21(2)	39(3)	80(4)	158(10)	74(10)	232(20)
1982	118(13)	20(1)	35 (6)	97(0)	167(12)	65 (2)	232(14)
1983	159(13)	23(2)	58 (7)	38(0)	154(14)	68 (9)	222(23)
1984	157 (0)	37(0)	63 (0)	97(0)	225 (0)	118 (0)	341 (0)
1985	174(30)	25(4)	53(10)	85(0)	191(20)	93(16)	284(36)
1986	192(79)	33(8)	57(14)	109(3)	223(31)	124(24)	347(55)

[a] These data were assembled by G. Holton, L. Shandruk, and B. Turner, from the original CWS flight reports. Since 1978, most surveys have included contiguous portions of British Columbia. Therefore, to aid between-year comparisons, the data since 1978 are presented in the format: Alberta survey results (British Columbia survey results).

[b] Incomplete survey.

Appendix VII. Summer census results for the Interior Canada flocks (excluding the Grande Prairie flock)[a].

Year	Yukon Toobally Lakes	NWT Nahanni	Saskatchewan Cypress Hills	Alberta Otter Lakes	Edson	Other[b]	British Columbia Fort Nelson	Fort St. John[c]
1979	ns[d]	ns	ns	3- 0	ns	ns	ns	ns
1980	66-19	ns	ns	ns	ns	ns	ns	ns
1981	68-26	ns	ns	ns	ns	2- 3	34-0	10- 0
1982	ns	ns	ns	8-12	ns	2- 0	ns	ns
1983	ns	ns	ns	9- 7	8-15	2- 4	ns	ns
1984	ns	26-10	5-0	7- 0	12-12	13- 8	ns	ns
1985	49-10[e]	51-24	4-2	8- 3	15- 8	8- 3	16-4	40-23
1986	ns	84-55	5-0	3- 4	13-12	19-21[f]	ns	ns

[a] Data are from McKelvey *et al.* 1983, McKelvey 1986, Shandruk 1986, McCormick 1985, Holton 1983-85; G. Holton and L. Shandruk, pers. comm.

[b] Includes Chinchaga River, Pincher Cr./Cardston, Howard Lake.

[c] Included Boudreau Lakes in 1981 and Boudreau Lakes and Dawson Creek area in 1985.

[d] ns = not surveyed.

[e] 49-10 observed in same area as surveyed previously. In addition, 21-8 were counted in the Ross/McEvoy area, and 12-2 were found in areas adjacent to the core Toobally Lakes area (McKelvey 1986).

[f] Chinchaga River area only.

241

Appendix VIII. Distribution and total number of nests at Red Rock Lakes National Wildlife Refuge, 1935-86[a].

Year	Upper Lake	Lower Lake	River Marsh	Swan Lake	Ponds	Total Refuge
1935	1	3	nd[b]	nd	0	4+
1936	5	3	2	nd	0	10+
1937	3+	5+	3 +	1+	0	12+
1939	4+	4+	5+	2+	0	15+
1940	2+	6+	4+	nd	0	12+
1941	nd	7	nd	nd	0	nd
1942	4+	6+	2+	1+	0	13+
1943	4+	4+	nd	3+	0	11
1944	nd	nd	nd	nd	0	23+
1945	6	7	nd	nd	0	22+
1946	9	10	nd	nd	0	19+
1947	8+	7+	nd	2+	0	17+
1948	nd	nd	nd	nd	0	30
1954	7	6	25+	7	0	45+
1955	6	8	23+	7	0	44+
1956	6	9	26+	9	0	50+
1957	6	10	19+	9	0	44+
1958	5	12	24+	12	0	60
1959	7	19	38	14	0	78
1960[c]	7	18	4	2	0	31
1961	3	11	15+	8	0	37+
1962	7	12	14	12	0	45
1963	8	14	18	11	0	51
1964	6	12		9	0	53
1965	8	11	23	9	0	51
1966	11	11	25	13	0	60
1967	9	13	19	10	2	53
1968	9	11	25	11	4	60
1969	8	8	16	8	7	47
1970	3	7	16	9	4	39
1971	4	5	13	9	5	38
1972	5	4	17	9	5	42
1973	7	5	20	8	7	48
1974	8	6	20	11	3	48
1975	4	5	12	9	6	36

[a] All data were assembled from RRLNWR Annual Narrative Report and Annual Trumpeter Swan Reports. Values followed by + are minimum estimates of numbers of nests based upon number of broods reported.

[b] nd = no data recorded in RRLNWR files.

[c] Personnel change, comparability of data with past years is unknown. Observers reported 62 pairs on territories but only 31 actual nests.

Appendix VIII, cont.

Year	Upper Lake	Lower Lake	River Marsh	Swan Lake	Ponds	Total Refuge
1976	9	6	10	7	7	39
1977	5	7	11	5	6	34
1978	10	5	13	9	6	43
1979	9	6	16	10	6	47
1980	7	5	11	11	7	41
1981	9	4	17	11	3	44
1982	3	1	7	8	5	24
1983	3	4	11	6	5	29
1984	3	4	12	7	5	31
1985	4	2	9	5	4	24
1986	1	3	10	8	3	25

Appendix IX. Comparison of reproduction parameters in Trumpeter Swan flocks.

Area	Year	Number of active nests	% Nest success[a]	Clutch size Mean	Clutch size Sample	% Hatch of eqqs
Red Rock Lakes NWR (RRLNWR files 1981-85)	1981-85	152	nd[b]	4.9	95	50
Red Rock Lakes NWR Page (1976)	1971-73	127	75	4.9	100	55
National Elk Refuge, WY (NER swan files 1985)	1944-84	43	72	3.9	15	50
Yellowstone and vicinity (Shea 1980)	1977-79	72	60	4.0	56	56
Lower elevation WY (Lockman et al. 1987)	1981-86	45	87	4.3	23	61
Targhee N.F., ID (IDFG files[c])	1984-85	15	100	3.6	6	91
Targhee N.F., ID (Maj 1983)	1979-81	21	95	4.4	21	84
Other Idaho (IDFG files[c])	1984-85	14	71	3.8	6	61
Grande Prairie, AB (Holton 1985)	1978-80	34	94	5.6	34	76
Grande Prairie, AB (Burgess 1972)	1970-72	19	100	5.8	6	81
Cypress Hills, SK (Nieman 1979)	1971-78	15	100	6.4	10	80
Lacreek NWR, SD (Leach 1977)	1963-76	60	nd	6.2	nd	64

[a] Percent nest success equals the percent of nests which hatch at least one egg.
[b] nd = No data available.
[c] Data on file at Region 6 Office, Idaho Falls, ID.

Appendix IX, cont.

| Area | Year | Active nests | | Successful nests | | Cygnet Survival to Fledging |
		Cygnets hatched per nest	Cygnets fledged per nest	Cygnets hatched per nest	Cygnets fledged per nest	
Red Rock Lakes NWR (RRLNWR files 1981-85)	1981-85	2.3	.7	nd	nd	30%.
Red Rock Lakes NWR (Page 1976)	1971-73	2.1	.6	2.8	.8	27%
National Elk Ref. WY (NER swan files 1985)	1944-84	2.1	1.6	2.9	2.3	77%
Lower elevation, WY (Lockman *et al.* 1987)	1981-86	2.7	1.9	3.1	2.2	73%
Yellowstone and (Shea 1980)	1977-79	1.8	.5	3.0	.9	29%
Targhee N.F., ID (IDFG files)	1984-85	2.9	1.4	2.9	1.4	48%
Targhee N.F. ID (Maj 1983)	1979-81	3.7	1.2	3.9	1.3	33%
Other Idaho (IDFG files)	1984-85	2.3	1.9	3.2	2.7	84%
Grande Prairie, AB (Holton 1985)	1978-80	4.3	2.4	4.5	2.5	56%
Grande Prairie, AB (Burgess 1972)	1970-72	nd	nd	4.9	3.1	77%
Cypress Hill, SK (Nieman 1979!	1971-78	4.4	2.9	4.4	2.9	65%
Lacreek NWR, SD (Leach 1977)	1963-76	4.0	2.3	nd	nd	58%

Appendix X. Annual production data, Red Rock Lakes National Wildlife Refuge.

Year	Number of nests·	Mean clutch size (n)	Estimated total eggs laid	Percent eggs (n) hatched	Estimated total hatched	Number of cygnets fledged
1935	4+	nd[a]	nd	nd	nd	16
1936	10+	nd	nd	nd	nd	26
1937	12+	nd	nd	nd	nd	51
1938		nd	nd	nd	nd	42
1939	15+	nd	nd	nd	nd	59
1940	12+	nd	nd	nd	nd	48
1941	nd	nd	nd	nd	nd	44
1942	13+	nd	nd	nd	nd	43
1943	11+	5.67 (3)	62+	88% (17)	55+	25
1944	23+	5.00 (6)	115+	nd	nd	60
1945	22+	5.00 (12)	110+	nd	nd	50
1946	19+	5.01 (14)	95+	nd	nd	46
1947	17+	6.00 (7)	102+	nd	nd	49
1948	30+	nd	nd	nd	nd	73
1949	nd	5.08 (12)	nd	50% (61)	nd	61
1950	nd	nd	nd	nd	nd	40
1951	nd	5.61 (13)	112+	66% (73)	nd	76
1952	nd	5.18 (17)	nd	nd	nd	55
1953	nd	nd	nd	nd	nd	38
1954	45+	4.88 (27)	220+	nd	nd	36
1955	44+	4.90 (32)	216+	63% (157)	136+	44
1956	50+	5.0+ estimate	250	nd	nd	39
1957	44+	nd	nd	nd	nd	45
1958	60	4.32 (38)	259	50% (164)	129	55
1959	78	5.00 (58)	390	59% (290)	230	40
1960	62	5.40 (5)	335[b]	50% (20)	157	33
1961	37+	nd	nd	nd	nd	32
1962	45	4.00 (11)	180	66%. (44)	119	50
1963	51	5.15 (13)	263[c]	70% (40)	163	122
1964	53	4.25 (16)	225[d]	18% (50)	37	24
1965	51	4.67 (12)	238[e]	75% (49)	172	16

[a] no data available
[b] 20 eggs removed
[c] 30 eggs removed
[d] 20 eggs removed
[e] 9 eggs removed
[f] Flooding probably lowered the % hatch substantially below 75%

Appendix X, cont.

Year	Number of nests·	Mean clutch size (n)	Estimated total eggs laid	Percent eggs (n) hatched	Estimated total hatched	Number of cygnets fledged
1966	60	5.32 (22)	319	74% (117)	236	65
1967	53	3.41 (32)	181	57% (109)	103	20
1968	60	4.70 (33)	282	78% (155)	220	90
1969.	47	4.33 (21)	204	53% (95)	108	9
1970	39	4.31 (26)	168	59% (112)	99	50
1971	38	4.34 (35)	161	53% (148)	85	13
1972	42	5.39 (33)	226	47% (178)	106	20
1973	48	4.84 (31)	232	66% (150)	153	39
1974	48	4.70 (28)	226	nd	86+	33
1975	36	3.50 (16)	126	71% (56)	90	22
1976	39	4.26 (23)	166	47% (98)	78	23
1977	34	4.25 (4)	145[g]	59% (17)	76	39
1978	43	4.64 (25)	200[h]	61% (116)	102	40
1979	47	5.00 (34)	235[i]	66% (148)	137	56
1980	41	4.80 (32)	197[j]	21% (114)	37	5
1981	44	5.33 (41)	235	58% (113)	136	39
1982	24	4.07 (15)	98[k]	29% (17)	23	4
1983	29	4.60 (23)	133[i]	55% (121)	67	19
1984	31	4.75 (12)	147	20% nd	29	4
1985	24	5.25 (4)	126	76% (21)	96	42
1986	25	4.25 (4)	106	53% (4)	56	15

[g] 17 eggs removed
[h] 32 eggs removed
[i] 27 eggs removed
[j] 21 eggs removed
[k] 16 eggs removed
[l] 12 eggs removed

Appendix XI. Hatching dates of Trumpeter Swans at Red Rock Lakes NWR[a].

Year	First Brood Noted	Peak of Hatch	Last New Brood Noted
1938	6/30	nd	nd
1939	6/10	6/10-15	nd
1941	6/7	6/7-15	6/15
1942	6/13	nd	nd
1943	6/10	nd	nd
1944	6/16	6/16-23	6/28
1945	6/25	nd	early July
1946	6/10	nd	6/24
1947	nd	nd	6/20
1949	6/15	6/17	6/30
1952	6/17	6/26	7/3
1954	6/11	nd	nd
1960	6/11	nd	6/30
1962	6/25	6/30	7/10
1963	5/31	6/10	7/14
1964	7/1	7/3-5	7/10
1965	nd	6/28-7/2	7/12
1966	6/7	6/15-21	nd
1967	6/17	6/25-7/1	7/7
1968	6/16	6/20	nd
1969	nd	6/20	nd
1970	nd	711	7/16
1971	nd	6/21	nd
1972	6/5	6/5-10	nd
1973	6/5	6/21	nd
1974	6/6	6/18	nd
1975	nd	7/2-5	7/8
1976	5/30	6/15-25	7/10
1977	5/30	6/10-13	nd
1978	5/31	6/10-17	6/20
1979	6/14	6/20-30	nd
1980	6/15	6/22-25	6/30
1981	6/10	6/22-7/2	7/3
1982	nd	nd	nd
1983	6/12	6/20-21	7/3
1984	6/19	6/27-28	7/5
1985	6/9	6/20	6/27
1986	nd	6/11	6/27

[a] All data were extracted from RRLNWR Annual Narrative Reports
and Annual Trumpeter Swan Reports.

[b] nd = no data reported in RRLNWR files.

Appendix XII. Mean annual number of broods and cygnets per·brood at fledging, by decade[a].

Mean annual number of broods at fledging

Decade	Red Rock Lakes	Centennial Valley and Other Montana	Idaho	Yellowstone and Wyoming	Grande Prairie
1930s	10.3	nd[b]	nd	5.2	nd
1940s	15.7	2.9	1.3	4.4	nd
1950s	16.2	4.4	5.0	9.7	nd
1960s	17.6	5.6	6.4	7.6	7.8
1970s	12.1	7.0	3.8	5.8	15.0
1980s	6.8	5.3	6.5	5.8	28.6

Mean number of cygnets per brood at fledging

Decade	Red Rock Lakes	Centennial Valley and Other Montana	Idaho	Yellowstone and Wyoming	Grande Prairie
1930s	3.7	nd	nd	3.1	nd
1940s	3.3	2.8	3.3	2.8	nd
1950s	2.7	3.0	2.5	2.6	nd
1960s	2.8	3.0	2.3	2.2	3.24
1970s	2.7	3.1	2.6	2.6	3.28
1980s	2.6	3.3	2.5	2.7	3.42

[a] All data were derived from the USFWS Tri-state Trumpeter Swan Surveys.
[b] No data

Appendix XIII. Approximate dates of freeze-up and ice-out for Red Rock Lakes National Wildlife Refuge, 1937-86[a].

Freeze-up	Ice-out
11/1/37	5/10/38
11/3/38	4/28/39
11/10/39	4/22/40
11/10/40	4/29/41
11/4/41	4/25/42
10/28/42	4/1/43
10/31/43	5/4/44
11/13/44	5/l/45
11/9/45	4/28/46
10/28/46	5/5/47
11/10/47	5/12/48
11/7/48	4/30/49
11/18/49	5/15/50
11/9/50	5/4/51
11/1/51	early May/52
11/24/52	early May/53
11/19/53	4/25/54
11/28/54	May/55
10/31/55	late April/56
10/28/56	May/57
11/10/57	4/58
11/15/58	4/59
11/1/59	5/10/60
11/7/60	4/30/61

[a] Data were assembled from numerous RRLNWR files, primarily Annual Narrative Reports and Annual Trumpeter Swan Reports.

Appendix XIII, cont.

Freeze-up	Ice-out
10/24/61	5/10/62
11/15/62	5/1/63
11/15/63	5/20/64
11/27/64	5/1/65
11/28/65	4/20/66
11/23/66	5/11/67
11/3/67	4/68
11/2/68	4/69
10/25/69	5/21/70
10/25/70	5/15/71
10/24/71	4/72
11/21/72	5/10/73
no data	4/30/74
11/23/74	5/28/75
11/3/75	5/14/76
11/4/77	5/3/78
11./14/78	5/10/79
10/31/79	5/5/80
11/12/80	4/28/81
11/20/81	5/10/82
11./11/82	5/15/83
11/20/83	5/10/84
10/21/84	5/3/85
10/8/85[b]	4/24/86

[b] Lakes initially froze hard on 10/8/85 then thawed and refroze on 11/9/85

Appendix XIV. Water level gauge readings at the lower structure, Red Rock Lakes National Wildlife Refuge[a]

	Monthly median gauge reading (Peak gauge reading)[b]					Peak Annual	Date of
Year	April	May	June	July	August	Reading	Peak Flow
1958	9.04 (9.23)	9.23 (9.23)	8.46 (8.64)	7.76 (7.93)	no data	9.23	5/27
1959	8.43 (8.61)	8.24 (8.38)	8.11 (8.12)	7.79 (8.00)	7.50 (7.58)	8.61	5/21
1960	8.57 (8.98)	8.36 (8.56)	7.88 (7.88)	7.43 (7.56)	7.22 (7.22)	8.98	4/12
1961	8.21 (8.40)	8.03 (8.08)	7.94 (8.08)	7.49 (7.70)	8.29 (8.32)	8.40	4/20
1962	9.90 (9.90)	8.77 (9.12)	8.54 (8.68)	8.07 (8.08)	7.93 (8.06)	9.90	4/27
1963	8.70 (8.70)	8.30 (8.38)	8.30 (8.30)	8.02 (8.26)	7.50 (7.54)	8.70	4/17
1964	no data	8.81 (8.94)	8.88 (9.05)	8.43 (8.75)	7.99 (8.10)	9.05	6/?
1965	no data	8.70 (8.80)	8.68 (8.80)	8.86 (8.96)	8.48 (8.60)	8.96	7/15
1966	8.93 (8.96)	8.74 (8.82)	8.16 (8.47)	7.50 (7.70)	7.29 (7.34)	8.96	4/20
1967	no data	8.97 (9.10)	8.98 (9.10)	8.80 (8.80)	8.64 (8.74)	9.10	5/11,6/30
1968	8.44 (8.86)	10.18(10.59)	8.57 (8.69)	8.06 (8.33)	7.71 (7.80)	10.59	5/6
1969	9.11 (9.11)	9.14 (9.14)	9.19 (9.32)	8.84 (9.06)	8.20 (8.60)	9.32	6/10
1970	no data	9.02 (9.02)	8.86 (8.96)	8.56 (8.56)	7.66 (7.48)	9.02	5/8
1971	no data	9.25 (9.25)	8.93 (9.00)	9.09 (9.33)	8.00 (8.20)	9.33	7/2
1972	8.76 (8.76)	8.60 (8.60)	8.87 (8.96)	8.58 (8.78)	8.06 (8.20)	8.96	6/29
1973	no data	9.35 (9.50)	no data[c]	no data	no data	no data	no data
1974	9.32 (9.32)	9.77 (9.77)	9.86 (9.96)	8.30 (8.58)	7.86 (7.96)	9.96	6/24
1975	no data	9.60 no data	9.39 (9.48)	9.07 (9.25)	8.47 (8.75)	9.60+	5/17-25
1976	9.27 (9.27)	9.25 (9.50)	8.80 (8.94)	8.47 (8.68)	8.45 (8.51)	9.50	5/7
1977	8.80 (8.86)	8.86 (8.90)	8.88 (9.04)	8.43 (8.60)	8.36 (8.43)	9.04	6/15
1978	9.21 (9.48)	8.98 (9.56)	8.42 (8.47)	8.46 (8.55)	8.12 (8.30)	9.56	5/4
1979	9.30 (9.60)	9.20 (9.65)	8.42 (8.75)	7.79 (8.08)	7.41 (7.53)	9.65	5/4
1980	8.92 (9.35)	9.19 (9.35)	9.04 (9.25)	8.47 (8,83)	7.97 (8.05)	9.35	5/1
1981	8.60 (9.14)	8.84 (9.04)	8.63 (9.02)	8.01 (8.10)	8.12 (8.12)	9.04	5/28
1982	9.36 (9.36)	9.47 (9.90)	8.90 (8.96)	8.68 (8.96)	8.14 (8.40)	9.90	5/2
1983	no data	9.12 (9.14)	9.27 (9.44)	9.45 (9.60)	9.11 (9.22)	9.60	7/11-18
1984	9.60 (9.60)	9.61 (9.80)	9.52 (9.60)	9.10 (9.34)	8.66 (8, 70)	9.80	5/18
1985	9.85 (9.85)	8.99 (9.12)	8.54 (8.84)	7.82 (8.00)	7.53 (7.55)	9.85	4/17
1986	9.00 (9.14)	8.72 (8.79	8.78 (8.99)	8.36 (8.52	8.12 (8.19)	8.99	6/10

[a] All data were assembled from RRLNWR files.
[b] Data represent the actual reading minus 6000 ft.
  Gauge "00" elevation   6606.00 ft
  Operational level   6607.00 ft
  Flow line elevation   6607.00 ft
  Crest elevation   6607.00 ft
[c] Gauge was damaged by high water.

Appendix XV. Summary of aquatic vegetation survey data, Red Rock Lakes National Wildlife Refuge, 1956-85[a].

	Species composition (% of total weight) of vegetation in Lower Lake, RRLNWR											
Species	1956	1967	1969	1971	1971*	1973	1975	1977	1979	1981	1983	1985
Elodea canadensis	60.5	44.1	53.8	2.8	0.6	0.2	0.1	0.4		Tr	13.4	3.7
Lemna trisulca	0.8	6.2	3.4	34.7	33.9	7.1	15.0	4.0	5.8	1.6	10.5	2.3
Myriophyllum spicatum	2.0	16.6	4.5	10.4	14.5	11.3	20.5	17.6	45.0	19.6	7.9	7.1
Ceratophyllum demersum			17.3	11.8	4.7	13.3	1.4	15.5		11.8	15.5	38.7
Chara vulgaris	0.9	2.9	0.7	3.1	1.2	1.8	3.5	0.8	4.2	1.7	10.2	12.0
Potamogeton pectinatus	3.4	2.9	0.9	2.7	8.4	6.9	1.7	2.5	1.6	2.0	3.1	0.6
E. zosteriformis	2.9	4.4	1.3	0.5	3.8	9.3	12.7	13.4	4.0	6.8	10.2	2.5
E. richardsonii	14.0	14.6	15.0	20.9	24.6	32.6	34.8	15.1	24.7	17.9	16.6	16.7
E. praelongus	4.0	4.7	0.7	9.6	4.8	12.3	9.0	11.0	8.4	12.0	9.5	14.9
E. pusillus		2.0	0.5	2.5	0.4	0.1				0.7		0.3
E. foliosus	3.1				0.1	0.4					Tr	
E. friesii					0.1	0.2		7.1	4.5	4.5	1.0	Tr
Sagittaria cuneata	7.7	0.9	0.2	0.4	1.3	0.6	0.3	3.1	1.1	2.4	1.6	
Najas flexilis	0.3	0.7	Tr			3.9	0.4	2.9		Tr	0.1	
Hippuris vulgaris					0.5							
Utricularia vulgaris				0.6	1.1							
Isoetes sp.						Tr		1.1	0.2			
Ranunculus aquatilis								5.3		18.9		1.1
Nitella flexilis							0.4					

[a] All data were compiled from RRLNWR Aquatic Vegetation Survey Files, except for 1971* were the results of surveys conducted by Paullin (1973). Blanks indicate that the species was not recorded.

Appendix XV, cont.

Species	Production (in tons) of major aquatic vegetation species in Lower Lake, RRLNWR								
	1967	1969	1971	1973	1975	1977	1979	1981	1983
Elodea canadensis	10400	12528	252	31	18	54		Tr	2156
Potamogeton richardsonii	3214	3260	1880	4699	8546	2142	4101	3292	2665
P. praelongus	1035	152	863	1780	2218	1561	1395	2218	1525
Myriophyllum exalbescens	4045	1083	935	1623	5036	2500	7450	3614	1275
Ceratophyllum demersum		4029	1061	1923	348	2200		2178	2486
Chara vulgaris	889	212	279	264	858	107	698	322	1646
P. zosteriformis	969	282	45	1346	3108	1905	657	1252	1632
P. pectinatus	730	224	243	997	425	358	268	376	505
Lemna trisulca	1755	950	3120	1019	3685	568	957	376	1686
Ranunculus aquatilis						751		3479	
Total Production	23901	23320	8993	14427	24534	14221	16609	18426	16046

Appendix XV, cont.

Species	Species composition (% of total weight) in Upper Lake, RRLNWR											
	1956	1966	1968	1970	1971*	1972	1974	1976	1978	1980	1982	1984
Chara vulgaris	25.3	24.0	33.2	58.9	20.5	56.7	60.0	37.3	53.8	45.8	50.5	70.5
Elodea canadensis	46.6	12.0	0.8	0.1	0.9	0.6		0.7	Tr	Tr	0.3	1.7
Myriophyllum spicatum	0.2	0.7	14.5	9.7	5.4	9.8	2.9	11.1	14.3	16.6	10.6	0.4
Potamogeton richardsonii	3.3	43.1	36.8	13.2	16.0	16.5	18.3	26.8	14.6	20.9	15.1	13.1
P. praelongus	3.9	6.7	4.9	8.4	25.8	6.5	3.0	12.8	10.4	9.9	7.1	6.6
P. pectinatus	4.8	8.7	4.7	4.3	27.1	4.2	2.0	3.7	2.7	2.1	2.6	1.1
P. friesii					Tr				Tr	Tr		
P. zosteriformis	3.2			Tr	0.8	2.5	0.4	1.3	Tr	0.6	0.8	3.4
P. foliosus	5.7						Tr		0.3	0.2		0.5
P. pusillus				0.7	0.8	Tr	0.1	2.0	Tr			Tr
P. filiformis				3.0								
Sagittaria cuneata	2.8	1.8	4.1	1.2	0.5	1.2	0.3	1.0	0.5	1.1	1.4	0.3
Najas flexilis	4.0	0.4	1.0	0.3	0.9	0.5	2.0	1.4	Tr	0.2	0.5	0.3
Utricularia vulgaris				0.3	Tr							
Ranunculus aquatilis		2.0		Tr	1.5	1.5		1.3	3.0	2.2	10.5	1.8
Ceratophyllum demersum							9.9					Tr
Isoetes sp.								0.4	Tr	0.4	0.4	0.3
Callitriche hermaphroditica							Tr				Tr	0.2

Appendix XV, cont.

| Species | Production (in tons) of major aquatic plant species in Upper Lake, RRLNWR | | | | | | | | | |
	1966	1968	1970	1972	1974	1976	1978	1980	1982	1984
Elodea canadensis	2605	164	31	220		191	T	8	59	780
Chara vulgaris	6773	8802	23834	20397	41238	10702	9385	9300	9804	32951
Myriophyllum exalbescens	157	3076	3109	3534	2004	3178	2491	3377	2063	178
Potamogeton richardsonii	8733	7052	3832	5932	12592	7673	2542	4241	2924	6118
P. praelongus	1551	939	2428	2330	2055	3673	1805	2017	1381	3076
P. pectinatus	2015	1029	1437	1517	1398	1055	466	419	496	513
P. zosteriformis			Tr	898	292	369	30	114	157	1614
P. foliosus					T		56	38	4	242
P. pusillus			234	T	85	568	34			4
Ranunculus aquatilis				525		364	525	441	2034	873
Ceratophyllum demersum					6809					
Najas flexilis	93	219	100	194	1381	398	8	51	97	140
Total Production	22841	22014	36585	35976	68782	28667	17435	20320	19401	46780

Appendix XV, cont.

Species	Species composition (% of total weight) of aquatic plants in Swan Lake, RRLNWR										
	1956	1966	1969	1971	1971•	1973	1975	1977	1979	1981	1983
Myriophyllum spicatum	48.9	65.6	67.2	41.1	41.9	25.7	30.2	58.2	45.0	58.6	49.2
Chara vulgaris	2.2	28.8	23.8	25.9	41.9	48.5	46.5	33.7	46.3	21.2	42.9
Potamogeton pectinatus	17.4	2.2	4.1	15.6	5.4	3.8	3.4	2.3	3.2	2.5	2.1
P. richardsonii	16.3	3.0	2.5	9.4	8.5	10.9	17.0	3.7	4.0	14.1	4.6
P. foliosus	9.7			0.5	0.3			0.7	1.2	1.7	
P. pusillus	5.6	0.2	1.0	4.5	1.9	8.9				4.0	0.1
P. friesii				1.7		1.5	0.9	0.5	0.3	0.3	0.9
Lemna trisulca		0.1	0.2	0.7	0.1	0.6	1.5	Tr	Tr	Tr	Tr
Ceratophyllum demersum							0.3				
Utricularia vulgaris			1.2	0.1							
Zannichellia palustris		0.2					0.4	Tr			
Elodea canadensis				Tr			Tr	Tr			
Ranunculus aquatilis	0.4			Tr			0.3		Tr		
Nitella flexilis				0.5					0.6		
Najas flexilis											0.1
Isoetes sp.							0.1				

Appendix XV, cont.

Species	Production (in tons) of the major aquatic plant species in Swan Lake, RRLNWR								
	1966	1969	1971	1973	1975	1977	1979	1981	1983
Myriophyllum spicatum	5977	5205	3551	2479	2209	2906	2697	4239	2690
Chara vulgaris	2629	1848	3551	4664	3397	1685	2760	1538	2346
Potamogeton richardsonii	272	200	720	1052	1226	185	37	1023	251
P. pectinatus	195	320	458	364	249	113	192	183	117
P. foliosus						36	72	126	
P. pusillus				858				63	5
Lemna trisulca				63	111	Tr	Tr	Tr	Tr
Total Production	9118	7761	8471	9627	7306	4996	5960	7238	5465

258

Appendix XV, cont.

Species	Species composition (% of total weight) of aquatic plants in River Marsh, RRLNWR								
	1967	1969	1971	1973	1975	1977	1979	1981	1983
Myriophyllum spicatum	31.5	49.3	10.6	24.7	30.5	25.9	7.2	32.4	31.7
Chara vulgaris	29.3	14.3	9.7	16.5	13.4	15.5	17.5	26.2	22.4
Elodea canadensis	0.4	0.2	0.2	1.4	2.3	1.0	7.5	4.3	15.7
Sagittaria cuneata	2.0	0.1	0.1	0.1		0.8	0.2	0.8	0.2
Potamogeton richardsonii	9.9	25.9	41.5	22.7	35.9	34.2	39.2	17.6	11.7
P. pectinatus	12.1	8.7	4.0	0.6	3.6	2.3	0.4	1.6	0.4
P. zosteriformis	5.5	0.5	15.1	11.2	2.2	1.7	1.4	2.5	2.0
P. pusillus	7.1	0.5	2.6	2.6	0.2			0.1	3.4
P. friesii						6.9	3.0	7.5	4.0
P. foliosus			5.2	14.1	10.8	3.4		5.6	
Nitella flexilis	1.9	0.4	0.2	Tr					
Utricularia vulgaris	0.4		9.5	6.1					
Lemna trisulca							Tr	Tr	
Ranunculus aquatilis						7.0		0.4	
Ceratophyllum demersum								0.7	7.0
Isoetes sp.				Tr	1.0	0.9	0.2	0.2	
Najas flexilis				T	T	0.3	0.2	T	T

Appendix XV, cont.

	Production (in tons) of major aquatic plant species in River Marsh, RRLNWR								
Species	1967	1969	1971	1973	1975	1977	1979	1981	1983
Myriophyllum spicatum	7626	16148	2847	6390	6220	4449	4182	8851	6453
Chara vulgaris	7094	4686	2605	4283	2829	2660	2509	7138	4554
Elodea canadensis	98	64	54	307	470	168	1074	1173	3200
Potamogeton richardsonii	2400	8489	11145	5875	7318	5866	5628	4815	2376
P. pectinatus	2934	2863	1074	166	738	401	215	441	81
P. foliosus				3659	2195	587		1522	
P. zosteriformis				2892	456	290	203	674	407
P. friesii						1179	430	2044	825
P. pusillus				1	46			29	703
Ranunculus aquatilis						1208		122	
Total Production	24247	32745	26802	25928	20474	17163	14340	27286	20328

Appendix XV, cont.

Species	Species composition (% of total weight) of aquatic plants in Widgeon Pd., RRLNWR									
	1966	1970	1971*	1972	1974	1976	1978	1980	1982	1984
Elodea canadensis	99	90	58	47	63	25	77	23	14	4
Myriophyllum spicatum	1	2	24	5		5	5	8	1	2
Ceratophyllum demersum		Tr	Tr	45	30	41	3	2	9	4
Chara vulgaris		Tr	Tr	2	Tr	1	3	9	1	Tr
Potamogeton pectinatus		Tr	9	1				Tr	Tr	Tr
P. zosteriformis			Tr		Tr	23	9	50	67	54
P. richardsonii		3	4	Tr	Tr	2	2	1	2	3
P. pusillus		3	3			2			Tr	Tr
P. praelongus								1	5	27
P. foliosus							Tr	119	4	
P. friesii			Tr				Tr			Tr
Ranunculus aquatilis			Tr				Tr			
Isoetes sp.						Tr				4

Appendix XV, cont.

Species	Production (in tons) of major aquatic plant species at Widgeon Pond, RRLNWR								
	1966	1970	1972	1974	1976	1978	1980	1982	1984
Elodea canadensis	2160	2126	1332	2733	230	1873	548	238	82
Myriophyllum spicatum	5	39	140		47	119	193	6	37
Ceratophyllum demersum		2	1275	640	382	74	40	152	89
Chara vulgaris		6	68	20	12	77	211	13	10
Potamogeton pectinatus		5	22		Tr		4	6	5
P. zosteriformis				1	215	218	1207	1108	1070
P. richardsonii		53	5	11	16	60	29	35	63
P. pusillus					22			3	Tr
P. praelongus							34	78	533
P. foliosus						Tr	119	4	
P. friesii						Tr			Tr
Ranunculus aquatilis						3			
Isoetes sp.					2				81
Total Production	2186	2360	2843	3406	928	2442	2396	1644	1972

Appendix XV, cont.

Species composition (% of total weight) of aquatic plants in Culver Pond, RRLNWR

Species	1966	1968	1970	1971*	1972	1974	1976	1978	1980	1982	1984
Elodea canadensis	53	65	62	25	33	26	29	60	12	25	46
Chara vulgaris	2	16	13	5	11	7	20	10	16	30	28
Ranunculus aquatilis	19	6	5	Tr	9		4	10	11	2	10
Myriophyllum spicatum	23	4	6	21	14	17	14	3	19	8	3
Potamogeton richardsonii		Tr	Tr	3	3	3	1	2	1	5	1
P. pectinatus	2	10	10	19	9	12		1	8		5
P. pusillus	Tr		4	22	19	14	10	1	4	6	4
P. foliosus							19	14	27	18	
Zannichellia palustris	Tr		Tr	4	2						2

Appendix XV, cont.

| Species | Production (in tons) of major aquatic vegetation species in Lower Lake, RRLNWR | | | | | | | | | |
	1967	1969	1971	1973	1975	1977	1979	1981	1983	1985
Elodea canadensis	10400	12528	252	31	18	54		Tr	2156	434
Potamogeton richardsonii	3214	3260	1880	4699	8546	2142	4101	3292	2665	1972
P. praelongus	1035	152	863	1780	2218	1561	1395	2218	1525	1767
Myriophyllum exalbescens	4045	1083	935	1623	5036	2500	7450	3614	1275	841
Ceratophyllum demersum		4029	1061	1923	348	2200		2178	2486	4575
Chara vulgaris	889	212	279	264	858	107	698	322	1646	1422
P. zosteriformis	969	282	45	1346	3108	1905	657	1252	1632	295
P. pectinatus	730	224	243	997	425	358	268	376	505	107
Lemna trisulca	1755	950	3120	1019	3685	568	957	376	1686	246
Ranunculus aquatilis						751		3479		125
Total Production	23901	23320	8993	14427	24534	14221	16609	18426	16046	11819

Appendix XV, cont.

Species	Production (in tons) of major aquatic plant species in Culver Pond, RRLNWR									
	1966	1968	1970	1972	1974	1976	1978	1980	1982	1984
Elodea canadensis	877	594	670	152	252	242	445	36	138	407
Chara vulgaris	40	144	146	48	69	164	72	48	163	247
Ranunculus aquatilis	321	52	51	42		31	73	33	13	85
Myriophyllum spicatum	383	33	66	62	165	114	25	57	42	23
Potamogeton pectinatus	28	95	100	40	120		11	26		48
P. richardsonii	3	12	26	10	13	2	26	12		
P. pusillus	9		46	88	103	79	9	12	31	36
P. foliosus						156	101	83	98	
Zannichellia palustris	4		5	42						19
Total Production	1669	920	1088	457	971	827	748	304	547	878

Appendix XVI. Winter feeding of swans at RRLNWR, 1935-87[a].

Winter	No. Swans counted (Low count/High count)					Quantity Fed (Bu)	Feeding Period	Feed Type	Method Used[c]	Feeding Frequency
	Nov.	Dec.	Jan.	Feb.	Mar.					
1935-36	10/ 43	24/ 29	0	12/ 22	33/ 43	66	ND	WH	SH	3X/WK
1936-37	ND[b]	0	5	ND	44	166	ND	WH	RB	1X/WK
1937-38	-------------- 20-80 -------------			58	78	233	12/15-3/1	WH	RB	2-3X/WK
1938-39	ND	25/ 40	45/ 66	12/ 76	30/ 83	202	12/12-3/30	WH	RB	2-3X/WK
1939-40	ND	ND	71 65	19/120	19/120	183	1/5 -3/27	WH	RB	3X/WK
1940-41	30/ 50	30/ 50	5/ 28	12/150	15/150	226	1/1 -3/31	WH	RB	2-3X/WK
1941-42	ND	2	2/ 44	28/ 69	6/ 61	173	12/29-4/3	WH	RB	2X/WK
1942-43	ND	16/ 37	8/ 44	0/ 62	15/ 64	110	12/21-3/31	WH	RB	1X/WK
1943-44	24/ 36	16/129	0/132	0/124	40/173	710	11/27-4/19	WH/BAR	RB	2X/WK
1944-45	35	0/168	0/170	2/150	59/150	816	11/14-4/14	WH/BAR	RB	2X/WK
1945-46	96	140	113/144	112/134	84/152	947	11/8 -4/1	WH/BAR	RB	2X/WK
1946-47	89/155	155	117/150	84/131	47/160	978	10/28-4/11	WH/BAR	RB	2X/WK
1947-48	117	87/150	44/147	84/170	67/172	1159	11/3 -4/12	WH	RB	2X/WK
1948-49	ND	0/140	1/160	9/132	30/150	934	12/10-3/31	WH	RB	2X/WK
1949-50	ND	105	01 80	0/200	25/150	612	12/14-4/7	WH	RB	2X/WK
1950-51	ND	6/ 31	0/ 47	0/100	0/120	488	12/22-4/5	WH	RB	2X/WK
1951-52	ND	60/ 75	1/ 50	21 50	25/150	480	12/20-4/17	WH	RB	2X/WK
1952-53	ND	30/ 35	1/ 61	3/ 75	5/100	587	12/27-4/17	WH/BAR	RB	2X/WK
1953-54	ND	12	3/ 90	85/146	65/250	441	1/5 -4/1	WH	RB	1-2X/WK
1954-55	200	25/150	0/150	3/135	15/250	1043	11/18-4/11	WH	RB	2X/WK
1955-56	ND	10/ 55	1/154	1/325	0/280	703	12/8 -4/4	WH	RB	2X/WK
1956-57	ND	2/ 24	2/ 95	40/209	150/230	1081	12/18-4/30	WH/BAR	RB	2X/WK
1957-58	20/ 31	69/224	147/197	127/274	130/311	965	11/22-4/24	WH	RB	2X/WK
1958-59	123/240	40/256	50/242	67/296	87/300	1497	11/17-4/9	WH/BAR	RB	3X/WK
1959-60	40/132	108/146	138/161	104/192	84/188	1039	11116-4/7	WH/BAR	RB	3X/WK
1960-61	23/ 61	65/122	39/247	138/266	134/267	1188	11/14-4/3	WH/BAR	RB	2X/WK
1961-62	50/ 75	134/150	87/197	154/257	66/302	1319	11/8 -4/9	WH/BAR	RB	3X/WK
1962-63	80/150	40/193	69/230	204/260	220/250	1344	11/19-4/25	WH/BAR	RB	2X/WK
1963-64	46/241	171/230	171/201	97/215	197/277	1410	11/19-4/30	WH/BAR	RB	2X/WK

[a] Data were compiled from RRLNWR Annual Narrative Reports, RRLNWR winter feeding files, RRLNWR Annual Trumpeter Swan Reports, McEneaney 1986b, and C. Mitchell, pers. comm.

[b] ND = no data found in Refuge files;

[c] WH = wheat; BAR = barley; SH = grain fed from shoreline; RB = rowboat; WF = over-water feeders; DF = dryland feeders; GF = Gravity flow from bin feeder; 2X/WK = grain was fed twice per week. WH = wheat; BAR = barley; SH = grain fed from shoreline; RB = rowboat;

Appendix XVI, cont.

Winter	No. Swans Counted (Low count/High count) Nov.	Dec.	Jan.	Feb.	Mar.	Quantity Fed(Bu)	Feeding Period	Feed Type	Method Used	Feeding Frequency
1964-65	3/126	118/328	170/293	125/350	191/401	1271	11/20-4/26	WH/BAR[b]	RB,WF	2X/WK
1965-66	150/200	55/253	79/240	184/294	99/331	1291	12/8 -4/21	WH/BAR	RB,WF	2X/WK
1966-67	141/360	95/283	278/322	260/323	197/297	1500	ND	WH/BAR	GF	2X/WK
1967-68	158/245	112/174	220/227	217/238	220/240	1500	11/24-ND	BAR	GF	1X/WK
1968-69	225/268	249/260	300/330	305/325	300/330	1500	11/18-ND	WH/BAR	GF	<1X/WK
!969-70	150/200	150/150	135/165	100/165	130/172	500	3/16 -4/30	WH	GF	ND
1970-71	90/100	120/150	200/200	225/250	250/300	850	12/18-5/5	WH	GF	1X/WK
1971-72	3/ 40	104/145	0/154	175/200	156/236	850	12/10-3/30	WH/BAR	GF	1X/WK
1972-73	234	122/270	190/257	116/258	81/257	1300	12/4 -4/13	WH/BAR	GF	2X/WK
1973-74	ND	150/205	150/275	250/266	218/321	1247	12/10-4/11	WH/BAR	GF	2X/WK
1974-75	ND	149/177	172/230	139/234	177/396	966	12/11-5/6	WH/BAR	GF	2X/WK
1975-76	ND	94/157	146/261	219/282	281/324	735+	12/3 -4/6	WH/BAR	GF	2X/WK
1976-77	ND	206/250	230/289	237/325	271/373	ND	1/4 -4/12	WH/BAR	GF	2X/WK
1977-78	155	198	224	245	255	565	12/16-4/7	WH/BAR	GF,WF	2X/WK
1978-79	200/221	209/225	249/266	212/273	273/310	500	11/15-3/20	BAR	GF,WF	2X/WK
1979-80	253/287	265/318	260/353	332/357	254/333	250+	12/12-4/15	WH	GF,WF	2X/WK
1980-81	261/284	242/275	225/290	295/321	215/288	250	1/8 -4/8	WH/TF	GF,DF	2X/WK
1981-82	118/396	251/329	228/294	259/321	270/282	300	12/12-4/28	WH/BAR/TF	GF,DF	2X/WK
1982-83	98/191	115/187	97/188	177/244	145/252	190	1/25 -4/9	WH/BAR/TF	GF,DF	2X/WK
1983-84	171/229	154/188	168/265	211/214	211/271	265	12/3 -4/28	WH/BAR/TF	GF,DF	2X/WK
1984-85	41/221	190/234	139/244	225/267	25/272	600	12/4 -4/7	WH/TF	GF,DF	2X/WK
1985-86	70/123	109/170	160/212	195/271	85/266	1065	11/20-3/20	WH/TF	GF,DF	2X/WK
1996-97	153/255	152/366	230/325	ND	ND	964[d]	11/10-4/7	WH/TF	GF, DF, RB	2X/WK

[d] 964 bu wheat plus 2600 lbs. of commercially formulated dry feeds. Observers noted that the swans ate only a small portion of the dry feed.

Appendix XVII. Removals of eggs, cygnets and adults from the RRLNWR area, 1938-1986. (M=male, F=female, PR=breeding pair, NB=nonbreeding adult or subadult). Items presented in parentheses were not counted as losses to the RRLNWR flock.

Date	Number removed			Comments	Destination
	Eggs	Cygnets	Adults		
10/24/38	0	4	0		Nat. Elk Ref.
10/1/39	0	3	0		Nat. Elk Ref.
10/16/39	0	3	0		Malheur NWR
9/19/41	0	6	1		Malheur NWR
9/23/41	0	3	0		Nat. Elk Ref.
5/44	5			(2 hatched, Cygnets sent on 9/4 to Malheur NWR)	
9/4/44	0	20	0		Malheur NWR
9/13/45	0	20	0		Malheur NWR
9/12/48	0	20	0	1 died en route	Malheur NWR
9/12/49	0	10	0	5M,5F	Ruby L. NWR
7/14/54	0	7	7	3PR, 1NB	Malheur NWR
7/28/54	0	1	6	3PR	Ruby L.. NWR
8/7/55	0	3	3	NB	Delta, Manitoba
9/26/55	0	16	0	8M, 8F	Ruby L.. NWR
9/27/55	0	17	0	12M, 5F	Malheur NWR
8/11/56	0	0	12	NB:6M,6F; 1died	Malheur NWR
10/10/56	0	12	0		Ruby L. NWR
8/2/57	0	0	19	NB: 10M, 9F	Ruby L. NWR
8/2/57	0	0	2	NB: 1M, 1F	Nat. Zool. Pk.
9/25/57	0	20	0		Malheur NWR
9/4/58	0	0	4	NB	Malheur NWR
9/22/58	0	20	0	8M, 12F	Ruby L. NWR
7/15/59	0	0	12	NB: 6M, 6F	zoos
5/25/60	(20)			(7 reared and release at RRLNWR, incubated at Bowdoin NWR)	
7/18/60	0	0	19	NB: 10M, 9F	zoos
9/12/60	0	20	0	7 from off-refuge	LaCreek NWR
7/14/61	0	1	18	NB: 9M, 9F	zoos
7/14/61	0	1	0	1F	Cornell zoo
7/20/61	0	0	2	NB	Malheur NWR
9/21/61	0	17	0	8M, 8F, 1 died	LaCreek NWR
7/62	0	0	16	NB: 9M, 7F	zoos
9/20/62	0	20	0		LaCreek NWR
5/63	30			(hatched into 16 handreared cygnets)	
		10		5M, 5F	zoos
		6		2M, 4F	Turnbull NWR
9/64	0	2		1M, 1F	zoo
5/64	(20)			(only 1 hatched and it died)	RRLNWR

Appendix XVII, cont.

Date	Number removed			Comments	Destination
	Eggs	Cygnets	Adults		
6/23/65	(9)	(7 hatched, all died by 9/17)			RRLNWR
9/21/65	0	1	0	6M, 5F	Turnbull NWR
		1			
9/24/65	0	2	0	1M,1F; 1died	Kellogg Sanctuary
7/29/65	0	0	7	NB:4M, 3F	zoos
8/29/66	0	0	2	1M,1F	Hennepin Pk. Res.
9/7/66	0	2	0	10M,10F	Turnbull NWR
		0			
9/7/66	0	0	13	4M, 9F	zoos
9/19/66	0	1	0		Nova Scotia
8/67	0	2		1M, IF	Belgium zoo
8/31/67	0	1	0		Hennepin Pk. Res.
		0			
8/67	0	0	11	5M, 6F	zoos
8/31/67	0	0	10		Hennepin Pk. Res.
9/10/68	0	1	0		Hennepin Pk. Res.
		0			
9/68	0	2	0		Portugal
7/68	0	0	14	8M, 6F	zoos
8/69	0	0	2	1M, IF	CM Russel NWR
8/69	0	2	13		zoos & priv.
8/17/70	0	0	4	NB:3M, IF	CMRNWR; Sandhills
9/9/70	0	1	0	5M, 5F	Hennepin Pk. Res.
		0			
8171	0	0	1	1F	priv. propagators
7174	0	0	1	1F	Tracy Aviary
9/74	0	0	1	1M	Logan zoo
1975	0	3	0	1M, 2F	Crane Cr.
6/10/77	17	0	0		priv. propagators
9/9/78	0	2	0	1M, 1F	Logan Utah Zoo
5-6/78	28	0	0	5 clutches	priv. propagators
6/11/78	4	0	0	1 clutch	Kellogg Sanctuary
6/6-12/79	22	0	0	4 clutches	priv. propagators
6/13/79	5	0	0	1 clutch	Kellogg Sanctuary
8/17/79	0	2	0	1M, IF	priv. propagators
8/27/79	0	2	0	1M, IF	priv. propagators
1980	21	0	0	4 clutches	priv. propagators
8/31/81	0	2	0	1M, IF	priv. propagators
6/82	16	0	0	4 clutches	priv .propagators
1983	12	0	0		Minnesota
Total Removed	160	323	208		

LITERATURE CITED

Adams, M. S. and M. D. McCracken, 1974. Seasonal production of the *Myriophyllum* component of the littoral of Lake Wingra, Wisconsin. J. Ecol. 62:457-465.

Aiken, S. G., P. R. Newroth, and I. Wile. 1979. The biology of Canadian weeds. 34. *Myriophyllum spicatum* L. Can. J. Plant Sci. 59:201-215.

Allen, E. D., and P. R. Gorham. 1973. Changes in the submerged macrophyte communities of Lake Wabamun as a result of thermal discharge. Proc. Symposium on the lakes of western Canada. University of Alberta, Edmonton. Pp. 313-324.

Allendorf, F. W. 1986. Genetic drift and loss of alleles versus heterozygosity. Zoo Biol. 5: 181-190.

Anderson, D. R., R. C. Herron, and B. Reiswig. 1986. Estimates of annual survival rates of Trumpeter Swans banded 1949-82 near Red Rock Lakes National Wildlife Refuge, Montana. J. Wildl. Manage. 50(2): 218-221.

Anderson, M. G. 1978. Distribution and production of sago pondweed (*Potamogeton pectinatus* L.) on a northern prairie marsh. Ecol. 59(1): 154-160.

Anderson, M. G., and J. B. Low. 1976. Use of sago pondweed by waterfowl on the Delta Marsh, Manitoba. J. Wildl. Manage. 40:233-242.

Anderson, R. R. 1969. Temperature and rooted aquatic plants. Chesapeake Sci. 10: 157-164.

Angradi, T. 1986. Draft – Henry's Fork Report. Idaho State Univ. Pocatello.

Ankney, C. D., and C. D. MacInnes. 1978. Nutrient reserves and reproductive performance of female lesser snow geese. Auk 95: 459-471.

Annear, J. T. 1969. Historical notes on the transfers of Trumpeter Swans. U.S. Fish Wildl. Serv., Red Rock Lakes Nat. Wildl. Ref. Unpub. Rept. 86 pp.

Archibald, G. W. 1977. Supplemental feeding and manipulation of feeding ecology of endangered birds – a review. Pages 131-134 *in* S.A. Temple, ed. Endangered birds – management techniques for preserving threatened species. The University of Wisconsin Press, Madison.

Avise, J. C., and C. F. Aquadro. 1982. A comparative summary of genetic distances in the vertebrates: patterns and correlations. Evol. Biol. 15:151-185.

Avise, J. C., J. C. Patton, and C. F. Aquadro. 1980. Evolutionary genetics of birds, II. Conservative protein evolution in North American sparrows and relatives. Syst. Zool. 29:323-334.

Bangs, E. E., V. D. Berns, and T. N. Bailey, 1981. Leech parasitism of Trumpeter Swans in Alaska. Murrelet 62(1): 24-26.

Banko, W. E. 1960. The Trumpeter Swan. N. Am. Fauna 63, U.S. Fish Wildl. Serv., Washington, D.C. 214 pp.

Banko, W. E., and R. H. Mackay. 1964. Our native swans. Pages 155-164 *in* J. P. Linduska, ed. Waterfowl tomorrow. U.S. Department of Interior, Washington, D.C.

Barko, J. W., D. G. Hardin, and M. S. Matthews. 1982. Growth and morphology of submerged freshwater macrophytes in relation on light and temperature. Can. J. Bot. 60:877-887.

Barrett, V. A. 1980. Genetic comparison of Trumpeter Swan (*Olor buccinator*) populations. M.S. Thesis. Montana State Univ., Bozeman. 101 pp.

Barrett, V. A., and E. R. Vyse. 1982. Comparative genetics of three Trumpeter Swan populations. Auk 99: 103-108.

Barrowclough, G. F., and S. L. Coats. 1985. The demography and population genetics of owls with special reference to conservation of the spotted owl (*Strix occidentalis*). Pages 74-85 *in* R. J. Gutierrez and A. B. Carey eds. Ecology and management of the spotted owl in the Pacific Northwest. U. S. Forest Service Gen. Tech. Rept. PNW-185.

Batt, B. D. J. 1971. Trumpeter nesting – Delta Waterfowl Research Station, Delta, Manitoba, Canada. Trumpeter Swan Soc. Newsletter No. 6:3.

Bennett, G. F., B. Turner, and G. Holton. 1981. Blood parasites of Trumpeter Swans, *Olor buccinator* (Richardson), from Alberta. J. Wildl. Diseases 17(2):213-215.

Berglund, B. E., K. Curry-Lindahl, H. Luther, V. Olsson, W. Rodhe, and C. Sellerberg. 1963. Ecological studies on the mute swan (*Cygnus olor*) in southern Sweden. *Acta vertebratica* 2:167-288.

Blackford, J. L. 1939. Vanishing glory? Nat. Mag. 32(7):397-402.

Bortner, J. B. 1985. Bioenergetics of wintering Tundra Swans in the Mattamuskeet region of North Carolina. M. S. Thesis. University of Maryland, College Park. 67 pp.

Botkin, D. B. and R. S. Miller. 1974. Mortality rates and survival of birds. Am. Nat. 10:181-192.

Boylen, C. W., and R. B. Sheldon. 1978. Submergent macrophytes: growth under winter ice cover. Science 194:841-842.

Brechtel, S. H. 1982. A management plan for Trumpeter Swans in Alberta-1982. Alberta Dept. Energy and Nat. Res., Fish and Wildl. Div., Edmonton. 135 pp.

Bristow, J. M. 1975. The structure and function of roots in aquatic vascular plants. Pages 221-236 in J. G. Torrey and d. T. Clarkson, eds. The development and function of roots. Academic Press, New York.

Brockelman, W. Y. and R. M. Fagen. 1972. On modeling density-independent population change. Ecol. 52(5):944-948.

Brown, J. M. A. 1975. Ecology of macrophytes. Pages 244-262 in V. H. Jolly and J. M. A. Brown, eds. New Zealand lakes. Auckland Univ. Press and Oxford Univ. Press, Auckland.

Brown, J. M. A. 1979. The Nelson Lakes and their aquatic weeds. New Zealand Dept. Sci. and Induct. Res. Info. Ser. No. 142.

Brussard, P. F., and M. E. Gilpin. 1987. Demographic and genetic problems associated with small or fragmented populations, with special reference to the black-footed ferret *Mustela nigripes*. In U. S. Seal, ed. The biology of the black-footed ferret. Yale Univ. Press, New Haven. (In press).

Bumby, M. J. 1977. Changes in submerged macrophytes in Green Lake, Wisconsin from 1921 to 1971. Trans. Wisc. Acad. Arts Sci. Lett. 65:120-151.

Burgess, H. 1986. Potential Trumpeter Swan restoration. Pages 97-111 in D. Compton, ed. Proc. And Papers Ninth Trumpeter Swan Soc., 1984. Omni Press, Madison, Wisconsin.

Burgess, T. E. 1972. Investigations into the breeding ecology of the Grande Prairie Trumpeter Swan population. Pages 16-18 in Proc. and Papers of the Second Trumpeter Swan Soc. Conf., Grande Prairie, 1972.

Chapman, V. J., J. M. A. Brown, F. I. Dromgolle, and B. T. Coffey. 1971. Submerged vegetation of the Rotorua and Waikato lakes. I. Lake Rotoiti. New Zealand J. Marine and Freshwater Res. 15:259-279.

Christian, J. J., J. A. Lloyd, D. E. Goldman, D. E. Davis. 1971. An empirical formula for the growth of some vertebrate populations. Curr. Mod. Biol. 4:26-34.

Churchill, B. P. 1987. Potential Trumpeter Swan nesting habitat in northeastern British

Columbia. Proc. and Papers Tenth Trumpeter Swan Soc. Conf. In press.

Coale, H. K. 1915. The present status of the Trumpeter Swan (*Olor buccinator*). Auk 32:82-90.

Conant, B., J. I. Hodges, J. G. King, and S. L. Cain. 1985. Alaska Trumpeter Swan status report – 1985. U. S. Fish Wildl. Ser., Juneau. 12 pp.

Condon, D. D. 1941. Preliminary report on the Trumpeter Swans of Yellowstone National Park. U. S. Natl. Pk. Serv. 27 pp.

Cooper, J. A. 1979. Trumpeter Swan nesting behavior. Wildfowl 30:55-71.

Corbin, K. W. 1978. Genetic diversity in avian populations. Pages 292-302 in S. A. Temple, ed. Endangered birds. University of Wisconsin Press, Madison.

Cornely, J. E., S. P. Thompson, E. L. McLaury, and L. D. Napier. 1985. A summary of Trumpeter Swan production at Malheur National Wildlife Refuge, Oregon. Murrelet 66: 50-55.

Dale, H. M. and G. E. Miller. 1978. Changes in the aquatic macrophyte flora of Whitewater Lake near Sudbury, Ontario from 1947 to 1977. Can. Field-Nat. 92(3):264-270.

Davis, G. J., and M. M. Brinson. 1980. Response of submersed vascular plant communities to environmental change. U. S. Fish Wildl. Serv., Washington, D.C. 69 pp.

Davis, G. J. and D. F. Carey, Jr. 1981. Trends in submersed macrophyte communities of the Currituck Sound 1977-79. J. Aquatic Plant Manage. 19:3-8.

Delacour, J., and E. Mayr. 1945. The family Anatidae. Wilson Bull. 57(1).

Dennington, M. 1987. Trumpeter Swan habitat in southern Yukon. Proc. And Papers of the 10th Trumpeter Swan Soc. Conf. Grande Prairie, 1986. In Press.

Denny, P. 1972. Sites of nutrient absorption in aquatic macrophytes. J. Ecol. 60:819-829.

Denson, E. P., Jr. 1970. The Trumpeter Swan, *Olor buccinator*: a conservation success and its lessons. Biol. Cons. 2(4):253-256.

De Vos, A. 1963. Observations on the behavior of captive Trumpeter Swans during the breeding season. *Ardea* 52:166-189.

Engel, S. 1985. Aquatic community interactions of submerged macrophytes. Tech. Bull. No. 56. Dept. Nat. Res. Madison, Wisconsin.

Erickson, R. 1969. Untitled paper presented at the 1st Trumpeter Swan Soc. Conf., 1969. Reprinted in The Trumpeter Swan Soc. Newsletter 5:7-12.

Fenchel, T. 1977. Aspects of the decomposition of seagrasses. Pages 123-145 in C. P. McRoy and C. Hefferich, eds. Seagrass ecosystems, a scientific perspective. Marine Sci. Vol. 4. Marvel Decker, Inc., New York.

Fjetland, C. A. 1974. Trumpeter Swan management in the National Wildlife Refuge system. Pages 136-141 in Trans. 39th N. Am. Wildl. And Nat. Res. Conf., Wildl. Mgmt. Inst., Washington, D.C.

Forest, H. S. 1977. Study of submerged aquatic vascular plants in northern glacial lakes (New York State, U.S.A.). Folia Geobot. Phytotax. Praha 12:329-341.

Forsberg, C. 1964. Phosphorus, a maximum factor in the growth of *Characeae*. Nature 201:517-518.

Frankel, O. H., and M. E. Soule. 1981. Conservation and evolution. Cambridge Univ. Press, Cambridge.

Franklin, I. R. 1980. Evolutionary change in small populations. Pages 135-149 in M. E. Soule and B. A. Wilcox, eds. Conservation Biology. Sinauer Assoc., Sunderland, Mass.

Franson, C. 1986. Immunosupressive effects of lead. Pages 106-109 in J. S. Feierabend and A. B. Russell, eds. Lead poisoning in wild waterfowl – a workshop. Natl. Wild. Fed.,

272

Washington, D.C. 139pp.

Fuerst, P. A., and T. Maruyama. 1986. Considerations on the conservation of alleles and genetic heterozygosity in small managed populations. Zoo Biol. 5:171-179.

Gauthier, G., J. Bedard, J. Huot, V. Bedard. 1984. Spring accumulation of fat by greater snow geese in two staging areas. Condor 86:192-199.

Gillette, L. 1985. Summary of the migration of Trumpeter Swans from Minnesota. Trumpeter Swan Soc. Newsletter 13(4)/14(1):6.

Goulder, R. and D. J. Boatman. 1971. Evidence that nitrogen supply influences the distribution of a freshwater macrophyte, Ceratophyllum demersum. J. Ecol. 59:783-791.

Gritman, R. B., and W. I. Jensen. 1965. Avian cholera in a Trumpeter Swan. Bull. Wildl. Dis. Assoc. 1(4):54-55.

Groves, C. R., and T. W. Clark. 1986. Determining minimum population size for recovery of the black-footed ferret. Great Basin Nat. Mem. 8:150-159.

Haag, R. w. 1979. The ecological significance of dormancy in some rooted plants. J. Ecol. 67:727-738.

Haines, A. L. 1977. The Yellowstone story. Vol. I. Yellowstone Library and Museum Assoc. Mammoth, Wyoming. 385 pp.

Hampton, P. D. 1981. The wintering and nesting behavior of the Trumpeter Swan. M.S. Thesis. University of Montana, Missoula. 185 pp.

Hansen, C. G. 1959. Report on the aquatic plants found in the Island Park area of Idaho during the fall and winter of 1958. U. S. Fish Wildl. Serv., Red Rock Lakes Nat. Wildl. Ref. Unpub. Rept. 3 pp.

Hansen, H. A 1973. Trumpeter Swan management. Wildfowl 24:27-32.

Hansen, H. A., P.E. K. Sheperd, J. G. King, and W. A. Troyer. 1971. The Trumpeter Swan in Alaska. Wildl. Mono. 26. 83 pp.

Harper, J. L., and J. White. 1971. The dynamics of plant populations. Pages 41-63 in J. den Boer and G. R. Gradwell, eds. Dynamics of populations. Proc. Of the Advanced Study Institute, Oosterbeek.

Hedrick, P. W. 1983. Genetics of populations. Van Nostrand Reinhold Co., New York.

Hedrick, P. W., P. F. Brussard, F. W. Allendorf, J. Beardmore, and S. Orzack. 1986. Protein variation, fitness, and captive propagation. Zoo Biol. 5:91-99.

Hochbaum, H. A. 1955. Travels and traditions of waterfowl. University of Minnesota Press, Minneapolis. 301 pp.

Hodges, J. I., B. Conant, and S. L. Cain. 1986. Alaska Trumpeter Swan status report. U. S. Fish Wildl. Serv., Juneau. 8 pp.

Holton, G. R. 1982. Habitat use by Trumpeter Swans in the Grande Prairie region of Alberta. M. S. Thesis. University of Calgary, Alberta. 164 pp.

Holton, G. R. 1983-85. Trumpeter Swans in the Peace River Region. Alberta Energy and Natural Resources. Unpub. Annual survey repts.

Holton, G. R. 1985. Biology of Trumpeter Swans breeding near Grande Prairie, Alberta. Unpub. Manuscript.

Holton, G. R. and L. J. Shandruk. 1987. Trumpeter Swans in the Peace River Region, 1986. Alberta Energy and Natural Resources and the Canadian Wildl. Serv. Unpub. Rept. 8 pp.

Hull, A. V. 1939. Trumpeter Swans, their management and preservation. Pages 378-382 in Trans. 4th N. Am. Wildl. Conf., Am. Wildl. Inst.

Jackson, H. O. and W. C. Starrett. 1959. Turbidity and sedimentation at Lake Chantaqua, Illinois. J. Wildl. Mgmt. 14:157-168.

Jeppson, P. 1977. Management of the Henry's Fork Fishery. Idaho Department Fish and Game. Unpub. Rept. 61 pp.

Jepson, W. L. 1925. Manual of flowering plants of California. Assoc. Students Store. Univ. of California, Berkeley. 1238 pp.

Johnsgard, P.A. 1973. Proximate and ultimate determinants of clutch size in *Anatidae*. Wildfowl 24: 144-149.

Johnsgard, P. a. 1978. The triumphant trumpeter. Nat. Hist. 77(9):72-77.

Jupp, B. P., and H. N. Spence. 1977. Limitations of macrophytes in a eutrophic lake, Loch Leven. II. Wave action, sediments, and waterfowl grazing. J. Ecol. 65:341-446.

Kaul, V. and K. K. Kaso. 1972. Production studies of some macrophytes of Srinagar lakes. Pages 725-731 in A. Jajak and A. Hillbrecht-Ilkowska, eds. Productivity problems of freshwaters. PWN-Polish Scientific Publishers, Warszawa, Krakow.

Kennington, G. S., R. M. Thomas, and J. E. Doerges. 1980. Report on alpha-emitting radioactivity in Trumpeter Swan (*Olor buccinator*) egg shells and feathers from the Yellowstone Park of Wyoming. University of Wyoming, Laramie. Unpub. Rept., 8 pp.

Killaby, M. 1987. Trumpeter Swan habitation and proposed management in Saskatchewan. Proc. And Papers 10th Trumpeter Swan Soc. Conf., Grande Prairie, 1986. In press.

King, J. G., and B. Conant. 1981. The 1980 census of Trumpeter Swans on Alaskan nesting habitats. American Birds 35:789-793.

King, J. R. 1973. Energetics of reproduction in birds. Pages 78-107 *in* D.S. Farner, ed. Breeding biology of birds. Nat. Acad. Sci. Washington, D.C.

Kiorboe, T. 1980. Distribution and production of submerged macrophytes in Tipper Grund (Ringkobing Fjord, Denmark) and the impact of waterfowl grazing. J. Appl. Ecol. 17:675-687.

Krapu, G. L. 1981. The role of nutrient reserves in mallard reproduction. The Auk 98:29-38.

Lande, R., and G. F. Barrowclough. 1986. Effective population size and its use in genetic management. In M. E. Soule, ed., Viable populations. Cambridge Univ. Press (in press).

Leach, J. T. 1977. Lacreek area Trumpeter Swan behavior and migration study. M. S. Thesis. Univ. of South Dakota, Vermillion. 50pp.

Lind, C. T. and G. Cottam. 1969. The submersed aquatics of University Bay: A study in eutrophication. Amer. Midl. Nat. 81:353-369.

Lockman, D. C. 1983. Rocky Mountain Trumpeter Swan Population – Wyoming flock. Progress report September 16, 1982 – September 15, 1983. Wyo. Game and Fish Department, Cheyenne. 9 pp.

Lockman, D. C. and M. Brandt. 1987. An evaluation of Trumpeter Swan winter habitat in the Snake River drainage, Wyoming. Special report draft manuscript. Wyoming Game and Fish Dept. Cheyenne.

Lockman, D. C., R. Wood, H. Smith, B. Raynes. 1985. Progress report on the Rocky Mountain Trumpeter Swan Population, Wyoming flock. Wyo. Game and Fish. Dept. Unpub. Rept. 16 pp.

Lockman, D.C., R. Wood, H. Smith, B. Smith, and H. Burgess. 1987. Rocky Mountain Trumpeter Swan Population – Wyoming flock, 1982-86. Progress report. Wyo. Game and Fish Dept., Cheyenne. 74 pp.

Lumsden, H. G. 1984. The pre-settlement breeding distribution of trumpeter, (*Cygnus buccinator*) and Tundra Swans (*C. columbianus*) in eastern Canada. Can. Field-Nat. 98(4):415-424.

Lumsden, H. G. 1987a. The food of Trumpeter Swan cygnets in Ontario. Proc. And papers 10th Trumpeter Swan Soc. Conf., Grande Prairie, 1986. In Press.

Lumsden, H. G. 1987b. Productivity of Trumpeter Swans in relation to condition. Proc. And papers 11th Trumpeter Swan Soc. Conf., Grande Prairie. In Press.

Mackay, R. H. 1957. Movements of Trumpeter Swans shown by band returns and observations. Condor 59 (5):339.

Mackay, R. H. 1978. Status report on Trumpeter Swan (*Olor buccinator*) in Canada, 1978. Status reports on endangered wildlife in Canada. Vol. 1. Committee on the Status of Endangered Wildlife in Canada. Can. Wildl. Serv., Ottawa. Unpub. 27 pp.

Mackay, R. H. 1987. Trumpeter Swan investigations, Grande Prairie area, Alberta, 1953-1975. Proc. And papers 10th Trumpeter Swan Society Conf., Grande Prairie, 1986. In press.

Maj, M. E. 1983. Analysis of Trumpeter Swan habitat on the Targhee National Forest of Idaho and Wyoming. M. S Thesis. Montana State University, Bozeman. 102 pp.

Mann, K. 1972. Macrophyte production and detritus food chains in coastal waters. Pp. 353-383 in Memorie dell Instituto Italiano di Idrobiologia Dolt. Marco de Marchi, 29 Suppl.

Marshall, D. B. Return of the Trumpeter Swan. Naturalist 19(1):2-10 [J. Nat. Hist. Soc. Minn.].

Martin, J. B., and L. B. Clements. 1968. Influence of lake bed sediment and reservoir water on nutrient uptake by a submerged aquatic plant, *Najas sp.* Agron. Abstr. S-6. Soil Sci. Div. P. 128.

McCormick, K. J. 1985. A survey of Trumpeter Swans and their habitat in Nahanni National Park Reserve and vicinity. Tech. Rept. No. 84-12. Can Wildl. Serv., Yellowknife. 35 pp.

McCormick, K. J., and L. J. Shandruk. 1986. A survey of Trumpeter Swans and their habitat in southern Mackenzie District, Northwest Territories. Tech. Rept.. No. 86-5. Can. Wildl. Serv., Yellowknife, Northwest Terr. 34 pp.

McEneaney, T. 1986a. Movements and habitat use patterns of the Centennial Valley Trumpeter Swan population (Montana) as determined by radio telemetry data. U. S. Fish Wildl. Serv., Red Rock Lakes NWR. Unpub. Rept. 31 pp.

McEneaney, T. 1986b. An analysis of the Red Rock Lakes National Wildlife Refuge Trumpeter Swan winter feeding program. U. S. Fish Wildl. Serv. Unpub. Rept. 42 pp.

McEneaney, T., and R. Sjostrom. 1983. Migration and movement of the Tri-state Trumpeter Swan Population, an analysis of neck-banding data. U. S. Fish Wildl. Serv., Red rock Lakes NWR. Unpub. Rept. 12 pp.

McEneaney, T., and R. Sjostrom. 1986. Trumpeter Swan movements and seasonal use, Centennial Valley (Montana). U. S. Fish Wildl. Serv., Red Rock Lakes NWR. Unpub. Rept. 20 pp.

McKelvey, R. W. 1986. The status of Trumpeter Swans in British Columbia and Yukon, summer, 1985. Tech. Rept. Series No. 8. Can. Wildl. Serv., Pacific and Yukon Region, B. C. 12 pp.

McKelvey, R. W., M. C. Dennington, and D. Mossop. 1983. The status and distribution of Trumpeter Swans (Cygnus buccinator) in the Yukon. Arctic 36(1):76-81.

McKelvey, R. W., K. McCormick, and L. Shandruk. 1986. The status of Trumpeter Swans, Cygnus buccinator, breeding in western Canada, 1985. Ms. Submitted to Canadian Field Naturalist.

McLandress, M. R. and D. G. Raveling. 1981. Changes in diet and body composition of Canada geese before spring migration. Auk 98:65-79.

Monnie, J. B. 1966. Reintroduction of the Trumpeter Swan to its former prairie breeding range. J. Wildl. Manage. 30(4):691-696.

Morris, R. F. 1963. Predictive equations based on key-factors. Mem. Entomol. Soc. Can. 32:16-21.

Moyle, J. B. 1945. Some chemical factors influencing the distribution of aquatic plants in Minnesota. Am. Midl. Nat. 34:402-420.

Nei, M. 1972. Genetic distance between populations. Am. Nat. 106:283-292.

Nichols, S. A. 1975. Identification and management of Eurasian milfoil in Wisconsin. Trans. Wis. Acad. Sci. Arts Lett. 63:116-128.

Nichols, S. A. and S. Mori. 1971. The littoral macrophyte species of Lake Wingra: an example of a Myriophyllum spicatum invasion in a southern Wisconsin lake. Trans. Wisc. Acad. Arts Sci. Lett. 59:107-119.

Nieman, D. 1976. Population status and management of Trumpeter Swans in Saskatchewan. Trumpeter Swan Soc. Newsletter No. 9.

Nieman, D. 1979. Population and productivity data for Saskatchewan Trumpeter Swans. The Trumpeter Swan Soc. Newsletter No. 17:12.

Nieman, D. J. and R. J. Isbister. 1974. Population status and management of Trumpeter Swans in Saskatchewan. Blue Jay 32(2): 97-101.

Ogden, E. C. 1974. Anatomical patterns of some aquatic vascular plants of New York. Bull. 424. New York St. Mus. Sci. Serv., St. Ed. Dept., Albany.

O'Rear, C. W., Jr. 1975. The effects of stream channelization on the distribution of nutrients and metals. Water Resource Res. Inst. Univ. No. Carolina, Rept. No. 108. Raleigh. 52 pp.

Page, R. 1976. The ecology of the Trumpeter Swans on Red Rock Lakes National Wildlife Refuge, Montana. Ph.D. Thesis, University of Montana, Missoula. 165 pp.

Palmer, R. S. (ed.) 1976. Handbook of North American birds. Vol. 2, 3. Yale University Press, New Haven.

Papike, R. 1971. Investigations of nonbreeding Trumpeter Swan range, movements, migration, and nesting territories. Wildlife Management Study RRL-1. U. S. Fish Wildl. Serv. Red Rock Lakes Nat. Wildl. Ref. Unpub. Rept. 10 pp.

Patten, B. C., Jr. 1956. Notes on the biology of Myriophyllum spicatum L. in a New Jersey lake. Bull. Of Torrey Bot. Club. 83(1):5-18.

Paullin, D. G. 1971. The ecology of submerged aquatic macrophytes of Red Rock Lakes National wildlife Refuge, Montana. M. S. Thesis. University of Montana, Missoula. 171 pp.

Peltier, W. H., and e. B. Welch. 1969. Factors affecting growth of rooted aquatics in a river. Weed Sci. 17:412-416.

Perverly, J. H., and R. Johnson. 1976. Changes in submersed plants at the south end of Cayuga Lake following tropical storm Agnes. Hydrobiol. 48:251-255.

Rabotnov, T. a. 1974. Differences between fluctuations and successions. Pages 19-25 in R. Knapp, ed. Vegetation dynamics. Handbook of vegetation science VIII. Junk, The Hague.

Randall, L. C. and L. C. Peterson. 1978. Red Rock Lakes National Wildlife Refuge: An aquatic history 1899-1977. Dept. of Interior, Fish and Wildl. Serv. Kalispell.

Raveling, D. G. 1979. The annual cycle of body composition of reproduction. Auk 96:234-252.

Reynolds, C. M. 1972. Mute swan weights in relation to breeding. Wildfowl 23:111-118.

Rich, P. H., R. G. Wetzel, and N. Van Thuy. 1971. Distribution, production, and role of aquatic macrophytes in a southern Michigan marl lake. Freshwater Biol. 1:3-21.

Richmond, C. J. 1981. Macrophytes in the Rotorua lakes: a review of distribution and nuisance. Dept. of Internal Affairs. Rotorua, New Zealand. Unpub. Rept.

Robel, R. J. 1961. Water depth and turbidity in relation to growth of sago pondweed. J. Wildl. Manage. 25:436-438.

Robel, R. J. 1962. Changes in submersed vegetation following a change in water level. J. Wildl. Manage. 26:221-224.

Rogers, J. s. 1972. Measures of genetic similarity and genetic distance. Pages 145-153 in Studies in Genetics IV. University of Texas Publ. No. 7203.

Rogers, P.M. and D. A. Hammer. 1978. Ancestral breeding and wintering ranges of the Trumpeter Swan (*Cygnus buccinator*) in the eastern United States. Tennessee Valley Auth. Unpub. 45 pp.

Romanoff, A. L. 1972. Pathogenesis of the avian embryo – an analysis of causes of malformations and prenatal death. Wiley Interscience. J. Wiley and Sons, Inc., New York. 476 pp.

Roscoe, J. W. 1976. Aquatic vegetation report, Red rock lakes National Wildlife Refuge. U. S. Fish Wildl. Serv. Red Rock Lakes NWR. Unpub. Rept. 18 pp.

Salter, R. J. 1954. The Trumpeter Swan in Idaho. Pages 254-257 in Proc. 34th Conf. Western Assoc. Fish and Game Comm.

Sallwasser, H., S. P. Mealey, and K. Johnson. 1984. Wildlife population viability – a question of risk. Trans. N. Am. Wildl. And Nat. Res. Conf. 49:421-439.

Schiemer, F. 1979. Submerged macrophytes in the open lake. Distribution pattern, production, and long term changes. In H. Loffler, ed. Neusiedlersee: The limnology of a shallow water lake in central Europe. Junk, The Hague.

Schiemer, F. and M. Prosser. 1976. Distribution and biomass of submerged macrophytes in Neusiedlersee. Aquat. Bot. 2: 289-307.

Schorger, A. W. 1964. The Trumpeter Swan as a breeding bird in Minnesota, Wisconsin, Illinois, and Indiana. Wilson Bull. 76(4):331-338.

Scott, P. and the Wildfowl Trust. 1972. The swans. Houghton Mifflin Co., Boston. 242 pp.

Scott, M. L. 1973. Nutrition in reproduction – direct effects and predictive functions. Pages 46-59 in D. S. Farner, ed. Breeding biology of birds. Natl. Acad. Sci., Washington, D.C. 515 pp.

Sculthorpe, C. D. 1967. The biology of aquatic vascular plants. Arnold, London.

Seddon, B. 1872. Aquatic macrophytes as limnological indicators. Freshwat. Biol. 2:107-130.

Shandruk, L. J. 1986. A survey of Trumpeter Swan breeding habitats in Alberta, Saskatchewan, and northeastern British Columbia. Tech. Rept. No. 86-6. Can. Wildl. Serv., Edmonton. 22 pp.

Shea, R. E. 1979. The ecology of Trumpeter Swan in Yellowstone National Park and vicinity. M.S. Thesis, University of Montana, Missoula. 132. pp.

Shea, R. E. 1980. Causes of pre-fledging mortality of Trumpeter Swans. U. S. Fish Wildl. Serv., Madison Wildl. Health Lab. Unpub. Rept. 32 pp.

Sheehan, B. 1987. Early history of Trumpeter Swans in the Grande Prairie area. Proc. And papers 10th Trumpeter Swan Soc. Conf., Grande Prairie, 1986. In press.

Sheldon, R. B. and C. W. Boylen. 1977. Maximum depth inhabited by aquatic vascular plants. Amer. Midl. Nat. 97 (1):248-254.

Singhal, P. K. and J. s. Singh. 1978. Ecology of Naini Tal Lakes: morphology and macrophytic vegetation. Trop. Ecol. 19(2):178-188.

Sjostrom, R. r. 1982. Trumpeter Swan collaring at Red Rock Lakes National Wildlife Refuge. Papers and proc. 8th Trumpeter Swan Society Conf., Hickory Corners, Mich., 1982.

Smith, J. H. 1985. Trumpeter Swan behavior report. Wyoming Dept. of Game and Fish. Unpub. Rept. 12 pp.

Southwick, C. H. and F. W. Pine. 1975. Abundance of submerged vascular vegetation in Rhode River from 1966 to 1973. Chesapeake Sci. 16:147-151.

Spence, D. H. N. 1964. The macrophytic vegetation of freshwater locks, swamps, and associated fens. Pages 306-425 in J. H. Burnett, ed. The vegetation of Scotland. Oliver and Boyd, Edinburgh.

Spence, D. H. N. 1967. Factors controlling the distribution of freshwater macrophytes with particular reference to the locks of Scotland. J. Ecol. 55:147-170.

Spence, D. H. N. 1970. Photosynthesis and zonation of freshwater macrophytes. I. Depth distribution and shade tolerance. New Phytol. 69:205-215.

Spence, D. H. N. 1982. The zonation of plants in freshwater lakes. Adv. Ecol. Res. 12:37-125.

Spence, D. H. N., R. M. Campbell, and J. Chrystal. 1973. Specific leaf areas and zonation of freshwater macrophytes. J. Ecol. 61:317-328.

Spence, D. H. N., and J. Chrystal. 1970. Photosynthesis and zonation of freshwater macrophytes: I. Depth distribution and shade tolerance. New Phytol. 69:205-215.

Spence, D. H. N., and J. Chrystal. 1970. Photosynthesis and zonation of freshwater macrophytes. II. Adaptability of species of deep and shallow water. New Phytol. 69: 217-227.

Steenis, J. H. 1947. Recent changes in the marsh and aquatic plant status at Reelfoot Lake. J. Tenn. Acad. Sci. 22:22-27.

Stuckey, R. L. 1971. Changes of vascular aquatic flowering plants during 70 years in Put-in-Bay Harbor, Lake Erie, Ohio. Ohio J. Sci. 71:321-342.

Templeton, A. R., H. Hemmer, G. Mace, U. S. Seal, W. M. Shields, and D. S. Woodruff. 1986. Local adaptation, coadaptation, and population boundaries. Zoo Biol. 5:115-125.

Thorne, T. T., E. S. Williams, S. L. Anderson, D. Lockman. 1985. Trumpeter Swan disease investigations, northwest Wyoming, southeast Idaho, and Red Rock Lakes National Wildlife Refuge, Montana. Pages 9-14 in Job Performance Rept. Wyo. Dept. Game and Fish., Cheyenne.

Trauger, D. L., and J. C. Bartonek. 1977. Leech parasitism of waterfowl in North America. Wildfowl 28:143-152.

Trumpeter Swan Society. 1977. A guideline for propagation of captive Trumpeter Swans. The Trumpeter Swan Soc. Maple Plain, Minnesota. 24 pp.

Trumpeter Swan Society. 1986. Declining productivity of Trumpeter Swans at Red Rock Lakes National Wildlife Refuge, Lima, Montana. Pages 124-131 in Proc. and Papers of the 9th Trumpeter Swan Soc. Conf., West Yellowstone, Montana, 1984.

Turner, B. 1978. Population dynamics of the Grande Prairie Trumpeter Swan population. Proc. Of the 9th Trumpeter Swan Society Conf., Anchorage, Alaska. 1978.

Turner, B. 1981. Status of the Grande Prairie Trumpeter Swan population and its habitat. Presented at the Trumpeter Swan Society 7th Conf.

Turner, B. 1982. The interior Trumpeter Swans of Canada. Presented at the Trumpeter Swan Soc. 8th Conf., Hickory Corners, Mich.

Turner, B. 1987. The Grande Prairie Trumpeter Swan neck-band program. Proc. and Papers of the 10th Trumpeter Swan Soc., Conf., Grande Prairie, 1986. In press.

278

Turner, B., and R. H. Mackay. 1981. The population dynamics of the Trumpeter Swans of Grande Prairie. Can. Wildl. Serv., Edmonton. Unpub. Rept. 17 pp.

U. S. Fish and Wildlife Service. 1984. North American management plan for Trumpeter Swans. Office of Migratory Bird Manage., Washington, D.C. 62 pp. plus appendices.

U. S. Fish and Wildlife Service. 1986a. Use of lead shot for hunting migratory birds n the United States. Off. of Migratory Bird Mgmt. Washington, D. C.

U. S. Fish and Wildlife Service. 1986b. Results of the workshop on grizzly bear population genetics. Manuscript submitted to the Interagency Grizzly Bear Committee by the Office of the Grizzly Bear Recovery Coordinator, University of Montana, Missoula.

Van der Valk, A. G. and C. B. Davis. 1976. Changes in the composition, structure, and production of plant communities along a perturbed wetland coenocline. Vegetation 32:87-96.

Verhoeven, J. T. a. 1980. Ecology of Ruppia-dominated communities in western Europe. II. Synecological classification. Structure and dynamics of the macroflora and macrofauna communities. Aquatic Botany 8:1-85.

Ward, J., and J. Talbot. 1984. Distribution of aquatic macrophytes in Lake Aluxandrina, New Zealand. New Zealand J. Marine and Freshwater Res. 18:211-220.

Whittaker, R. H. 1975. The design and stability of plant communities. Pages 169-181 in W. H. van Dobben and R. H. Lowe-McConnell, eds. Unifying concepts in ecology. Junk, The Hague.

Wobeser, G. 1986. Interactions between lead and other disease agents. Pages 109-112 in J. S. Feierabend and a. B. Russell, eds. Lead poisoning in wild waterfowl – a workshop. Natl. Wild. Fed. Washington, D.C. 139 pp.

Wright, S. 1931. Evolution in Mendelian populations. Genetics 16:97-159.

Wypkema, R. C., and C. D. Ankney. 1979. Nutrient reserve dynamics of lesser snow geese staging at James Bay, Ontario, Can. J. Zool. 57:213-219.